Research Reports in Physics

Research Reports in Physics

Nuclear Structure of the Zirconium Region
Editors: J. Eberth, R. A. Meyer, and K. Sistemich

Ecodynamics Contributions to Theoretical Ecology
Editors: W. Wolff, C.-J. Soeder, and F. R. Drepper

Nonlinear Waves 1 Dynamics and Evolution
Editors: A. V. Gaponov-Grekhov, M. I. Rabinovich, and J. Engelbrecht

Nonlinear Waves 2 Dynamics and Evolution
Editors: A. V. Gaponov-Grekhov, M. I. Rabinovich, and J. Engelbrecht

Nuclear Astrophysics
Editors: M. Lozano, M. I. Gallardo, and J. M. Arias

Optimized LCAO Method and the Electronic Structure of Extended Systems   By H. Eschrig

Nonlinear Waves in Active Media
Editor: J. Engelbrecht

Problems of Modern Quantum Field Theory
Editors: A.A. Belavin, A.U. Klimyk, and A.B. Zamolodchikov

Fluctuational Superconductivity of Magnetic Systems
By M.A. Savchenko and A.V. Stefanovich

Nonlinear Evolution Equations and Dynamical Systems
Editors: S. Carillo and O. Ragnisco

Nonlinear Physics
Editors: Gu Chaohao, Li Yishen, and Tu Guizhang

Gu Chaohao Li Yishen
Tu Guizhang (Eds.)

Nonlinear Physics

Proceedings of the International Conference,
Shanghai, People's Rep. of China, April 24–30, 1989

With 47 Figures

Springer-Verlag
Berlin Heidelberg New York London
Paris Tokyo Hong Kong Barcelona

Professor Gu Chaohao
Professor Li Yishen
University of Science and Technology of China,
Hefei, Anhui 230026, People's Rep. of China

Professor Tu Guizhang
Computing Center of Academia Sinica, Beijing 100080, People's Rep. of China

Executive Editor
Associate Professor Zeng Yunbo
Department of Mathematics, University of Science and Technology of China,
Hefei, Anhui 230026, People's Rep. of China

Advisory Committee

Chairman
Yang, C. N. (USA)

Members
Ablowitz, M. J. (USA)	Flato, M. (France)
Araki, H. (Japan)	Gu, C. H. (China)
Calogero, F. (Italy)	Kruskal, M. (USA)
Degasperis, A. (Italy)	Sato, M. (Japan)
Faddeev, L. D. (USSR)	

Organizing Committee

Chairman
Gu Chaohao

Members
Cao Cewen	Hu Hesheng
Guo Benyu	Li Yishen
Ge Molin	Ni Guangjiong
Hou Boyu	Tu Guizhang

ISBN 3-540-52389-8 Springer-Verlag Berlin Heidelberg New York
ISBN 0-387-52389-8 Springer-Verlag New York Berlin Heidelberg

Library of Congress Cataloging-in-Publication Data. Nonlinear physics / Gu Chaohao ... [et al.], eds. p. cm. -- (Research reports in physics) A selection of refereed papers presented at the International Conference on Nonlinear Physics held Apr. 24-30, 1989, in Shanghai. Includes bibliographical references. ISBN 0-387-52389-8 (U.S. : alk. paper) 1. Solitons--Congresses. 2. Nonlinear theories--Congresses. I. Gu, Ch'ao-hao. II. International Conference on Nonlinear Physics (1989 : Shanghai, China) III. Series. QC 174.26.W28N66 1990 530.1'4--dc20 90-34467

This work is subject to copyright. All rights are reserved, whether the whole or part of the material is concerned, specifically the rights of translation, reprinting, reuse of illustrations, recitation, broadcasting, reproduction on microfilms or in other ways, and storage in data banks. Duplication of this publication or parts thereof is only permitted under the provisions of the German Copyright Law of September 9, 1965, in its current version, and a copyright fee must always be paid. Violations fall under the prosecution act of the German Copyright Law.

© Springer-Verlag Berlin Heidelberg 1990
Printed in Germany

The use of registered names, trademarks, etc. in this publication does not imply, even in the absence of a specific statement, that such names are exempt from the relevant protectiv laws and regulations and therefore free for general use.

Printing and binding: Weihert-Druck GmbH, D-6100 Darmstadt
2157 / 3150-543210 – Printed on acid-free paper

Preface*

The International Conference on Nonlinear Physics took place in Shanghai during 24–30 April, 1989. About one-fifth of the 108 participants came from regions outside China and Hong Kong.

The main topic of the conference was soliton theory. These proceedings contain 30 of the 83 contributions and reflect the most recent and significant developments in the field, including new methods and new applications. The selected contributions have all undergone referee review. Some of the papers are rather broad surveys of the recent research while others are narrow scope research articles.

Because of the wide range of the covered topics, the contributions are divided into five groups. The first part is devoted to the general theory of the integrable system, such as Hamiltonian structure, symmetries, Bäcklund and Darboux transformations. In the second part entitled "Finite Dimensional Dynamical Systems", one paper is concerned with the symmetries and integrability of coupled nonlinear oscillators, others concentrate on finite dimensional integrable systems reduced from infinite dimensional integrable Hamiltonian systems. The third part "Quantum Aspects and Statistical Mechanics" includes knot theory, braid groups and the R-matrix method. The fourth part, "Physical Phenomena", deals with nonlinear evolution equations, such as the K-P equation applied to water waves, soliton phenomena in porous media, and lattice models. In addition, findings related to chaos and the cellular automata are presented. The final part comprises contributions addressing various questions ranging from the most fundamental ones to those of the exciting theory of strings.

We are particularly pleased to thank the International Union for Pure and Applied Physics, the International Center for Theoretical Physics, the National Commission of Education of China and the National Natural Science Foundation of China for providing financial support. We are indebted to the authors for the hard work they invested in their contributions. We take this opportunity to express our thanks to Executive Editor Zeng Yunbo for the hard work and direct help in preparing the proceedings.

Hefei
January 1990

Gu Chaohao
Li Yishen
Tu Guizhang

---

*The complete manuscript was received by Springer-Verlag on February 14, 1990

Contents

Part I	Integrable Systems: Hamiltonian Structure, Symmetries, Bäcklund and Darboux Transformations

Liouville Integrability of Zero Curvature Equations
By Tu Guizhang 2

2+1 Dimensional Integrable Hierarchies, Lax Operators and Relevant Algebraic Structures
By Cheng Yi 12

Determination of Nondegenerate Darboux Operators of First Order in 1+2 Dimensions
By Zhou Zixiang 23

A Series of New Exact Solutions to the Nonlinear Equation $y_t+y_{xxx}-6y^2 y_x+6\lambda y_x=0$
By Au Chi (With 1 Figure) 29

Bäcklund Transformations for the Isospectral and Non-Isospectral KdV Hierarchies
By Tian Chou and Zhang Youjin 35

Multiple Darboux Transformations and Multiple Pole Solutions for AKNS Hierarchy
By Gu Xinshen 42

A Lie Algebraic Structure of G.J. and Its Gauge Equivalent Yang Hierarchies
By Li Yishen, Cheng Yi, and Zeng Yunbo 47

Part II	Finite Dimensional Dynamical Systems

Coupled Nonlinear Oscillators: Symmetries and Integrability
By M. Lakshmanan 54

Classical Integrable Systems Generated Through Nonlinearization of Eigenvalue Problems
By Cao Cewen and Geng Xianguo 68

The Confocal Involutive System and the Integrability of the Nonlinearized
Lax Systems of AKNS Hierarchy
By Ma Wenxiu 79

Two Kinds of Finite-Dimensional Systems Related to the Generalized
Schrödinger Equation
By Zeng Yunbo and Li Yishen 85

Nonlinearization of the Lax Pair for the KdV Equation and Integrable
Hamiltonian Systems
By Zhuang Dawei and Lin Yuanqu 92

Part III Quantum Aspects and Statistical Mechanics

Quantum and Classical Statistical Mechanics of the Integrable Models in
1+1 Dimensions
By R.K. Bullough, D.J. Pilling, J. Timonen, Yi Cheng, and
Yu-Zhong Chen 98

Link Polynomials and Exactly Solvable Models
By M. Wadati, Y. Akutsu, and T. Deguchi (With 17 Figures) 111

R-Matrices and Higher Poisson Brackets for Integrable Systems
By W. Oevel 136

Classical R-Matrix and Semi-Simple Lie Algebras
By Liu Zhangju and Qian Min 146

Witten's Approach, Braid Group Representations and X-Deformations
By M.L. Ge, F. Piao, L.Y. Wang, and K. Xue 152

Part IV Physical Phenomena

Nonlinear Evolution Equations, Solitons, Chaos and Cellular Automata
By M.J. Ablowitz, B.M. Herbst, and J.M. Keiser (With 14 Figures) 166

Kadomtsev-Petviashvili Equations in the Description of Water Waves
By D. Levi (With 5 Figures) 190

Three-Dimensional Lattice Model Based on Soliton Theory
By N. Saitoh 205

Soliton Phenomena in a Porous Medium
By D. Takahashi, J.R. Sachs, and J. Satsuma (With 5 Figures) 214

Two-Dimensional Chiral Gauge Theories on a Lattice
By Ma Zhongshui and Guo Shuohong 221

Transformation for the Solutions of the Two-Dimensional Toda Lattice
By Liu Qiming 227

Part V	Other Topics

Some Ideas on Nonlinear Evolution Equations
By F. Calogero 232

Some Problems of the Generalized Kuramoto-Sivashinsky Type Equations with Dispersive Effects
By Guo Boling 236

Standard Nonlinearities Associated with KdV-like Two-Soliton Solutions
By F. Lambert and R. Willox 242

Complex Singularities and the Riemann Surface for the Burgers Equation
By D. Bessis and J.D. Fournier (With 5 Figures) 252

From Soliton Theory to String Theory
By S. Saito and H. Kato 258

Non-Linear Equations from a String-Theoretical Point of View
By H. Kato and S. Saito 266

Painlevé Analysis and Integrability of the Evolution Equation $u_t = u_{xxx} + u^2 u_{xx} + 3u u_x^2 + 1/3 u^4 u_x$
By M. Daniel and R. Sahadevan 273

Subject Index 281

List of Participants............................... 283

Index of Contributors............................. 287

Part I

**Integrable Systems:
Hamiltonian Structure, Symmetries,
Bäcklund and Darboux Transformations**

Liouville Integrability of Zero Curvature Equations

*Tu Guizhang**

Computing Centre of Academia Sinica, Beijing 100080, People's Rep. of China

1. INTRODUCTION

Let $S = S(\infty, -\infty)$ be the Schwartz space, and

$$u_t = K(u) \tag{1}$$

be a nonlinear evolution equation (NLEE), where $u = (u_1, \ldots, u_p) \in S^p$. There are different definitions on integrability of the equation (1), we shall adopt the following two definitions:

(A.) We call the equation (1) Lax integrable if it can be written as a zero curvature equation

$$U_t - V_x + [U, V] = 0,$$

where U=U(u), V=V(u) are two matrices which contain u as the 'potential', and [U,V]=UV-VU.

(B.) We call the equation (1) Liouville integrable if (1) it can be written as a generalized Hamiltonian equation

$$u_t = J\delta H/\delta u$$

where J is a Hamiltonian operator; and (2) it possesses an infinite number of conserved densities $\{H_n\}$ that are involution in pairs:

$$\{H_n, H_m\} = 0 \ (mod \ D),$$

where

$$\{f, g\} = (J\delta f/\delta u).(\delta g/\delta u)$$

and f=0 (mod D) means $f = (d/dx)h$ for some $h \in S^p$.

Example. The well-known KdV and AKNS hierarchies of NLEEs are both Lax integrable and Liouville integrable.

In the past decades the theory of generalized Hamiltonian ayatems has undergone a rapid development (see [1]-[17]). A central and very important problem in the theory of integrable systems is to search for the nonlinear evolution equations that are both Lax and Liouville integrable. The aim of the present report is the following:

(1.) First, we propose a scheme for generating Lax integrable hierarchies of equations

$$U_t - V_x^{(n)} + [U, V^{(n)}] = 0;$$

---

*Supported by National Natural Science Foundation through Nankai Institute of Math.

(2.) Second, we give sufficient conditions for the above hierarchies of equations to be Liouville integrable.

The key result is an explicit formula for the Poisson brackets

$$\lambda^k \{H(\lambda), H(\mu)\} = (d/dx) <V(\lambda), \lambda V(\mu)/(\lambda-\mu) + \Delta(\lambda)>.$$

We shall explain below the meaning of this formula.

2. A Scheme for Generating Lax Integrable Systems

Consider the isospectral problem

$$\psi_x = U(u, \lambda)\psi$$

where

$$U = U(u, \lambda) = e_0(\lambda) + u_1 e_1(\lambda) + \ldots + u_p e_p(\lambda),$$

$$u = (u_1, \ldots, u_p) \in S^p,$$

$$e_0, \ldots, e_p \in \check{G} = G \otimes C(\lambda, \lambda^{-1}),$$

and G is a semi-simple matrix Lie algebra with $\check{G}$ being its loop algebra. We have the following theorem.

Theorem 1. Let the matrix $U = U(u, \lambda)$ be defined above and $V = V(\lambda)$ be a solution of the equation

$$V_x(\lambda) = [U(\lambda), V(\lambda)]. \tag{2}$$

If there exists a matrix $\Delta = \Delta(\lambda, \mu)$ such that

$$[\mu(U(l) - U(\mu))/(l-\mu), V(\mu)] + \Delta_x(l, \mu) - [U(l), \Delta(l, \mu)] = \sum_{i=1}^{p} f_i(\mu, u) e_i(\lambda),$$

where $f_1(\mu, u), \ldots, f_p(\mu, u)$ are p independent functions, then we can relate the isospectral problem $\psi_x = U\psi$ with the following hierarchy of equations:

$$u_{t_n} = (f_{1n}, \ldots, f_{pn})^T, \tag{3}$$

where f_{in} are defined by

$$f_i(\mu, u) = \sum_{i \geq 0} f_{in}(u)\mu^{-n}.$$

Moreover the equation (3) are Lax integrable, the corresponding zero curvature representations are

$$U_{t_n} - V_x^{(n)} + [U, V^{(n)}] = 0,$$

where

$$V^{(n)} = (\lambda^n V)_+ + \Delta_n, \quad \Delta(\lambda, \mu) = \sum_{i \geq 0} \Delta_i(\lambda)\mu^{-i}.$$

$$V(\lambda, u) = \sum_{i \geq 0} V_n(u)\lambda^{-n}, \quad (\lambda^n V)_+ = \sum_{i=0}^{n} V_i(u)\lambda^{n-i}.$$

Example: Jaulent-Miodek (JM) Hierarchy. Consider the isospectral problem
$$y_{xx} + (\lambda^2 - \lambda q - r)y = 0,$$
where $u_1 = q$, $u_2 = r$ are two potentials. This problem can be reduced to the above standard form $\psi_x = U\psi$ with
$$U = (r + q\lambda - \lambda^2)f + e, \quad \psi = (y, y_x)^T,$$
where
$$\left\{ e = \begin{pmatrix} 0 & 1 \\ 0 & 0 \end{pmatrix}, f = \begin{pmatrix} 0 & 0 \\ 1 & 0 \end{pmatrix}, h = \begin{pmatrix} 1 & 0 \\ 0 & -1 \end{pmatrix} \right\}$$
represents a basis of the Lie algebra $sl(2, C)$. Let
$$V(\lambda) = ah + bc + ef$$
be a solution of the equation $V_x = [U, V]$, we have
$$[\mu(U(l) - U(\mu))/(l - \mu), V(\mu)] = \mu(\lambda + \mu - q)(\bar{b}_x f + \bar{b}h),$$
hereafter we write $a = a(\lambda)$, $\bar{a} = a(\mu)$ and so on. Since $U_t = (q_t\lambda + r_t)f$ involves only f, we have to cancel the tear $\bar{b}h$. To this end we introduce
$$\Delta(\lambda, \mu) = (\bar{b}\mu^2 - \bar{b}\mu q + \lambda\mu\bar{b})f,$$
for which it is easy to verify that
$$[\mu(U(l) - U(\mu))/(l - \mu), V(\mu)] + \Delta_x(l, \mu) - [U(l), \Delta(l, \mu)] = (f_1\lambda + f_2)f,$$
with
$$f_1(\mu, u) = 2\mu\bar{b}_x,$$
$$f_2(\mu, u) = 2\mu^2 \bar{b}_x - \mu((\bar{b}q)_x + q\bar{b}_x).$$
Therefore the desired JM hierarchy reads
$$q_t = 2\mu\bar{b}_x,$$
$$r_t = 2\mu^2 \bar{b}_x - \mu(2q\bar{b}_x + \bar{b}q_x).$$
It is a concise writing for the hierarchy of equations
$$q_{t_n} = 2b_{n+1\,x},$$
$$r_{t_n} = 2b_{n+2\,x} - 2qb_{n+1\,x} - q_x b_{n+1}.$$

3. A Formula for Poisson Brackets

The following theorem gives an explicit expression for Poisson brackets:
Theorem 2. Let the matrices U, V, Δ and (f_i) be mentional as above. If there exists a skewsymmetric operator $J : S^p \to S^p$ such that
$$J\lambda^k (\ldots, <V, \partial U/\partial u_i>, \ldots)^T = (\ldots, f_i, \ldots)^T,$$

where
$$<A,B> = tr(AB),$$
then
$$\lambda^k \{H(\lambda), H(\mu)\} = (d/dx) <V(\lambda), \lambda V(\mu)/(\lambda - \mu) + \Delta(\lambda)>,$$
where
$$\{f, g\} = (J\delta f/\delta u).(\delta g/\delta u).$$

A direct consequence of the above formula is
$$\{H_n, H_m\} = 0 \ (mod\ D),$$
where $H'_n s$ are defined by the expansion
$$H(\lambda) = \sum H_n \lambda^{-n}.$$

In other words, if in addition J is a Hamiltonian operator, then the infinite number of functions $\{H_n\}$ are in involution in pairs with respect to the above mentioned Poisson bracket.

4. Liouville Integrability of Zero Curvature Equations

Combining the above two theorems we deduce the following theorem:

Theorem 3. Let the isospectral problem be given by $\psi_x = U\psi$ with $U(\lambda) = e_0(\lambda) + u_1 e_1(\lambda) + \ldots + u_p e_p(\lambda)$, and V is a solution of the equation (2). Suppose that
(A.) the equation
$$[\mu(U(l) - U(\mu))/(l - \mu), V(\mu)] + \Delta_x(l, \mu) - [U(l), \Delta(l, \mu)] = \sum_{i=1}^{p} f_i(\mu, u) e_i(\lambda),$$
holds for a matrix $\Delta(\lambda, \mu)$ and p independent functions $f_1(\mu, u), \ldots, f_p(\mu, u)$;
(B.) The operator J defined by
$$J\lambda^k(\ldots, <V, \partial U/\partial u_i>, \ldots)^T = (\ldots, f_i, \ldots)^T,$$
is Hamiltonian.
Then the hierarchy of equations
$$u_{t_n} = f_n$$
admit zero curvature representations and they can be written as Hamiltonian equations
$$u_{t_n} = J\delta H_{n+k}/\delta u,$$
where H_n is defined by
$$(\lambda^{-\gamma}(\partial/\partial \lambda)\lambda^\gamma)H = <V, \partial U/\partial \lambda>, \ (\gamma = const.),$$
and for which it holds that
$$\delta H/\delta u_i = <V, \partial U/\partial u_i>.$$

Moreover the set $\{H_n\}$ constitutes the common set of infinite number of conserved densities which are in involution in pairs. Therefore the above mentioned hierarchy of Hamiltonian equations are Liouville integrable.

Remark The above couple of equations

$$(\lambda^{-\gamma}(\partial/\partial\lambda)\lambda^{\gamma})H = <V, \partial U/\partial\lambda>,$$

$$\delta H/\delta u_i = <V, \partial U/\partial u_i>.$$

is a consequence of the following trace identity [19,20]

$$(\delta H/\delta u_i)<V, \partial U/\partial\lambda> = (\lambda^{-\gamma}(\partial/\partial\lambda)\lambda^{\gamma})<V, \partial U/\partial u_i>,$$

where γ is a constant that can be easily fixed each time.

Example. Consider again the JM hierarchy. The corresponding matrix $U = (r + q\lambda - \lambda^2)f + e$. We have found that $\Delta(\lambda, \mu) = (\bar{b}\mu^2 - \bar{b}\mu q + \lambda\mu\bar{b})f$ and the resulting hierarchy of equations is

$$q_{t_n} = 2b_{n+1\,x},$$
$$r_{t_n} = 2b_{n+2\,x} - 2qb_{n+1\,x} - q_x b_{n+1}.$$

where a,b,c are defined by the matrix V=ah+be+ef. To write the above hierarchy in its Hamiltonian form we note that

$$<V, \partial U/\partial\lambda> = (q - 2\lambda)b,$$
$$<V, \partial U/\partial q> = b\lambda,$$
$$<V, \partial U/\partial r> = b,$$

Thus the trace identity mentioned above implies that

$$(\delta/\delta q, \delta/\delta r)((q - 2\lambda)b) = (\lambda^{-\gamma}(\partial/\partial\lambda)\lambda^{\gamma}(b\lambda, \beta),$$

or

$$(\delta/\delta q, \delta/\delta r)((qb_{n+1} - 2b_{n+2}) = (\gamma - n)(b_{n+1}, b_n).$$

This equation suggests us to search for the recursion relation among (b_{n+1}, b_n). It is easy to derive such a relation as

$$b_{n+2} = R(q)b_{n+1} + (R(r) - \partial^2/4)b_n,$$

Where

$$R(f) = f - (\frac{1}{2})\partial^{-1}f_x.$$

We have $b_0 = b_1 = 0$, $b_2 = \beta = const$. Setting n=2 in the above equation following the trace identity, we obtain

$$(\gamma - 1)(\beta, 0) = 0,$$

thus $\gamma = 1$ and we conclude that

$$(\delta/\delta u)H_n = (b_{n+1}, b_n)^T$$

where $u = (q, r)$, $H_1 = \beta q$, and

$$H_n = (2b_{n+2} - qb_{n+1})/(n-1), \quad n > 1.$$

Therefore the JM hierarchy takes the following Hamiltonian form
$$u_{t_n} = J(b_{n+2}, b_{n+1})^T = J\delta H_{n+1}/\delta u,$$
where
$$J = \begin{pmatrix} 0 & 2\partial \\ 2\partial & -2\partial R(q) \end{pmatrix}$$

By theorem 3 we see that JM hierarchy consists of Liouville integrable Hamiltonian systems and the corresponding Poisson bracket is given by
$$\{H(\lambda), H(\mu)\} = (d/dx) < V(\lambda), V(\mu)/(\lambda - \mu) + \Delta(\lambda)/\lambda >$$
$$= (d/dx)(2a\bar{a} + b\bar{c} + \bar{b}c)/(\lambda - \mu) + b\bar{b}(\lambda + \mu - q)).$$

Example: a nonlinear reduction of the Glachette-Johnson (GJ) hierarchy

Consider the following isospectral problem
$$U = (s - \lambda)h + qe + rf = \begin{pmatrix} s - \lambda & q \\ r & \lambda - s \end{pmatrix}.$$

Taking $\Delta(\mu) = \bar{\delta}h$ we have, as easily verified, that
$$[\mu(U(l) - U(\mu))/(l - \mu), V(\mu)] + \Delta_x(l, \mu) - [U(l), \Delta(l, \mu)]$$
$$= (-2\mu\bar{b} + 2q\bar{\delta})e + (2\mu\bar{c} - 2r\bar{\delta})f + \bar{\delta}_x h.$$

Thus we are led to the hierarchy of equations
$$q_{t_n} = -2b_{n+1} + 2q\delta_n,$$
$$r_{t_n} = 2c_{n+1} - 2r\delta_n, \quad s_{t_n} = \delta_{nx}$$

We emphasize here that the function δ remains arbitrary at this step, this will allow us to make several reductions. We now search for the following nonlinear reduction
$$s = \beta qr,$$
where $\beta = const$. Then
$$\bar{\delta}_x = s_t = \beta(q_t r + qr_t) = 2\beta\mu(q\bar{c} - r\bar{b}) = 2\beta\mu\bar{a}_x,$$
thus we have to choose that
$$\bar{\delta} = 2\beta\mu\bar{a}.$$

Then the above hierarchy of equations reduces to
$$q_t = -2\mu\bar{b} + 4\beta q\mu\bar{a},$$
$$r_t = 2\mu\bar{c} - 4\beta r\mu\bar{a},$$
or
$$q_{t_n} = -2b_{n+1} + 4\beta qa_{n+1},$$
$$r_{t_n} = 2c_{n+1} - 4\beta ra_{n+1}.$$

To write it to its Hamiltonian form we observe that

$$< V, \partial U/\partial \lambda > = -2a$$
$$< V, \partial U/\partial q > = c + 2\beta ra \equiv f,$$
$$< V, \partial U/\partial r > = b + 2\beta qa \equiv g.$$

By applying the trace identity we obtain

$$(\delta/\delta q, \delta/\delta r)(-2a) = (\lambda^{-\gamma}(\partial/\partial\lambda)\lambda^\gamma)(f,g),$$

or

$$(\delta/\delta q, \delta/\delta r)(-2a_{n+2}) = (\gamma - n - 1)(f_{n+1}, g_{n+1})$$

Setting n=0 in the above equation we find that

$$(r,q) = (\gamma - 1)(-r, -q),$$

thus $\gamma = 0$ and we have

$$(\delta/\delta q, \delta/\delta r) H_{n+1} = (f_{n+1}, g_{n+1})$$

with

$$H_{n+1} = 2a_{n+2}/(n+1).$$

Therefore we obtain the following GJ hierarchy:

$$\begin{pmatrix} q \\ r \end{pmatrix}_{t_n} = J \delta H_{n+1}/\delta u$$

where

$$J = 2 \begin{pmatrix} 4\beta q \partial^{-1} q & -1 - 4\beta q \partial^{-1} r \\ 1 - 4\beta r \partial^{-1} q & 4\beta r \partial^{-1} r \end{pmatrix}$$

We observe that

$$J = 2J_1 + 8\beta J_2.$$

where

$$J_1 = \begin{pmatrix} 0 & -1 \\ 1 & 0 \end{pmatrix},$$

$$J_2 = \begin{pmatrix} q \\ -r \end{pmatrix} \partial^{-1} (q, -r).$$

It is easy to check that J_1 and J_2 form a pair of Hamiltonian operators, thus J is a HAmiltonian operator, and by Theorem 3 we conclude that the above GJ hierarchy of Hamiltonian equations are Liouville integrable.

It is interesting to observe that since J_1 and J_2 are compatible, thus the operator $L = J_1^{-1} J_2$ is a hereditary symmetry, which is exactly the recursion operator for the well-known AKNS hierarchy.

A typical equation of this hierarchy is, after a further reduction $q = \pm r^*$,

$$q_{t_2} = i(q_{xx} - 2|q|^2 q + 4\gamma^2 |q|^4 q) + 4\gamma(|q|^2)_x q,$$

which can be viewed as a higher order correction of the well-known Nonlinear Schrodinger (NLS) equation

$$q_{t_2} = i(q_{xx} - 2|q|^2 q).$$

5. LIOUVILLE INTEGRABILITY OF THE STATIONARY ZERO CURVATURE EQUATION $V_x = [U, V]$

We have seen from the above sections that the equation (2) plays an essential role in the constructing of integrable Hamiltonian equations starting from the isospectral problem $\psi_x = U\psi$, we have established a general theorem under a broad assumption the equation (2) can be reduced to a classical Hamiltonian system of finite dimensions

$$q_i' = \partial h/\partial p_i, \quad p_i' = -\partial h/\partial q_i, \quad i = 1, \ldots, N,$$

that is Liouville integrable. It is worth mentioning the fact that the formula on Poisson brackets mentioned above (Theorem 2) provides us with a constructive algorithm for generating the N first integrals as needed by the classical Liouville theorem on complete integrability.

To save the space we give in the following only an illustrative example.

It is well-known that the AKNS hierarchy reads [15]

$$u_t = J\delta H_n/\delta u$$

with

$$u = (v, w)^T, \quad J = 2(f - e),$$

and

$$H_n = 2a_{n+1}/n,$$

where the sequence $\{a_n\}$ can be derived from the above GJ hierarchy by taking $s = 0$.

As an example we take n=4 then

$$H_4 = 2a_5/4$$
$$= (vw_{xxx} - wv_{xxx} - v_x w_{xx} + w_x v_{xx} - 6v^2 ww_x + 6w^2 vv_x)$$
$$= H \equiv (v_{xx}w_x + 3vv_x w^2) \pmod{D}.$$

Thus the corresponding stationary equation is

$$\delta H/\delta u = 0.$$

that can be reduced to a classical Hamiltonian system by performing the corresponding Jacobi-Ostrogradskii transformation

$$q_1 = v, \quad q_2 = v_x, \quad q_3 = w,$$
$$p_1 = \delta H/\delta v_x = 3vw^2 - w_{xx}, \quad p_2 = \delta H/\delta v_{xx} = w_x, \quad p_3 = \delta H/\delta w_x = v_{xx}$$

or equivalently

$$v = q_1, \quad v_x = q_2, \quad v_{xx} = p_3$$
$$w = q_3, \quad w_x = p_2, \quad w_{xx} = 3q_1 q_3^2.$$

The corresponding Hamiltonian h is given by

$$h = q_1' p_1 + q_2' p_2 + q_3' p_3 - H$$
$$= p_2 p_3 + q_2 p_1 - 3q_1 q_2 q_3^2.$$

The resulting Hamiltonian system
$$q'_i = \partial h/\partial p_i, \quad p'_i = -\partial h/\partial q_i, \quad i = 1,2,3.$$
is
$$q_{1x} = q_2, \quad q_{2x} = p_3, \quad q_{3x} = p_2$$
$$p_{1x} = 3q_2 q_3^2, \quad p_{2x} = -p + 3q_1 q_3^2, \quad p_{3x} = 6q_1 q_2 q_3.$$

Moreover by the general theorem we have the following set of first integrals:
$$h_i = \partial^{-1}\{H_i, H_4\}.$$

By applying the formula on Poisson brackets we have
$$h_1 = c_3 b_1 + c_1 b_3 + a_2^2 + b_2 c_2$$
$$= p_3 q_2 - p_1 q_1 - q_2 p_2,$$
$$h_2 = (2a_2 a_3 + b_3 c_2 + c_3 b_2) = h,$$
$$h_3 = (a_3^2 + b_3 c_3)$$
$$= q_1^2 p_2^2 + q_3^2 p_2^2 - 2q_1 q_2 q_3 p_2 - p_1 p_3 + q_1 q_3^2 p_3 + 2q_1^2 p_1 q_3 - 2q_1^3 q_3^3,$$

for which we have
$$\{h_i, h_j\} = 0,$$
where
$$\{f, g\} \equiv \sum((\partial f/\partial p_i)(\partial g/\partial q_i) - (\partial f/\partial q_i)(\partial g/\partial p_i)).$$

Therefore we conclude that the above Hamiltonian system is Liouville integrable in the strict sense.

REFERENCES

1 Ablowitz M.J. and Segur H., Solitons and the inverse scattering transform, SIAM Philadelphia PA 1981.
2 Newell A.C., Solitons in mathematics and physics, SIAM Philadelphia PA 1985.
3 Faddeev L.D. and Takhtajan L.A., Hamiltonian method in the theory of solitons, Springer-Verlag, Berlin 1987.
4 Magri F., J. Math. Phys. 19(1978), 1156.
5 Gel'fand I.M. and Dorfman I.G., Funk. Anal. Pril. 13 (1979), 13.
6 Fuchssteiner B., Nonl. Anal. Theor. Meth. Appl. 3(1979), 849.
7 Chen H.H., Lee Y.C. and Liu C.S., Phys. Script. 20 (1979), 490.
8 Olver P.J., Math. Proc. Camb. Philos. Soc. 88(1980),71.
9 Fuchssteiner B. and Fokas A.S., Physica D4 (1981), 47.
10 Wilson G., Ergod. Theor. Dynam. Syst. 1(1981), 361.
11 Tu G.Z., Nuovo. Cimento B73 (1983), 15.
12 Boiti M. and Tu G.Z., ibid 75(1983), 145.
13 Boiti M., Pempinelli F. and Tu G.Z., ibid 79(1984), 231.
14 Tu G.Z. Scientia Sinica 31A (1988), No. 12, 28.
15 Tu G.Z. J. Math. Phys. 30 (1989), 330.
16 Tu G.Z. and Meng D.Z., Acta Math. Appl. Sinica 5 (1989), No.1, 101.

17 Tu G.Z. and Xu B.Z., The trace identity, a powerful tool for constructing the Hamiltonian structrue of integrable systems (III), preprint CTR-1989-T2.
18 Tu G.Z.,On Liouville integrability of zero curvature equations and the Yang hierarchy, J. Phys. A. Math. Gen. (in press).
19 Tu G.Z., Scientia Sinica 24A(1986), 138.
20 Tu G.Z., Advan. Sci. China (ser. Math.) 2(1988), 45.
21 Tu G.Z., On complete integrability of a large class of Hamiltonian systems of finite dimensions, submitted for publication.

2 + 1 Dimensional Integrable Hierarchies, Lax Operators and Relevant Algebraic Structures

Cheng Yi

Department of Mathematics, University of Science and Technology of China, Hefei 230026, People's Rep. of China

An approach is described to construct integrable hierarchies, their Lax operators and relevant algebraic structures for the given nonlinear integrable equations in 2+1 dimensions. The hierarchies consist of isospectral and non-isospectral flows and their Lax operators form an infinite dimensional Lie algebra—-called the Lax algebra which in turn gives rise to the commutability and non-commutability of the flows and the algebraic structures for the symmetries and conserved quantities of each equations in the hierarchies. The so-called mastersymmetries of the given equations can also be derived straightforwards through the approach.

1. Introduction.

Associated with a given nonlinear evolution equation (NEE) exactly solved by the inverse scattering transform (IST), one of the problem of fundamental interest is *to construct hierarchies of NEEs which have similar property to the given equation (i.e., they are solved by IST of the same spectral problem, possess infinitely many conserved quantities and symmetries, etc.) and investigate the relevant algebraic and geometrical properties*. As is known, for NEEs in 1+1 dimensions, the general theory concerning with the above problem has been progressed spectacularly since the work of P.D. Lax on the Korteweg-de Vries (KdV) equation [1]. A variety of methods have been achieved in the theory, which mainly include the method in terms of the recursion opeartor (e.g. see [2-4]), generalized Wronskian technique [5] and the approach based on the theory of loop, or Kac-Moody algebras (e.g. see [2, 6-8] and references therein).

Recently, various interesting studies have been made for NEEs in 2+1 dimensions. The Sato's theory gievs rise to a general form of commuting flows by using of the microdifferential operator and the further development of this theory shed light to the algebraic structure from group theoretical veiw-point (e.g. see [9, 10]). The so-called mastersymmetry (MS) in the sense of Fuchssteiner [11], the works of Chen et.al., and Orlof and Shulman (e.g. see [12, 13]), and the extended resursion operator recently achieved by Fokas and Santini [14-16] provide methods to study the algebraic and geometrical properties analogue to these found in 1+1 dimensions. However, for a given equation in 2+1, to construct a nontrivial mastersymmetry and an extended recursion operator with canonical property (i.e., the operator admits the factorization of two Hamiltonian operators, or satisfies the hereditary property) is not so easy. One needs to derive them firstly and then verify the conditions they should satisfy by, somewhat, a lengthy computations.

In this article, I hope to describe a novel approach to construct hierarchies of NEEs and relevant algebraic structures in 2+1 dimensions. A convenient starting point for building these hierarchies of NEEs is to look for hierarchies of the Lax operators associated with the given multidimensional spectral problem. It is shown that these Lax operators are derived by an algebraically recursive procedure rather than the standard method in 1+1 dimensions. Consequently, by using them and the well studied IST in multidimensions, the initial value problem of hierarchies of NEEs can be solved, from which one finds that the hierarchies consist of a familar family of isospectral flows, and a hierarchies of non-isospectral flows depending explicitly on the spatial variables x, y. These results are analogue to that found in 1+1 (e.g. see [5]).

The further study shows that the Lax operators themselves form an infinite dimensional Lie algebra—called the Lax algebra for the sake of convenience. This Lax algebra then naturally gives rise to the commutability and non-commutability of the isospectral and non-isospectral flows respectively, and the algebraic structures for the symmetries and conserved quantities of each equation in the hierarchies. The other bonus from our treatment is how to derive the mastersymmetries systematically and straightforwards for the given equations in 2+1.

To explain the basic idea of the approach, I will take the Kadomtsev-Petviashvilli (KP) equation as an typical example and sketch the main results for others including the Benjamin-Ono (BO), the 2+1 dimensional Caudrey-Dodd-Gibbon-Kotera-Sawada (CDGKS) equations and the example associated with the Schrodinger spectral problem in the plane. The last example seems to be more interesting, since the extended recursion operator in the case is no longer canonical [22], however it still shares the same algebraic property with others. The presence of the example with non-canonical recursion operator in 1+1 and its further study were shown by Li, Cheng and Zeng [23]. Interested readers can find detail there.

Some results presented here have been shown in [17-21], however, in this article, the general theory on the approach and some novel insights in the algebraic structure of hierarchies of NEEs are emphasized.

2. Integrable Hierarchies and Lax Operators for the KP Equation.

Take the KP equation in the form

$$u_t = K_2(u) = \partial_x^{-1} u_{yy} - 6uu_x - u_{xxx}, \tag{2.1}$$

where $u = u(x, y, t)$, $u_t = \partial u/\partial t$ etc. and the usual boundary conditions (i.e. u and its derivatives vanish rapidly enough as x, $y \to \infty$ and $\int u dx = 0$) are always assumed.

As is known, the KP equation can be written as the compatibility conditions of the linear system [24, 25]

$$L\varphi = 0, \quad L = \alpha \partial_y + \partial_x^2 + u, \quad \alpha = \frac{i}{3}\sqrt{3}, \tag{2.2}$$

$$\varphi_t = A_2(u, \partial_x)\varphi, \quad A_2 = -4\partial_x^3 - 6u\partial_x - 3u_x + i\sqrt{3}(\partial_x^{-1} u_y), \tag{2.3}$$

i.e., it is equivalent to its Lax representation $L_t = [A_2, L]$, where $[\ ,\ ]$ denotes the usual bracket of operators. Notice that as the polynomial in ∂_x with coefficients being functional of u, A_2 satisfies $[A_2,\ L] = K_2(u)$ identically. In this sense, we recall the definition of Lax operator as follows:

A is called a Lax operator, if A is a polynomial in ∂_x :

$$A(u,\partial_x) = \sum_{j=0}^{N} a_j(u)\partial_x^j, \qquad (2.4)$$

such that $[A, L] = a(u)$ *be a multiplication operator (i.e. a zeroth order polynomial in ∂_x), where L is given in (2.2)*. By equating coefficients of powers ∂_x^k in $[A, L] = a$, the Lax operator can obtained by solving

$$a_{Nx} = 0,$$

$$\sum_{j=k+1}^{N} \binom{j}{k} a_j u_{j-k} - \alpha a_{ky} - a_{kxx} - 2a_{(k-1)x} = 0, \qquad (2.5)$$

for $k = N, N-1, \ldots 1$, and so $a(u)$ is given by

$$a(u) = \sum_{j=1}^{N} a_j u_j - \alpha a_{0y} - a_{0xx}, \qquad (2.6)$$

here $u_j = \partial^j u / \partial^j x$ is used for short. According to the boundary conditions assumed previously, an observation is that two Lax operators A, $\tilde{A}$ equals to each other iff $A - \tilde{A} \to 0$ as $x, y \to 0$, namely, the Lax operator is uniquely determined by its asymptotic behaviour at infinity. Equations (2.5) also imply that the leading coefficients of Lax operator may be an arbitrary function of y. However, in this paper, I will restrict it only in two choices: a_N is either constant, or proportional to y. The case that a_N depends on powers of y is discussed in the forthcoming paper [26]. Under this restriction, some specific Lax operators and the correspondent multiplication operators can be calculated explicitly, they are

$$A_0 = \frac{1}{3}\partial_x, \quad K_0 = \frac{1}{3}u_x; \quad A_1 = 2\alpha(\partial_x^2 + u), \quad K_1 = \frac{2}{3}u_y;$$

A_2, K_2 are in (2.3), (2.4);

$$A_3 = -24\alpha\partial_x^4 - 46\alpha u\partial_x^2 - [48\alpha u_x + 8(\partial_x^{-1}u_y)]\partial_x \qquad (2.7)$$
$$- (24\alpha u^2 + 24\alpha u_{xx} + 4u_y - 4\alpha\partial_x^{-1}u_{yy}),$$

$$K_3 = -4u_{xxy} + \frac{4}{3}(\partial_x^{-1}u_{yyy}) - 8u_x\partial_x^{-1}u_y + 6uu_y;$$

and

$$B_1 = \frac{1}{3}yA_0 - \frac{1}{2}\alpha x, \quad \sigma_1 = yK_0;$$

$$B_2 = yA_1 + xA_0 + \frac{1}{3}, \quad \sigma_2 = yK_1 + xK_0 + \frac{2}{3}u;$$

$$B_3 = yA_2 + xA_1 + 2\alpha\partial_x + \alpha(\partial_x^{-1}u), \quad \sigma_3 = yK_2 + xK_1 + \frac{4}{3}(\partial_x^{-1}u_y); \qquad (2.8)$$

$$B_4 = yA_3 + xA_2 - 6\partial_x^2 + 2\alpha(\partial_x^{-1}u_y) + u,$$

$$\sigma_4 = yK_3 + xK_2 - 8u^2 - 4u_{xx} - 2u_x(\partial_x^{-1}u) + 2(\partial_x^{-2}u_{yy}).$$

where $K_i = [A_i, L]$, $\sigma_i = [B_i, L]$, etc. To proceed further the same way would be impossible tedious. However, two hierarchies of Lax operators can be constructed recursively as follows: if the Lax operator A_n and B_n is known respectively, then the next one of each A_n and B_n is given by

$$A_{n+1} = \frac{3}{n+1} \|[A_n, \ B_3]\|, \quad n \geq 0; \tag{2.9}$$

$$B_{n+1} = \frac{3}{n-3} \|[B_n, \ B_3]\|, \quad n \geq 4, \tag{2.10}$$

where the bracket of any two Lax operators A and $\check{A}$ are defined as

$$\|[A, \ \check{A}]\| = A'[\check{a}] - \check{A}'[a] + [A, \ \check{A}], \tag{2.11}$$

$$A'[\check{a}] = \lim_{\varepsilon \to 0} \frac{\partial}{\partial \varepsilon} A(u + \varepsilon \check{a}), \quad \text{etc.}$$

and a, $\check{a}$ are multiplication operators of A, $\check{A}$ respectively. The proof of the above statement can be found in [17-20]. Recall that

$$K_n = [A_n, \ L], \quad \sigma_n = [B_n, \ L] \tag{2.12}$$

as the respective multiplication operators of A_n, B_n, and then by letting $x, y \to \infty$, the following asymptotics can be calculated from (2.9), (2.10)

$$A_n \to \overline{A}_n = \frac{1}{3}(6\alpha)^n \partial_x^{n+1}, \quad K_n \to 0, \quad n \geq 0; \tag{2.13}$$

$$B_n \to \overline{B}_n = \frac{1}{3} y (6\alpha)^{n-1} \partial_x^n + \frac{1}{3} x (6\alpha)^{n-1} \partial_x^{n-1}$$
$$+ \frac{1}{6}(n-1)(6\alpha)^{n-2} \partial_x^{n-2}, \tag{2.14}$$

$$\sigma_n \to 0, \quad n \geq 1.$$

Therefore, (2.9) and (2.10) generate Lax operators nontrivially, i.e. (2.9) ((2.10)) maps the (n+1)th (nth) order of Lax operator to the (n+2)th ((n+1)th) one. Furthermore, one can check that both B_n and σ_n explicitly, but linearly depend on x, y.

We, now, consider the hierarchies of equations

$$u_t = K_n(u), \quad n \geq 0; \tag{2.15}$$

$$u_t = \sigma_n(u), \quad n \geq 1. \tag{2.16}$$

The first hierarchy includes the KP itself ($n = 2$) and is the usual family of the higher-order KP equations, while the second one is the hierarchy of equations depending on x, y explicitly. Since $L_t = u_t$ and so both flows (2.15) and (2.16) admit the Lax representations

$$L_t = [A_n, \ L], \quad n \geq 0; \tag{2.17}$$

$$L_t = [B_n, \ L], \quad n \geq 1, \tag{2.18}$$

namely (2.15) and (2.16) are, respectively, the compatibility conditions of the spectral problem (2.2) with $\varphi_t = A_n \varphi$, and with $\varphi_t = B_n \varphi$.

3. Solvability of the KP Hierarchies.

The IST of spectral problem (2.2) has been studied recently. Here we follow the work of Fokas and Ablowitz [25] and wish to solve the initial value problem of both hierarchies of (2.15) and (2.16). The complete set of scattering data necessary to carry out the IST of (2.2) are $f(k,l,t)$; and $k_j^\pm$, $\gamma_j^\pm$, $j = 1,...M$, they are defined by

$$\psi^+(k) = \psi^-(k) + \int_{-\infty}^{+\infty} \psi^-(l)f(k,l,t)dl \tag{3.1}$$

for $f(k,l,t)$, where $\psi^\pm$ solve (2.2) and satisfy $exp[-i(kx - \sqrt{3}k^2y)]\psi^\pm = \mu^\pm \to 1$ as $x,y \to \infty$, $\mu^\pm$ satisfy a Fredholm equation of the second type and is meromorphic in the upper (+) and lower (-) half plane respectively. The $k_j^\pm$ are simple poles of $\mu^\pm(k)$ and $\gamma_j^\pm$ are defined through

$$\lim_{k \to k_j^\pm}(\mu^\pm - \frac{i\psi_j^\pm}{k - k_j^\pm}) = (x - 2\sqrt{3}k_j^\pm y + \gamma_j^\pm)\psi_j^\pm \tag{3.2}$$

where the residues $\psi_j^\pm$ are homogeneous solutions of the Fradholm equation normalized by $(x - 2\sqrt{3}k_j^\pm)\psi_j^\pm \to 1$ as $x,y \to \infty$.

The detail of the IST of (2.2) can be found in [25]. Our purpose here is to derive the time evolution of the scattering data corresponding to (2.15) and (2.16). For (2.15), we set

$$L_n(k)\psi = (\partial_t - A_n + \alpha_n(k))\psi, \tag{3.3}$$

since $\psi^\pm = (exp[i(kx - \sqrt{3}k^2y)])\mu^\pm$, and $\mu^\pm \to 1$, $A_n \to \overline{A}_n$ as $x,y \to \infty$, so $\alpha_n(k) = \frac{i}{6\sqrt{3}}(-2\sqrt{3}k)^{n+1}$ which is then essentially the linearized dispersion relations of the equations in (2.15). By applying the operator $L_n(k)$ to (3.1) and (3.2), the time evolutions of the scattering data can be obtained [16]

$$f_t(k,l) = \frac{i}{6\sqrt{3}}(\xi^{n+1} - \eta^{n+1})f(k,l); \tag{3.4}$$

$$(k_j^\pm)_t = 0, \quad (\gamma_j^\pm)_t = \frac{1}{3}(n+1)(\xi_j^\pm)^n, \tag{3.5}$$

where we have used $\xi = -2\sqrt{3}k$, $\eta = -2\sqrt{3}l$, etc. Results in (3.4) and (3.5) coincide, for $n = 2$ with those derived for the $KP - I$ [25].

For the second hierarchy (2.16), the $L_n(k)$ are defined through

$$L_n(k)\psi = (\partial_t - B_n + \beta_n(k))\psi = 0. \tag{3.6}$$

From the asymptotics of $\psi^\pm$ and B_n, we then find

$$k_t = -\frac{i}{6\sqrt{3}}(-2\sqrt{3}k)^{n-1} \tag{3.7}$$

for all k in the k plane, while $\beta_n(k) = -\frac{1}{6}(n-1)(-2\sqrt{3}k)^{n-2}$. The correspondent time evolutions of the scattering data can be derived similarly, which are [16]

$$df/dt = \frac{1}{6}(n-1)(\xi^{n-2} - \eta^{n-2})f(k,l) \tag{3.8}$$

$$(\gamma_j^\pm)_t + \frac{1}{3}(n-1)(\xi_j^\pm)^{n-2}\gamma_j^\pm = -\frac{1}{3}(n-1)(n-2)(\xi_j^\pm)^{n-3} \qquad (3.9)$$

and $k_j^\pm$ also satisfy (3.7), where $df/dt = f_t + f_k k_t + f_l l_t$, k_t is given by (3.7) and l_t is (3.7) with l replacing k; $\xi, \eta, \xi_j^\pm$ are as before.

In summary one finds the equations (2.15) are the isospectral flows, while (2.16) are non-isospectral ones, i.e., when u satisfies (2.16), the correspondent discrete spectrum of (2.2) is no longer fixed as the time evolves. The presence of both isospectral and non-isospectral flows have been already studied in 1+1 dimensions (e.g. see [5]).

4. Relevant Algebraic Structures.

Denote $\mathcal{A}$ the linear space finitely spanned by all A_n, B_n. According to the definition of the Lax operators described in section 2, $\mathcal{A}$ is a set of Lax operators, but not the set of whole ones. It does not include these depending explicitly, but nonlinearly on y. However, the space $\mathcal{A}$ defined here still forms an infinite dimensional Lie algebra for its kind of Lax operators. The Lie bracket on $\mathcal{A}$ is defined by (2.11) for any $A, \check{A} \in \mathcal{A}$. To check its Jacobi identity and the closeness of $\mathcal{A}$ under the bracket, one only needs to verify the brackets among the basis A_n and B_n. These brackets are

$$\|[A_m, A_n]\| = 0, \quad \|[A_m, B_n]\| = \frac{m+1}{3} A_{m+n-2},$$
$$\|[B_m, B_n]\| = \frac{m-n}{3} B_{m+n-2}. \qquad (4.1)$$

They immediately imply the closeness of $\mathcal{A}$ and the Jacobi identity among themselves and then among elements of $\mathcal{A}$.

The proof of (4.1) is as follows, we firstly check that, for instance, $\|[A_m, B_n]\|$ is still a polynomial in ∂_x, in terms of the definition of the bracket in (2.11), then by direct calculation, we have $[\|[A_m, B_n]\|, L] = \|[K_m, \sigma_n]\|$ which is the multiplication operator, where for any two functionals $a(u)$ and $\check{a}(u)$, the bracket is defined as

$$\|[a, \check{a}]\| = a'[\check{a}] - \check{a}'[a], \qquad (4.2)$$

$$a'[\check{a}] = \lim_{\epsilon \to 0} \frac{\partial}{\partial \epsilon} a(u + \epsilon \check{a}), \quad \text{etc.}$$

Therefore $\|[A_m, B_n]\|$ is a Lax operator. Finally, we have the asymptotics $\|[A_m, B_n]\| \to [\overline{A}_m, \overline{B}_n] = \frac{m+1}{3} \overline{A}_{m+n-2}$ as $x, y \to \infty$. By the argument of uniqueness of the Lax operator, the second one in (4.1) can be proved and similarly to prove others. Detail can be found in [17, 20].

As the result, the multiplication operators K_n and σ_n corresponding to A_n and B_n respectively, satisfy the same algebraic structure as in (4.1), i.e.

$$\|[K_m, K_n]\| = 0, \quad \|[K_m, \sigma_n]\| = \frac{m+1}{3} K_{m+n-2},$$
$$\|[\sigma_m, \sigma_n]\| = \frac{m-n}{3} \sigma_{m+n-2}, \qquad (4.3)$$

but with respect to $\|[\,,\,]\|$ defined in (4.2). Brackets in (4.3) are then the starting point to construct symmetries for each member of our hierarchies of equations (2.15) and (2.16). An example is for the KP equation $u_t = K_2(u)$. Its symmetries are K_n and $\tau_n = tK_n + \sigma_n$, namely, they satisfy the linearized equation of KP: $\tau_t(u) = K_2'[\tau]$. Simply calculation shows that the symmetries of KP, K_n and τ_n also satisfy the realation (4.3) among themselves, which coincides with that found in [12, 27]. The results for other members in the higher-order KP equations can be found similarly [12, 27, 28] and for these in the second hierarchy (2.16), was given by Cheng and Li [29].

It is shown that both K_n and σ_n correspond to integrals I_n and J_n, i.e. $K_n = \partial_x(\delta I_n/\delta u)$; $\sigma_n = \partial_x(\delta J_n/\delta u)$ [12, 27, 28, 30], which is contrary to that in 1+1 dimensions. One reason I think is both hierarchies of equations (2.15) and (2.16) have the exactly Lax representations. Other argument on this can be found in [14-16, 31]. Through the correspondence $\|[f, g]\| = \partial_x \delta\{F, G\}/\delta u$ [12, 28, 30], where $\{F, G\} = \iint (\delta F/\delta u)\partial_x(\delta G/\delta u)dxdy$ and $f = \partial_x(\delta F/\delta u)$ etc., one finds that I_n and J_n also satisfy the same structure as in (4.3), with respect to their own Poisson bracket, upon this structure, infinitely many conserved quantities for equations (2.15) and (2.16) can be derived [28, 29]. Inparticular for the KP, they are I_n and $T_n = tI_n + J_n$.

Equations in (4.3) also indicate the commutability among flows (2.15) and (2.16), namely, (2.15) commute among themselves and (2.16) are the non-commuting flows. Notice that the non-commuting flows in 2+1 dimensions have also been discussed by Nijhoff [32]. It is not clear to me at the present whether there exists correspondence of non-commuting flows obtained here and there.

Return to the commuting flows, by using the argument that the basical field u can be considered as the function of infinitely many temporal variables t_0, t_1, t_2, ... i.e. $u = u(t_0, t_1, ...)$ then the flows (2.15) can be written as $u_{t_n} = K_n$ and so t_0, t_1, t_2 are respectively proportional to x, y, t. The commutability means $\partial^2 u/(\partial t_m \partial t_n) = \partial^2 u/(\partial t_n \partial t_m)$ and the first equation in (4.1) becomes

$$(A_m)_{t_n} - (A_n)_{t_m} + [A_m, A_n] \tag{4.4}$$

which is the formal Zakharov-Shabat equation.

As we have see in this section, the essential step is to construct the Lax algebra. The Lax algebra including all possible Lax operators is discussed in [26], and then more symmetries and conserved quantities can be obtained. Inparticular, results in parts for the KP coincide with these found in [30].

5. Derivation of the Mastersymmetries.

The mastersymmetry (MS) argument was proposed by Fuchssteiner [11] and its significance can be found in [11, 33]. Recall the deifinition of MS [11]:

For the set of commuting flows (2.15), let $K = span\{K_n(u), \; n = 0, 1, 2, ...\}$, a functional $\sigma(u)$ is said to be a MS of degree 1, if $\|[K, \; \sigma]\| \subset K$; inductively, $\sigma(u)$ is said a MS of degree n if each one in $\|[K, \; \sigma]\|$ is the MS of degree of $n - 1$.

From equations in (4.3), one finds that all $\sigma_n(u)$ are actually the MSs of degree 1, inparticular $\sigma_3(u)$ is the first "non-trivial" one (i.e. it generates symmetries non-trivially). Therefore, through our analysis, the problem to find MSs is equivant to derive the Lax operators or to solve the linear equations (2.5). and the Lax algebra presents an guarantee of this equivalence. Examples have already been given in (2.8).

Observe that the MS of degree 1 corresponds to the Lax operator with leading coefficient linearly depending on y. The argument is then that to derive MS of degree s, one needs to solve (2.5) with the leading coefficient proportionally to y^s (see [26] for detail). For examples the first few MSs of degree 2 of the KP are

$$\sigma_1^{(2)} = y^2 u_y + xy u_x + 2yu$$

$$\sigma_2^{(2)} = y^2 K_2 + 2xy K_1 + x^2 K_0 + \frac{8}{3} y(\partial_x^{-1} u_y) + \frac{4}{3} xu - 4\alpha y u_x + \frac{2}{3}(\partial_x^{-1} u)$$

where K_i and α are given before.

In this sense, the approach presented in this article gives rise to a pratical and straightforward method to derive the mastersymmetries for a variety of equations in 2+1 dimensions, even for these possessing non-canonical recursion operators (see examples in the next section).

6. Other Examples.

In the previous sections, we have seen how our approach works for the KP equation. The crucial step one finds is to derive the special Lax operator $B = B_3$ in (2.7) which is used to generate others, then the constructions of hierarchies of equations, their Lax representations, mastersymmetries and relevant algebraic structures, etc. can be achieved. We call such Lax operator $B = B_3$ the generator of Lax operator (GLO). Corresponding to the GLO $B = B_3$, the multiplication operator $\sigma = \sigma_3$ is the nontrivial MS, it can generate symmetries independently. The existence of GLO for the KP is not a fluke, it can be derived for many other equations as long as the associated spectral problems are known. Therefore, in this section, I only list the GLOs and the correspondent MSs of other examples.

1) The BO Equation

$$u_t(x,t) = 2uu_x + Hu_{xx}, \quad Hf(x) = \frac{P}{\pi}\int_{-\infty}^{+\infty} \frac{f(x)}{z-x} dz \qquad (6.1)$$

where P denotes the principal value. The BO equation is an 1+1 dimensional integro-differential equation, but its features are mostly closed to equations in 2+1 dimensions (see [16, 34]). The associated spectral problem reads [34]

$$L\varphi^+ = \lambda \varphi^-, \quad L = iD_x + u + \lambda, \qquad (6.2)$$

where $\varphi^\pm$ represent the analytic functions in the upper (+) and lower (-) half λ planes, the GLO in this case is

$$B^+ = -ix(D_x - i\lambda)^2 - \frac{1}{2}(D_x - i\lambda) + ixP_+u_x + iP_+u$$
$$B^- = -ix(D_x - i\lambda)^2 - \frac{3}{2}(D_x - i\lambda) + ixP_-u_x + iP_-u \qquad (6.3)$$

and the nontrivial MS of degree 1 is

$$\sigma = B^- L - LB^+ = x(2uu_x + Hu_{xx}) + u^2 + \frac{3}{2}Hu_x, \qquad (6.4)$$

where $P_{\pm} = \pm I + iH$, and I is the identity operator. Therefore the results similar to these for the KP can be obtained, inparticular, the application of IST of the spectral problem (6.2) to solve the correspondent hierarchies of NEEs were shown in [18].

2) The 2+1 Dimensional CDGKS Equation

This equation was given by Konopelchenko and Dubrovsky [35], it has the form

$$u_t = K(u) = \partial_x^{-1} u_{yy} - \frac{1}{5} u_{xxxxx} - u_x u_{xx} - u u_{xxx} \\ - u^2 u_x - u_{xxy} - u u_y - u_x \partial_x^{-1} u_y, \qquad (6.5)$$

and its spectral problem reads [35]

$$L = \partial_x^3 + u\partial_x + \partial_y \qquad (6.6)$$

which is cubic in ∂_x. The GLO in this example is

$$B = \frac{4}{5} y \partial_x^5 + (3u - \frac{3}{5}x)\partial_x^3 + 3y u_x \partial_x^2 \\ + [y(2u_{xx} + u^2 - \partial_x^{-1} u_y) - \frac{3}{5} xu - \frac{1}{5}(\partial_x^{-1} u)]\partial_x, \qquad (6.7)$$

and the corresponding MS is

$$\sigma = [B, \ L]\partial_x^{-1} = yK(u) + \frac{3}{5}xu_y - \frac{1}{5}u_x\partial_x^{-1}u + \frac{4}{5}u_{xx} - \frac{1}{5}u^2 + \frac{6}{5}(\partial_x^{-1}u_y), \qquad (6.8)$$

where K(u) is in (6.5). The IST of (6.6) can be shown in terms of the method proposed by Caudrey [36] so that the solvability of hierarchies of equations in this example can be carried out.

3) The Example Associated With the Schrodinger Spectral Problem in the Plane

Such spectral problem is usually written as [37, 38]

$$L\varphi = \lambda\varphi, \quad L = \partial_\xi^2 - \partial_\eta^2 + u(\xi, \eta, t). \qquad (6.9)$$

It is convenient for us, however, to write down (6.9) in the form

$$L\varphi = \lambda\varphi, \quad L = \partial_x \partial_y + u(x, y, t) \qquad (6.10)$$

since an apparent transformation on variables between (6.9) and (6.10).

Associated with (6.10), some interseting nonlinear equations have been obtained [22, 37, 38], which include a 2+1 dimensional analogue of the KdV equation. The extended recursion operator was derived by Boiti, Leon and Pempinelli [22], but it is a non-canonical (in the sense of Fokas and Santini) one. Another difficulty is (6.10) does not admit the usual non-trivial Lax operators [22, 38]. Here, following Manakov's idea [39], we generalize the notation of the Lax operators, we say a pair of polynomials $A^{\pm}(u)$ in ∂_x and ∂_y are Lax operator if $A^+ L - LA^-$ is a multiplication operator (i.e. the zeroth order polynomial in ∂_x and ∂_y). The standard way leads us to reach a class of linear equations for coefficients of $A^{\pm}$, their solutions, in turn, give exactly the Lax operators and the correspondent multiplication operators. For Lax operators only including powers of ∂_x (called "in x-direction" for short), the first few are

$$A_0^{\pm x} = \partial_x, \quad a_0^x = u_x;$$
$$A_1^{+x} = \partial_x^3 + (3\partial_y^{-1} u_x)\partial_x + (3\partial_y^{-1} u_{xx}), \quad A_1^{-x} = \partial_x^3 + (3\partial_y^{-1} u_x)\partial_x, \quad (6.11)$$
$$a_1^x = A_1^{+x} L - L A_1^{-x} = u_{xxx} + 3(u \partial_y^{-1} u_x)_x$$

etc. The GLO in the x-direction reads

$$B^{+x} = x\partial_x^3 + 2\partial_x^2 + (3x\partial_y^{-1} u_x + \partial_y^{-1} u)\partial_x + (4\partial_y^{-1} u_x + 3x\partial_y^{-1} u_{xx})$$
$$B^{-x} = x\partial_x^3 + \partial_x^2 + (3x\partial_y^{-1} u_x + \partial_y^{-1} u)\partial_x, \quad (6.12)$$

i.e. $B^{\pm x}$ generate two hierarchies of ($\pm$) Lax operators in x-direction respectively as usual. The non-trivial mastersymmetry is

$$\begin{aligned} b^x =& B^{+x} L - L B^{-x} = x(u_{xxx} + 3(u\partial_y^{-1} u_x)_x) \\ &+ 2u_{xx} + (u\partial_y^{-1} u)_x + 3u\partial_y^{-1} u_x. \end{aligned} \quad (6.13)$$

Using the symmetric property of (6.10), the Lax operator in y-direction (i.e. the polynomial only in ∂_y) can be obtained through replacing x and y by y and x, respectively, in all their positions apparently occupied (i.e. replace ∂_y by ∂_x, $x\partial_x^3$ by $y\partial_y^3$, $x\partial_y^{-1} u_x$ by $y\partial_x^{-1} u_y$, etc.). We denote these Lax operators as $A_i^{\pm y}$, the GLO as $B^{\pm y}$, and the MS as b^y etc. For example a_1^y is $a_1^y = u_{yyy} + 3(u\partial_x^{-1} u_y)_y$. In general case, the Lax operators, etc. can always be expressed as combinations of these in x-direction and these in y-direction. The example is $A_1^{\pm} = A_1^{\pm x} + A_1^{\pm y}$, it corresponds to $a_1 = A_1^+ L - L A_1^- = a_1^x + a_1^y$ and the following flow

$$u_t = a_1(u) = u_{xxx} + u_{yyy} + 3(u\partial_y^{-1} u_x)_x + 3(u\partial_x^{-1} u_y)_y. \quad (6.14)$$

Equation (6.14) is a natural generalization of the KdV [22, 37, 38]. The GLO and its correspondent MS are $B^{\pm} = B^{\pm x} + B^{\pm y}$, $b = b^x + b^y$, and so the hierarchies of equations, the Lax algebra and symmetries with their algebraic structure etc. can be obtained. The detail for the present example is shown in [21].

There are some other examples, such as the 2+1 dimensional N-wave interaction including the Davey-Stewarton equation, and the modified KP equation. Results for them have been shown in [19, 20].

Acknowledgement.

I am greatly indebted to Professors R.K. Bullough and Yi-shen Li for their constant encouragement and fruitful discussion, some of results reviewed in this article are due to collaboration with them.

The work was supported by the Fok Ying-Tung Education Foundation in China.

References.

1. P.D. Lax, Commun. Pure Appl. Math. 21(1968), 467.
2. A.C Newell, "Solitons in Mathematics and Physics" (SIAM, Philadelphia 1985).

3. F. Magri, in "Nonlinear Evolution Equations and Dynamical Systems", M. Boiti, F. Peminelli and G. Soliani eds., Lect. Notes Phys. vol.120 (Springer, New York 1980) 233.
4. A.S. Fokas and B. Fuchssteiner, Physica 4D(1981), 47.
5. F. Calogero and A. Degasperis, "The Spectral Transform and Solitons I", (North-Holland, Amsterdam, 1982).
6. V.G. Drinfel'd and V.V. Sokolov, Dokl. Akad. SSSR 258(1981), 11.
7. G. Wilson, Ergodic Theory and Dynamical System 1(1981), 361.
8. H. Flashka, A.C. Newell and T. Ratiu, Physica 9D(1983), 300; 324.
9. Y. Ohta, J. Satsuma, D. Takahashi and T. Tokihiro, Progre. Theor. Phys. (Suppl.) 94(1988), 210.
10. M. Jimbo and T. Miwa, Publ. RIMS. Kyoto Univ. 19(1983), 943.
11. B. Fuchssteiner, Progre. Theor. Phys. 70(1983), 1508.
12. H.H. Chen, Y.C. Lee and J.E. Lin, Physica 9D(1983), 493.
13. A.Yu. Orlof and E.I. Shulman, Lett. Math. Phys.12(1986), 171.
14. P.M. Santini and A.S. Fokas, Commun. Math. Phys.
15. A.S. Fokas and P.M. Santini, Commun. Math. Phys.
16. A.S. Fokas and P.M. Santini, J. Math. Phys. 29(1988), 604.
17. Y. Cheng, Yi-shen Li and R.K. Bullough, J. Phys. A: Math. Gen. 21(1988), L443.
18. Y. Cheng, Phys. Lett. 127A(1988), 205.
19. Y. Cheng, Physica 34D(1989), 277.
20. Y. Cheng, "Hierarchies of equations and their symmetries for the 2+1 dimensional nonlinear evolution equations", to be published in Physica D.
21. Y. Cheng, "Integrable systems associated with the Schrodinger spectral problem in the plane", preprint, USTC, Hefei, China.
22. M. Boiti, J.J-P. Leon and F. Pempinelli, Stud. Appl. Math. 78(1988), 1.
23. Yi-shen Li, Y. Cheng and Y.B. Zeng, "A Lie algebraic structure of G.J. equation and its gauge equivalent Yang hierarchy", in this volume.
24. V. Dryuma, Sov. Phys. JEPT Lett. 19(1974), 387.
25. A.S. Fokas and M.J. Ablowitz, Stud. Appl. Math. 69(1983), 211.
26. Y. Cheng and G.C. Zhu, "The Lax algebra for the KP equation and its applications", preprint, USTC, Hefei, China.
27. W. Oevel and B. Fuchssteiner, Phys. Lett. 88A(1982), 323.
28. M. Case, J. Math. Phys. 26(1985), 1158.
29. Y. Cheng and Yi-shen Li, Physica 28D(1987), 189.
30. H.H. Chen and J.E. Lin, Physica 26D(1987), 171.
31. A.S. Fokas, Stud. Appl. Math. 77(1987),253.
32. F.W. Nijhoff, Physica 31D(1988), 339.
33. W. Oevel, in "Topics in soliton theory and exactly solable nonlinear equations", M. Ablowitz, B. Fuchssteiner and M. Kruskal eds. (World Scientific, Singapore 1987), 108.
34. A.S. Fokas and M.J. Ablowitz, Stud. Appl. Math. 68(1986), 271.
35. B.G. Konopelchenko and V.G. Dubrovsky, Phys. Lett. 102A(1984), 15.
36. P.J. Caudrey, "Two-dimensional spectral transforms", preprint, UMIST, Manchester.
37. A.P. Veselov and S.P. Novikov, Dokl. Acad. Nauk. SSSR 279(1984)
38. M. Boiti, J.J-P. Leon, M. Manna and F. Pempinelli, Inverse Problem 2(1986), 271.
39. S.V. Manakov, Uspekhi. Mat. Nauk. 31(1976), 245.

Determination of Nondegenerate Darboux Operators of First Order in 1 + 2 Dimensions

Zhou Zixiang

Institute of Mathematics, Fudan University,
200433 Shanghai, People's Rep. of China

In 1+2 dimensions, some Darboux operators have been constructed before. In this paper, all the nondegenerate Darboux operators of first order are given for quite general Lax pairs without reduction. They take the form which is already known. The Darboux operators or Darboux matrices in 1+1 dimensions are discussed as special cases.

1. Introduction

Darboux transformation method is an effective method to get explicit solutions of some nonlinear partial differential equations. For 1+1 dimensional problems, Darboux matrix has been known quite clearly (eg. [5,10]). In 1+2 dimensions, the spectral parameter in 1+1 dimensions is usually replaced by a derivative with respect to one variable. Thus the fundamental Darboux transformations are given by differential operators (Darboux operators(DOs)) rather than polynomials of the spectral parameter in 1+1 dimensions.

Let m_N be the set of all N×N complex matrices. Ω is a simply connected domain in $\mathbb{R}^3$ with coordinates x, y, t. Denote $\partial = \partial/\partial x$,

$$\mathcal{D}_N(\Omega) = \left\{ \sum_{j=0}^{r} A_j \partial^j \;\Big|\; A_j \in C^\infty(\Omega, m_N),\; r \geq 0 \right\}.$$

All the functions are assumed to be infinitely differentiable.

Now we consider an equation (or a system of equations)

$$F(x,y,t,u,u_x,u_y,u_t,u_{xx},\ldots) = 0 \tag{1}$$

of unknowns $u = (u_1, \ldots, u_s)$ in Ω which admits a Lax pair

$$\begin{cases} \Phi_y = U(\partial)\Phi \\ \Phi_t = V(\partial)\Phi \end{cases} \tag{2}$$

Here

$$\begin{cases} U(\partial) = U(x,y,t,u,u_x,\ldots,\partial) = \sum_{j=0}^{m} U_{m-j}(x,y,t,u,u_x,\ldots)\partial^j \\ V(\partial) = V(x,y,t,u,u_x,\ldots,\partial) = \sum_{j=0}^{n} V_{n-j}(x,y,t,u,u_x,\ldots)\partial^j \end{cases} \tag{3}$$

$U_j, V_j \in C^\infty(\Omega, m_N)$ if u is given.

(2) is a Lax pair of (1) implies that (1) is equivalent to

$$U_t(\partial) - V_y(\partial) + [U(\partial), V(\partial)] = 0 \tag{4}$$

which is the integrability condition of (2). Here we assume (2) is integrable in the sense that (2) is locally solvable for any initial data defined at $x=x_0$. [12]

The Bäcklund transformation in the form of differential equations or integro-differential equations (eg.[3,6,7,8]) as well as the inverse scattering transformation (eg.[1,2,4]) for a lot of equations or systems contained in (2) have already been known. As for Darboux transformation, the DO for the equations possessing scalar Lax pair as KP equation has also been known (eg.[11]). For general unreduced Lax pair (2), [12] showed that any nondegenerate matrix solution H of (2) generated a DO $\partial - H_x H^{-1}$. The DO for Davey-Stewartson equation was obtained in this way.

In this paper, we shall show that these are all the possible DOs in the form $\partial - S(x,y,t)$. Also, by reducing to 1+1 dimensions, we shall give the corresponding conclusions for 1+1 dimensional problems.

2. Darboux operators for 1+2 dimensional Lax pairs

For equation (1) with Lax pair (2), a differential operator $G(x,y,t,\partial) \in \mathcal{D}_N(\Omega)$ is called a DO if there exists $\tilde{u}$ such that for any solution Φ of (2), $\tilde{\Phi} = G(\partial)\Phi$ satisfies

$$\begin{cases} \tilde{\Phi}_y = \tilde{U}(\partial)\tilde{\Phi} \\ \tilde{\Phi}_t = \tilde{V}(\partial)\tilde{\Phi} \end{cases} \tag{5}$$

where $\tilde{U}(\partial) = U(x,y,t,\tilde{u},\tilde{u}_x,\ldots,\partial)$, $\tilde{V}(\partial) = V(x,y,t,\tilde{u},\tilde{u}_x,\ldots,\partial)$.

Obviously, $\tilde{U}(\partial)$, $\tilde{V}(\partial)$ satisfy

$$\begin{cases} \tilde{U}(\partial)G(\partial) = G(\partial)U(\partial) + G_y(\partial) \\ \tilde{V}(\partial)G(\partial) = G(\partial)V(\partial) + G_t(\partial) \end{cases} \tag{6}$$

$$\tilde{U}_t(\partial) - \tilde{V}_y(\partial) + [\tilde{U}(\partial), \tilde{V}(\partial)] = 0. \tag{7}$$

Therefore, we obtain a new solution $\tilde{u}$ of (1) by the action of the DO.

This section is devoted to the equation (4) without reduction, i.e. the entries of U_j, V_j are independent unknowns. Then $G(x,y,t,\partial) \in \mathcal{D}_N(\Omega)$ is a DO if and only if there exist $\tilde{U}(\partial)$, $\tilde{V}(\partial) \in \mathcal{D}_N(\Omega)$ such that (6) holds.

A nondegenerate DO of first order is a DO $G(x,y,t,\partial) = R(x,y,t) \cdot (\partial - S(x,y,t))$ with R nondegenerate. Since a matrix R is a trivial DO if we do not consider reduction, we can always choose R=I.

Nondegenerate DOs of first order can be constructed explicitly as follows.

Theorem 1. $\partial - S(x,y,t)$ is a DO of (2) if and only if $S = H_x H^{-1}$ for some N×N nondegenerate matrix solution H of (2).

Before the proof, we have some preparations.

For any $M \in C^\infty(\Omega, m_N)$, let M_j be defined inductively by

$$\begin{cases} M_0 = I \\ M_{j+1} = M_{j,x} + M_j M \quad (j \geqslant 0). \end{cases} \quad (8)$$

Let

$$U(M) = \sum_{j=0}^{m} U_{m-j} M_j. \quad (9)$$

Then, for any Ψ satisfying $\Psi_x = M\Psi$, we have

$$U(\partial)\Psi = U(M)\Psi. \quad (10)$$

Lemma. $\partial - S$ is a DO of (2) if and only if S satisfies

$$\begin{cases} S_y + [S, U(S)] = (U(S))_x \\ S_t + [S, V(S)] = (V(S))_x. \end{cases} \quad (11)$$

Proof. First suppose $\partial - S$ is a DO of (2). Choose a fundamental solution matrix Ψ of $\Psi_x = S\Psi$, then (6) implies

$$S_y \Psi = (\partial - S)U(S)\Psi = (U(S))_x \Psi - [S, U(S)]\Psi,$$

which leads to (11).

Conversely, suppose S is a solution of (11). Let

$$\tilde{U}(\partial) = \sum_{j=0}^{m} \tilde{U}_{m-j} \partial^j \quad (12)$$

where $\tilde{U}_j$'s are defined inductively by

$$\begin{cases} \tilde{U}_0 = U_0 \\ \tilde{U}_{j+1} = U_{j+1} + U_{j,x} - SU_j + \sum_{k=0}^{j} C_{m-k}^{m-j} \tilde{U}_k \partial^{j-k} S. \end{cases} \quad (13)$$

Then,

$$D(\partial) \equiv S_y - (\partial - S)U(\partial) + \tilde{U}(\partial)(\partial - S) \in C^\infty(\Omega, m_N).$$

However, for the fundamental solution matrix Ψ of $\Psi_x = S\Psi$, (11) gives $D(\partial)\Psi = 0$. This means $D(\partial) = 0$ as a matrix. QED.

Proof of Theorem 1. Suppose H is an N×N nondegenerate matrix solution of (2), $S = H_x H^{-1}$. Then (2) leads to (11) immediately.

Conversely, suppose $G(\partial) = \partial - S(x,y,t)$ is a DO of (2), we need to find a solution H of (2) such that $S = H_x H^{-1}$, or equivalently, we need to solve

$$\begin{cases} H_x = SH \\ H_y = U(\partial)H \\ H_t = V(\partial)H. \end{cases} \quad (14)$$

25

Again, this is equivalent to

$$\begin{cases} H_x = SH \\ H_y = U(S)H \\ H_t = V(S)H \end{cases} \qquad (15)$$

by (10). Therefore, we only need to verify the integrability condition of (15).

Let Ψ be a fundamental solution matrix of $\Psi_x = S\Psi$. From (6),

$$(\Psi_y - U(\partial)\Psi)_x = (S\Psi)_y - \partial U(\partial)\Psi = S(\Psi_y - U(\partial)\Psi),$$

thus,

$$(V_y(\partial) + V(\partial)U(\partial))\Psi = (V(\partial)\Psi)_y - V(\partial)(\Psi_y - U(\partial)\Psi)$$
$$= (V(S)\Psi)_y - V(S)(\Psi_y - U(S)\Psi) = V(S)_y \Psi + V(S)U(S)\Psi.$$

We have a similar equation by changing U and V. These lead to

$$U(S)_t - V(S)_y + [U(S), V(S)] = 0 \qquad (16)$$

by the integrability condition (4), since $\det \Psi \neq 0$.

The lemma implies that other two integrability conditions $H_{xy} = H_{yx}$, $H_{xt} = H_{tx}$ hold. Therefore, (15) has an N×N nondegenerate matrix solution. QED.

This theorem implies that any nondegenerate DO of first order can be determined only by an N×N matrix solution of the Lax pair. Thus, we obtain infinite number of solutions of (4) in the usually way.[5]

3. Application to 1+1 dimensional problems

We consider the equation (or the system of equations)

$$F(x, t, u, u_x, u_t, u_{xx}, \ldots) = 0 \qquad (17)$$

defined in Ω (a simply connected domain in $\mathbb{R}^2$) which possesses the Lax pair

$$\begin{cases} \lambda \Phi = U(\partial)\Phi \\ \Phi_t = V(\partial)\Phi. \end{cases} \qquad (18)$$

Here

$$\begin{cases} U(\partial) = U(x,t,u,u_x,\ldots,\partial) = \sum_{j=0}^{m} U_{m-j}(x,t,u,u_x,\ldots) \partial^j \\ V(\partial) = V(x,t,u,u_x,\ldots,\partial) = \sum_{j=0}^{n} V_{n-j}(x,t,u,u_x,\ldots) \partial^j, \end{cases} \qquad (19)$$

U_0 is nondegenerate.

Also, we assume that (18) is integrable, in the sense that for any $\lambda \in \mathbb{C}$, $(x_0, t_0) \in \Omega$ and $\Phi_0, \ldots, \Phi_{m-1} \in \mathfrak{m}_N$, there exists a local solution Φ of (18) such that $\partial^j \Phi(x_o, t_o) = \Phi_j$ ($j=0,1,\ldots,m-1$). The integrability condition (necessary) of (18) is

$$U_t(\partial) + [U(\partial), V(\partial)] = 0. \tag{20}$$

The simple examples of (20) are KdV equation and Boussinesq equation, the DO for the latter is given in [9].

For an unreduced equation (20), the entries of U_j, V_j are independent unknowns, then $G(x,t,\partial) \in \mathfrak{D}_N(\Omega)$ is a DO of (18) if and only if there exist $\tilde{U}(\partial), \tilde{V}(\partial) \in \mathfrak{D}_N(\Omega)$ such that for any solution Φ of (18), $\tilde{\Phi} = G(\partial)\Phi$ satisfies

$$\begin{cases} \lambda \tilde{\Phi} = \tilde{U}(\partial)\tilde{\Phi} \\ \tilde{\Phi}_t = \tilde{V}(\partial)\tilde{\Phi}. \end{cases} \tag{21}$$

The nondegenerate DO of first order $G(x,t,\partial) = \partial - S(x,t)$ is given as follows, using the conclusions in 1+2 dimensions.

Theorem 2. $\partial - S(x,t)$ is a DO of (18) if and only if $S = H_x H^{-1}$ where H is an N×N nondegenerate matrix solution of

$$\begin{cases} H\Lambda = U(\partial)H \\ H_t = V(\partial)H, \end{cases} \tag{22}$$

and Λ is a constant matrix.

Proof. Suppose $\partial - S(x,t)$ is a DO of (18). Let

$$\Delta(\partial) = (\partial - S)U(\partial) - \tilde{U}(\partial)(\partial - S).$$

For any solution Φ of (18), $\Delta(\partial)\Phi = 0$ by (21). This implies

$$\lambda(\tilde{U}_o U_o^{-1} - 1)\Phi_x + \lambda(\tilde{U}_1 - \tilde{U}_o U_o^{-1}(U_1 + U_{o,x}) + SU_o - \tilde{U}_o S)U_o^{-1}\Phi + M(\partial)\tilde{\Phi} = 0$$

by (18), where $M(\partial)$ is independent of λ, and the order is less than m. From the integrability of (18),

$$\tilde{U}_o = U_o, \quad \tilde{U}_1 = U_1 + U_{o,x} + [U_o, S].$$

Now it is easy to check that the order of $\Delta(\partial)$ is less than m. Hence $\Delta(\partial) = 0$ since it annihilates any solution Φ of (18). Thus

$$\begin{cases} 0 = (\partial - S)U(\partial) - \tilde{U}(\partial)(\partial - S) \\ S_t = (\partial - S)V(\partial) - \tilde{V}(\partial)(\partial - S). \end{cases} \tag{23}$$

(The second one is obtained from (21) directly.)

Now consider the equation

$$\begin{cases} \Psi_y = U(\partial)\Psi \\ \Psi_t = V(\partial)\Psi \end{cases} \tag{24}$$

for $(x,t) \in \Omega$, $y \in \mathbb{R}$. From (23), $\partial - S$ is a DO of (24). According to

Theorem 1, there exists a solution $K(x,y,t)$ of (24) such that $S=K_x K^{-1}$. Let $\Lambda = K^{-1} K_y$, then we can check that Λ is indeed a constant matrix by (11) and (16). Hence

$$K(x,y,t) = H(x,t)\exp(\Lambda y)$$

where H satisfies (22) and $S = H_x H^{-1}$.

Conversely, if H is a solution of (22), $S = H_x H^{-1}$, then it is easy to see that S satisfies (23), or equivalently $\partial - S$ is a DO of (18). QED.

If $N=1$, H must be a solution of the Lax pair (18). This gives the DO for Gelfand-Dikij system.

For the Lax pair

$$\begin{cases} \Phi_y = U(y,t,\lambda)\Phi \\ \Phi_t = V(y,t,\lambda)\Phi \end{cases} \qquad (25)$$

where U,V are two polynomials of λ, we can get the similar conclusions as Theorem 2, which are the partial results in [13].

Acknowledgement

This paper is supported by the Chinese Fund of Natural Sciences and the Chinese Fund of Doctor Program. The author would like to express his gratitude to Prof. Gu Chaohao and Prof. Hu Hesheng for encouragement and many helpful suggestions. He also would like to thank Prof. Tu Guizhang for helpful suggestions.

References

[1] Ablowitz M.J., Bar Yaacov D. and Fokas A.S., Stud. Appl. Math. 69 (1983), 135.
[2] Beals R. and Coifman R.R., Proc. Sym. Pure Math. 43 (1985), 45.
[3] Boiti M., Konopelchenko B.G. and Pempinelli F., Inverse Problem 1 (1985), 33.
[4] Fokas A.S. and Ablowitz M.J., J. Math. Phys. 25 (1984), 2494.
[5] Gu C.H., On the Darboux Form of Bäcklund Transformations, Proc. Nankai Symposium on Intrgrable Systems (1987), to be published.
[6] Hirota R. and Satsuma J., J. Phys. Soc. Japan 45 (1978), 1741.
[7] Konopelchenko B.G. and Dubrovsky V.G., Physica 16D (1985), 79.
[8] Levi D., Pilloni L. and Santini P.M., Phys. Lett. A81 (1981), 419.
[9] Li Y.S. and Gu X.S., Ann. Diff. Eqs. 2 (1986), 419.
[10] Sattinger D.H. and Zurkowski V.D., Physica 26D (1987), 225.
[11] Tian C., Generalized KP equation and Miura Transformation, preprint (1986).
[12] Zhou Z.X., Lett. Math. Phys. 16 (1988), 9.
[13] Zhou Z.X., General Form of Nondegenerate Darboux Matrices of First Order for 1+1 Dimensional Unreduced Lax Pairs, preprint (1989).

A Series of New Exact Solutions to the Nonlinear Equation $y_t + y_{xxx} - 6y^2 y_x + 6\lambda y_x = 0$

Au Chi

Department of Mathematical Studies, Hong Kong Polytechnic, Hong Kong

Following the Bäcklund transformation and using the theorems proved by the same authors, previously we obtained sets of new solutions to the KdV equation and the nonlinear equation $y_t + y_{xxx} - 6y^2 y_x + 6\lambda y_x = 0$ which transforms into the modified KdV equation when $\lambda = 0$. In this paper we present another new series of solutions to the above nonlinear equation.

I. Introduction

The same authors have employed the Bäcklund transformation[1,2] approach to study the mathematical and physical properties of a number of nonlinear equations. Using our methodology, via Bäcklund transformation we put in a "seed" (a KdV solution) and obtained previously a set of solutions to the KdV equation. In particular, using $u = b$ as our "seed", we have found[2] certain new soliton solutions to the KdV equation

$$u_t + u_{xxx} + 12uu_x = 0, \qquad (1)$$

and have discovered that the solutions contain the vacuum parameter b (in the KdV case, b is the asymptotic value of the solution as $x \to \pm \infty$) which has improtant physical significance; the soliton velocity, amplitude, and width are all functions of b. In the second paper of our series[3], we have used rather powerful theorems to relate the solution of the KdV equation and nonlinear equation

$$y_t + y_{xxx} - 6y^2 y_x + 6\lambda y_x = 0, \qquad (2)$$

which transforms into the modified KdV equation if $\lambda = 0$. One set of our solutions to Eq. (2) is a kink-antikink solution[3], which tends to $\pm\sqrt{\lambda - 2b}$ as $x \to \pm \infty$, instead of simply b in the case of KdV equation. It is interesting to note that the solutions so far found for the KdV equation are nontopoligical[3,4], while the solutions[3] to Eq. (2) are topological.

In the third and fourth of our papers[4,5], using the KdV solution $u = -1/x^2$ as the seed, we have derived other new solutions to both the KdV equation and Eq. (2) using our theorems[3] which bridge these

Research Reports in Physics Nonlinear Physics
Editors: Gu Chaohao · Li Yishen · Tu Guizhang
© Springer-Verlag Berlin, Heidelberg 1990

solutions. While these new soliton solutions to the KdV equation are close to the nature of the conventional one-soliton solutions (nontopological), the solutions to Eq. (2) are kink-antikink solutions (topological). We have discovered that the new soliton solutions to (1) and (2) change their velocities, amplitudes, and widths as they travel from remote distances. Such properties lead us to discover that soliton solutions can show "annihilation and creation" phenomena as they propagate.[6] It is worth noting that these two equations [(1) and (2)] are classical wave equations.

In this paper five theorems concerning the Bäcklund transformation which bridges the solutions of the KdV equation

$$u_t + 12uu_x + u_{xxx} = 0,$$

and the nonlinear equation (the λ MKdV equation)

$$y_t + y_{xxx} - 6y^2 y_x + 6\lambda y_x = 0,$$

are presented.[2-6] The reverse transformation of the Miura transformation is also given.

A number of sets of new solutions to the λMKdV equation are obtained via such a Bäcklund transformation.

Since the equation has been applied to many physical situations, new solutions to the equation are of both physical and mathematical significance.

II. Bridge Between the Solutions of the Korteweg-de Vries Equations and the λMKdV equation

Bäcklund (1882) transformation (BT) is a system of mathematical relations which connects a solution to a nonlinear differential equation with another solution.

There are two kinds of BT, the auto BT which connects a solution for a nonlinear differential equation with another solution of the same equation; the generalized BT which connects a solution with another solution of another equation. In the BT procedure, if one of the solutions of a certain nonlinear differential equation is known, by using the BT relation we can find out another solution to the differential equation.

Wahlquist and Estabrook[1] (1975) obtained the Bäcklund transformation for the KdV equation by using their prologation treatment. They found that if the function $u(x,t)$ is a solution to the KdV equation and the function $Y(x,t)$ satisfies

$$Y_x = \lambda - 2u - Y^2, \quad (3)$$
$$Y_t = -4(\lambda + u)Y_x + 2u_{xx} - 4u_x Y, \quad (4)$$

then the function $u*(x,t)$ defined as

$$u* = \lambda - u(x,t) - Y^2, \tag{5}$$

is a solution to the KdV equation (1); where λ is a constant.

Miura[7] (1968) discovered that if $y(x,t)$ is a solution to the modified KdV equation (MKdV for short)

$$y_t + y_{xxx} - 6y^2 y_x = 0, \tag{6}$$

then $u(x,t)$, defined by

$$u(x,t) = -\frac{1}{2}(y^2 + y_x), \tag{7}$$

satisfies the KdV equation (1).

It is well known that the Miura transformation enables us to obtain solutions to the KdV equation. It is a generalized BT. In fact as pointed out by Miura (1978)[8] himself: "However, here the transformation is between two nonlinear equations, neither of which can be solved." Thus, one can not obtain a solution via the Miura transformation directly. The most significant use of the Miura transformation is the development of the inverse scattering method for the KdV equation. (Gardner, Greene, Kruskal, and Miura 1967, 1974)[9,10].

In this section, according to our investigation on the relation between the function $u(x,t)$, $y(x,t)$, $u*(x,t)$ and $Y(x,t)$ expressed by equation (1), (2), (5) and (3), (4), we present five relevant theorems. Based on these theorems we establish the relations for the solutions of three different equations, the KdV equation (1), the prologation differential equations (3), (4) and λMKdV equation (2).

<u>Theorem 1</u> If $y(x,t)$ is a solution of the λMKdV equation (2)

$$y_t + y_{xxx} - 6y^2 y_x + 6\lambda y_x = 0,$$

then the function

$$u(x,t) = -\frac{1}{2}(y^2 + y_x - \lambda), \tag{8}$$

is a solution of the KdV equation.

We observe that the Miura transformation (7) becomes the special case of our relation (8) when $\lambda = 0$.

And we note that $-y$ is also a solution to the λMkDV equation (2). Substituting $-y$ into (8), the solution to u is shown to be identical to the Bäcklund transformation $u*$ in (5).

<u>Theorem 2</u> If $u(x,t)$ is a solution to the KdV equation (1), and $Y(x,t)$ is a solution to equations (3) and (4)

$$Y_x = \lambda - 2u - Y^2,$$
$$Y_t = -4(\lambda + u)Y_x + 2u_{xx} - 4u_x Y,$$

then $Y(x,t)$ is a solution to the λMKdV equation (2).

This is the reverse transformation of the above (generalized Miura) transformation.

If and only if $u(x,t)$ satisfies KdV equation (1), then the integrability condition

$$Y_{xt} = Y_{tx}$$

holds.

In fact previously one could not find the inverse transformation because the explicit expression for Y_t is not given in Miura's transformation. Our transformation (3), (4) and (5) bridge the solutions to the KdV and the λMKdV equation (which includes the MKdV equation as a special case) in both forward and backward directions. One can use them to find the exact solutions to KdV equation and λMKdV equation[2-6].

Theorem 3 If $y(x,t)$ is a solution to λMKdV equation:

$$y_t + y_{xxx} - 6y^2 y_x + 6\lambda y_x = 0,$$

and $u(x,t)$ is given by (8)

$$u(x,t) = -\frac{1}{2}(y^2 + y_x - \lambda),$$

then, $y(x,t)$ also satisfies (4):

$$Y_t = -4(\lambda + u)Y_x + 2u_{xx} - 4u_x Y.$$

It is worth noting that the λMKdV equation (2) can be transformed to the MKdV equation, via the following coordinates' transformation

$$t' = t,$$
$$x' = x - 6\lambda t.$$

Theorem 4 If $y(x,t)$ is a solution to the λMKdV Eq. (2), then the function $y(x + 6\lambda t, t)$ is a solution to the MKdV equation (6).

The reverse statement is also true:

Theorem 5 If $y(x,t)$ is a function to the MKdV equation (6), then the function $y(x - 6\lambda t, t)$ is a solution to the λMKdV equation (2).

We get new solutions to (2) via our Bäcklund transformation (3), (4) which contains a rather general parameter λ. If we set $\lambda = 0$ at the beginning (namely attempting to solve MKdV instead of (2)), we would lose some solutions for equation (2) and the MKdV equation[5].

III. Several Sets of New Solutions to the Nonlinear Equation $y_t + y_{xxx} - 6y^2 y_x + 6\lambda y_x = 0$, Using Bäcklund Transformation

We have already obtained a one-soliton KdV solution with nonzero vacuum parameter

$$u = (\mu-b) - (\mu-2b)\left(\frac{Ce^{(\mu-2b)^{1/2}r} - e^{-(\mu-2b)^{1/2}r}}{Ce^{(\mu-2b)^{1/2}r} + e^{-(\mu-2b)^{1/2}r}}\right)^2. \tag{9}$$

[where $r = x - 4(b+\mu)t$ and C, μ are constants and b is the vacuum parameter[2].] Now we use this solution as our seed, to find other new analytical solutions to the λMKdV equation (2), we find that Eq.(4) can be expressed as

$$y_t + 4(\lambda + b + \mu' - L^2)y_x = -4{\mu'}^2 + 16\mu' L^2 - 12L^4 + 8(\mu' L - L^3)y, \qquad (10)$$

The parameter μ' cannot be equal to zero, otherwise seed (9) will become $u(x,t) = b$, leading to solutions which have already been found[2].

The characteristic equations of (10) are

$$dt = \frac{dx}{4(\lambda+b+\mu'-L^2)} = \frac{dy}{-4{\mu'}^2+16\mu' L^2-12L^4+8(\mu' L-L^3)y}. \qquad (11)$$

We now solve Eqs. (11) and (3), and obtain the following real solutions to the λMKdV equation (2)

$$y_i = \frac{L(L^2-\mu')}{\lambda'-L^2} + \frac{\sqrt{\lambda'}(\lambda'-\mu')}{\lambda'-L^2} T(\sqrt{\lambda'}\,\xi_i), \qquad i = 1, 2, 3 \qquad (12)$$

where

$$T(\alpha\eta) \equiv \frac{ce^{\alpha\eta} - e^{-\alpha\eta}}{ce^{\alpha\eta} + e^{-\alpha\eta}}, \qquad (13)$$

$$L \equiv \sqrt{\mu'}\, T(\sqrt{\mu'}\, r), \qquad r = x - 4(b + \mu)t, \qquad (14)$$

$$\mu' = \mu - 2b, \qquad \lambda' = \lambda - 2b,$$

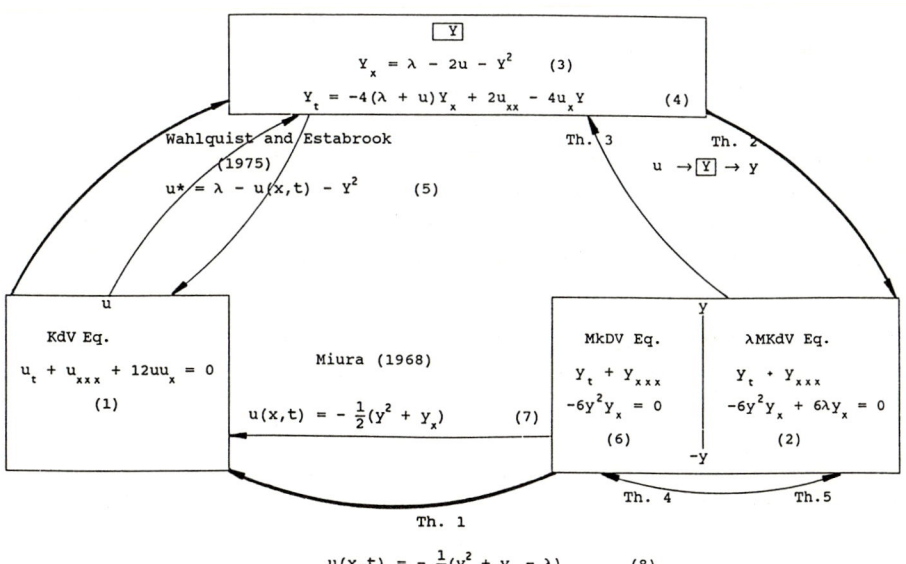

A bridge between the KdV and λMKdV equation

33

$$\xi_1 = \frac{-1}{2\sqrt{\lambda'}}\ln\frac{|\sqrt{\lambda'}+L|}{|\sqrt{\lambda'}-L|} - \frac{1}{\sqrt{|\mu'|}}\text{arctg}\frac{L}{\sqrt{|\mu'|}} - 4(\lambda'-\mu')t, \quad (\text{for } \lambda' > 0, \mu' < 0) \tag{15}$$

$$\xi_2 = \frac{1}{2\sqrt{\mu'}}\ln\frac{|\sqrt{\mu'}+L|}{|\sqrt{\mu'}-L|} + \frac{1}{\sqrt{|\lambda'|}}\text{arctg}\frac{L}{\sqrt{|\lambda'|}} - 4(\lambda'-\mu')t, \quad (\text{for } \lambda' < 0, \mu' > 0) \tag{16}$$

$$\xi_3 = \frac{1}{\sqrt{|\lambda'|}}\text{arctg}\frac{L}{\sqrt{|\lambda'|}} - \frac{1}{\sqrt{|\mu'|}}\text{arctg}\frac{L}{\sqrt{|\mu'|}} - 4(\lambda'-\mu')t, (\text{for}\lambda'<0,\mu'<0,\lambda'\neq\mu') \tag{17}$$

$$Y_4 = -L + \frac{\lambda'}{\lambda'-L^2}\frac{1}{\frac{L}{2(\lambda'-L^2)} - \frac{1}{2\sqrt{|\lambda'|}}\text{arctg}\frac{1}{\sqrt{|\lambda'|}} - 4\lambda't+D}, \quad (\text{for } \lambda' = \mu' < 0) \tag{18}$$

$$Y_5 = \frac{\mu'-L^2}{L} - \frac{\mu'}{L^2}\frac{1}{\frac{1}{L} + \frac{1}{\sqrt{|\mu'|}}\text{arctg}\frac{1}{\sqrt{|\mu'|}} - 4\mu't + D}. \quad (\text{for } \lambda' = 0, \mu' < 0) \tag{19}$$

References

1. H.D. Walhquist and F.B. Estabrook, J. Math. Phys. 16, 1(1975).
2. C. Au and P.C.W. Fung, Phys. Rev. B 25, 6460(1982).
3. P.C.W. Fung and C. Au, Phys. Rev. B 26, 4035(1982).
4. C. Au and P.C.W. Fung, J. Math. Phys. 25, 1364(1984).
5. P.C.W. Fung and C. Au, J. Phys. A 16, 1611(1983).
6. C. Au and P.C.W. Fung, Phys. Rev.,B 30, 1797(1984).
7. R.M. Miura, J. Math Phys. 9, 1902(1968).
8 R.M. Miura," Solitons in Action", ed. K. Lonngren and A. Scott, Academic Press, New York. (1978)
9. C.S. Gardner, J.M. Greene, M.D. Kruskal, and R.M. Miura, Phys. Rev. Lett, 19, 1095 (1967).
10. ... ..., Commun. Pure Appl. Math. 27, 91 (1974).

Bäcklund Transformations for the Isospectral and Non-Isospectral KdV Hierarchies*

Tian Chou and Zhang Youjin

Department of Mathematics, University of Science and Technology of China, Hefei, Anhui, People's Rep. of China

1. Introduction.

It is well known that Bäcklund transformation is a powerful means in the study of nonlinear evolution equations. Up to now, a large number of works have been devoted to this aspect [1,2,3,4,5,6]. Recently, it is noticed that a unified explicit form of Bäcklund transformations can be obtained for some hierarchies of isospectral equations, such as Kdv, MKdv, Sine-Gordon and AKNS hierarchy [7-12]. The approach to the study is to construct the Darboux matrices first, and then prove the gauge equivalence of the related Lax pairs. However, to demonstrate the t part is quite difficult in this approach, and it is also hard to employ this method to study hierarchies of non-isospectral evolution equations.

In this paper, we firstly convert the usual Lax pairs for the isospectral and non-isospectral Kdv hierarchies into Lax pairs in Riccati form, then by using an obvious invariability for the t part of the new Lax pair, we obtain concisely a unified explicit form of Bäcklund transformations for the isospectral and non-isospectral Kdv hierarchies, and we also obtain a superposition law. It should be emphasized here that the above transformation is an auto-Bäcklund transformation for the isospectral Kdv hierarchy, but this does not true for the non-isospectral Kdv hierarchy. In fact, it transforms one non-isospectral Kdv hierarchy to another one.

2. Isospectral Kdv hierarchy.

In the following we assume that $u(x,t)$ is a smooth function of x and t, and $u(x,t)$ and its derivatives of any order with respect to x tend to zero rapidly as $x \to -\infty$.

Consider the isospectral kdv hierarchy

$$u_t = \Phi^n u_x, \quad n = 0, 1, 2, ..., \tag{2.1}$$

where

$$\Phi = \Phi(u) = D^2 + 4u + 2u_x D^{-1}, \quad D = \frac{\partial}{\partial x}, \quad D^{-1} = \int_{-\infty}^{x} dx$$

It is known [12] that equation (2.1) has the following Lax pair:

$$\begin{pmatrix} v_1 \\ v_2 \end{pmatrix}_x = \begin{pmatrix} 0 & 1 \\ \lambda - u & 0 \end{pmatrix} \begin{pmatrix} v_1 \\ v_2 \end{pmatrix} = M_n \begin{pmatrix} v_1 \\ v_2 \end{pmatrix}, \begin{pmatrix} v_1 \\ v_2 \end{pmatrix}_t = \begin{pmatrix} A_n & B_n \\ C_n & -A_n \end{pmatrix} \begin{pmatrix} v_1 \\ v_2 \end{pmatrix} = N_n \begin{pmatrix} v_1 \\ v_2 \end{pmatrix} \tag{2.2}$$

here λ is the spectral parameter, $\lambda_t = 0$, and A_n, B_n, C_n, D_n are defined by

---

*Project supported by the National Science Fund

$$A_n = -\frac{1}{2}B_{nx}, \quad B_n = \sum_{j=0}^{n} b_{n-j}(4\lambda)^j, \quad C_n = -\frac{1}{2}B_{n,xx} + (\lambda - u)B_n \qquad (2.3)$$

$$b_0 = 1, \quad b_1 = 2u, \quad b_k = D^{-1}\Phi D b_{k-1}, k = 2, 3, ..., n. \qquad (2.4)$$

Let $\varphi = v_2/v_1$, then from (2.2) we have the Lax pair in Riccati form for (2.1):

$$\varphi_x = \lambda - u - \varphi^2, \qquad (2.5a)$$

$$\varphi_t = C_n - 2A_n\varphi - B_n\varphi^2. \qquad (2.5b)$$

Define
$$\tilde{\Phi} = \tilde{\Phi}(u,\lambda) = D^2 - 4\varphi^2 - 4\varphi_x D^{-1}\varphi + 4\lambda$$

Due to our assumption on u, we see that if φ is defined by $\varphi = v_2/v_1$, then we can choose φ so that it tend to $\sqrt{\lambda}$ or $-\sqrt{\lambda}$, and it's any order derivatives with respect to x tend to zero rapidly when x tend to negative infinite. The boundary conditions of φ insure that operations of $\tilde{\Phi}$ in the following are all well defined.

It is easy to prove that
$$(D + 2\varphi) \cdot \tilde{\Phi} = \Phi(D + 2\varphi) \qquad (2.6)$$

Using (2.6) we can obtain the following lemma:

Lemma 1. If u and φ are related by (2.5a), then
$$C_n - 2A_n\varphi - B_n\varphi^2 = \tilde{\Phi}^n \varphi_x$$

Proof. From (2.3)-(2.5a) we have

$$C_n - 2A_n\varphi - B_n\varphi^2 = -\frac{1}{2}D(D - 2\varphi)B_n = -\frac{1}{2}\sum_{j=0}^{n}(4\lambda)^j D(D - 2\varphi)b_{n-j}$$

$$= \sum_{j=0}^{n-2}(4\lambda)^j D(D - 2\varphi)D^{-1}\Phi^{n-j-1}(D + 2\varphi)\varphi_x + (4\lambda)^{n-1}(\varphi_{xxx} - 6\varphi^2\varphi_x + 6\lambda\varphi_x)$$

$$= \sum_{j=0}^{n-2}(4\lambda)^j D(D - 2\varphi)D^{-1}(D + 2\varphi)\tilde{\Phi}^{n-j-1}\varphi_x + (4\lambda)^{n-1}\tilde{\Phi}\varphi_x$$

$$= \sum_{j=0}^{n-2}(4\lambda)^j [\tilde{\Phi} - 4\lambda]\tilde{\Phi}^{n-j-1}\varphi_x + (4\lambda)^{n-1}\tilde{\Phi}\varphi_x = \tilde{\Phi}^n\varphi_x.$$

Remark 1. In the proof of the above lemma, we must pay much attention on the boundary conditions of φ, Since φ^2 tends to λ as $x \to -\infty$, we have

$$\tilde{\Phi}\varphi_x = (D^2 - 4\varphi^2 - 4\varphi_x D^{-1}\varphi + 4\lambda)\varphi_x = \varphi_{xxx} - 4\varphi^2\varphi_x - 4\varphi_x \frac{1}{2}\varphi^2|_{-\infty}^{x} + 4\lambda\varphi_x$$

$$= \varphi_{xxx} - 4\varphi^2\varphi_x - 2\varphi^2\varphi_x = \varphi_{xxx} - 6\varphi^2\varphi_x + 6\lambda\varphi_x$$

Lemma 2. If u and φ satisfy (2.5a) and (2.5b), then u satisfies the n-th order Kdv equation (2.1).

Proof. From lemma 1 we know that (2.5b) can be written as $\varphi_t = \tilde{\Phi}^n \varphi_x$, From (2.5a) we have $u_t = -(D + 2\varphi)\varphi_t$. Since

$$\Phi^n u_x = -\Phi^n(D + 2\varphi)\varphi_x = -(D + 2\varphi)\tilde{\Phi}^n \varphi_x$$

we get

$$u_t - \Phi^n u_x = -(D + 2\varphi)(\varphi_t - \tilde{\Phi}^n \varphi_x)$$

which proves the lemma.

Theorem 1. If $u(x,t)$ is a solution of the n-th order Kdv equation (2.1), $V(x,t,\lambda)$ satisfies equation (2.2), and $\varphi(x,t,\lambda) = \frac{v_2}{v_1}$, then

$$\bar{u} = u + 2\varphi(x,t,\lambda_0)_x = u + 2\varphi_{0x} \tag{2.7}$$

is also a solution of the equation (2.1) where λ_0 is any constant.

Proof. From the assumption of theorem we know that u and φ_0 satisfies the equations (2.5a) and (2.5b), therefore, from lemma 1, φ_0 satisfies

$$\varphi_{0t} = \tilde{\Phi}(\varphi_0, \lambda_0)^n \varphi_{0x} = (D^2 - 4\varphi_0^2 - 4\varphi_{0x}D^{-1}\varphi_0 + 4\lambda_0)\varphi_{0x}$$

Since $\tilde{\Phi}$ is invariant under the transformation $\varphi_0 \to -\varphi_0$, we know that $-\varphi_0$ also satisfies equation (2.5b). So if we define $\bar{u}$ by

$$-\varphi_{0x} = \lambda - \bar{u} - (-\varphi_0)^2 \tag{2.8}$$

then from lemma 2 we know that $\bar{u}$ is also a solution of the equation (2.1). Now substitute $\lambda = \varphi_{0x} + u + \varphi_0^2$ into (2.8) we have $\bar{u} = u + 2\varphi_{0x}$.

Furthermore, we can obtain the following superposition law.

Theorem 2. Let u and φ satisfy the assumption of theorem1, and $\varphi_0 = \varphi(x,t,\lambda_0)$, assume $Re\lambda \neq 0$, $\lim_{x \to -\infty} \varphi = \sqrt{\lambda}$, then

$$\bar{\varphi}(x,t,\lambda) = \frac{\lambda - \lambda_0}{\varphi - \varphi_0} - \varphi_0 \quad \lambda \neq \lambda_0 \tag{2.9}$$

satisfies the equation (2.5a) and (2.3b) with $\bar{u} = u + 2\varphi_{0x}$.

Proof. From $\varphi_x = \lambda - u - \varphi^2$, $\varphi_{0x} = \lambda_0 - u - \varphi_0^2$ we have

$$(\varphi - \varphi_0)_x + (\varphi - \varphi_0)^2 + 2\varphi_0(\varphi - \varphi_0) = \lambda - \lambda_0 \tag{2.10}$$

Let $\bar{\varphi}_x + \bar{\varphi}^2 = \lambda - \bar{u}$, then

$$\bar{\varphi}_x + \bar{\varphi}^2 = \lambda - (u + 2\varphi_{0x}) = \varphi_x - 2\varphi_{0x} + \varphi^2$$

So we have

$$(\bar{\varphi} + \varphi_0)_x + (\bar{\varphi} + \varphi_0)^2 - 2\varphi_0(\bar{\varphi} + \varphi_0) = \lambda - \lambda_0 \tag{2.11}$$

So from equation (2.10) it is obvious that

$$\frac{\bar{\varphi} + \varphi_0}{\lambda - \lambda_0} = \frac{1}{\varphi - \varphi_0} \quad \text{or} \quad \bar{\varphi} = \frac{\lambda - \lambda_0}{\varphi - \varphi_0} - \varphi_0$$

Since $\bar{u}$ satisfies (2.1) and $\bar{\varphi}$ satisfies (2.5a), from lemma 2 we have

$$0 = \overline{u}_t - \Phi(\overline{u})^n \overline{u}_x = -(D + 2\overline{\varphi})(\overline{\varphi}_t - \tilde{\Phi}(\overline{\varphi},\lambda)^n \overline{\varphi}_x)$$

So
$$\overline{\varphi}_t = \tilde{\Phi}(\overline{\varphi},\lambda)^n \overline{\varphi}_x + f(t,\lambda)e^{-2D^{-1}(\overline{\varphi} - \sqrt{\lambda}) - 2\sqrt{\lambda}x}$$

From $\lim_{x \to -\infty} \varphi = \sqrt{\lambda}$, it is obvious that $\lim_{x \to -\infty} \overline{\varphi} = \sqrt{\lambda}$, and $\lim_{x \to -\infty} \overline{\varphi}_t = \lim_{x \to -\infty} \tilde{\Phi}(\overline{\varphi},\lambda)^n$ $\overline{\varphi}_x = 0$ and we get $f(t,\lambda) = 0$, which proves the theorem.

From theorem 1 and theorem 2 we see that (2.7) and (2.9) constitute a Darboux transformation for (2.5), and (2.9) makes the further process of the auto–Bäcklund transformation merely an algebraic calculation.

Remark 2. Our above Darboux transformation is equivalent to the Darboux transformation given in references [6] and [7]. But as we can see, our approach is much more simple and clear, and more importantly, our approach can be employed to study the non–isospectral Kdv hierarchy and other hierarchies of non–isospectral equations. This is the main purpose of our paper.

3. Non–isospectral Kdv hierarchy.

Consider the non–isospectral Kdv hierarchy

$$u_t = \Phi^n(xu_x + 2u), \quad n = 0, 1, 2, \ldots \tag{3.1}$$

Equation (3.1) has the following Lax pair [2]

$$V_x = \begin{pmatrix} 0 & 1 \\ \lambda - u & 0 \end{pmatrix} V, \quad V_t = \begin{pmatrix} A_n & B_n \\ C_n & -A_n \end{pmatrix} V \tag{3.2}$$

where $\lambda_t = 2^{n+1}\lambda^{n+1}$, A_n, B_n, C_n are defined by

$$A_n = -\frac{1}{2}B_{nx}, \quad B_n = \sum_{j=0}^{n} b_{n-j}(4\lambda)^j, \quad C_n = -\frac{1}{2}B_{n,xx} + (\lambda - u)B_n, \tag{3.3}$$

$$b_0 = x, \quad b_1 = D^{-1}(4u + 2xu_x), \quad b_k = D^{-1}\Phi D b_{k-1}, \quad k = 2, 3, \ldots, n. \tag{3.4}$$

As in section 2, if V satisfies (3.2), and $\varphi = v_2/v_1$, then φ satisfies (2.5a) and (2.5b).

Lemma 3. If u and φ satisfy (2.5), and A_n, B_n, C_n are defined by (3.3),(3.4) then

$$C_n - 2A_n\varphi - B_n\varphi^2 = \tilde{\Phi}^n(x\varphi_x) - \sum_{j=0}^{n-1}(4\lambda)^j \tilde{\Phi}^{n-1-j}E + (4\lambda)^n\varphi$$

where

$$E = E(\varphi, \lambda) = D(D - 2\varphi)D^{-1}(2\lambda - \varphi_x - 2\varphi^2)$$
$$= -\varphi_{xx} - 4\varphi(\lambda - \varphi^2) - 2\varphi\varphi_x + 2\varphi_x D^{-1}\varphi_x - 2\varphi_x D^{-1}(2\lambda - 2\varphi^2)$$

Similar to the proof of lemma 1, we can prove this lemma by using

$$(D + 2\varphi)E = \Phi(2\lambda - \varphi_x - 2\varphi^2) - 4\lambda(2\lambda - \varphi_x - 2\varphi^2),$$

which can be checked easily.
Lemma 4. If u and φ satisfies (2.7a), then

$$\Phi^n(xu_x + 2u) = -(D + 2\varphi)[\tilde{\Phi}^n(x\varphi_x) - \sum_{j=0}^{n-1}(4\lambda)^j \tilde{\Phi}^{n-1-j}E + (4\lambda)^n\varphi] + 2^{2n+1}\lambda^{n+1}$$

Theorem 3. Let $u(x,t)$ be a solution of the equation (3.1). If $V(x,t,\lambda)$ satisfies (3.2) and $\varphi = \frac{v_2}{v_1}$, then $\overline{u} = u + 2\varphi_{0x}$ satisfies the equation

$$\overline{u}_t = \Phi(\overline{u})^n(x\overline{u}_x + 2\overline{u}) + 4\varphi_0(-\infty)\sum_{j=0}^{n-1}(4\lambda_0)^j \Phi(\overline{u})^{n-1-j}\overline{u}_x \qquad (3.5)$$

Here $\varphi_0 = \varphi(x,t,\lambda_0)$ and λ_0 is any constant with respect to x, and $\lambda_{0t} = 2^{2n+1}\lambda_0^{n+1}$, $\varphi_0(-\infty) = \lim_{x \to -\infty} \varphi_0$

Proof. From lemma 3 we have

$$\varphi_{0t} = \tilde{\Phi}(\varphi_0,\lambda_0)^n x\varphi_{0x} - \sum_{j=0}^{n-1}(4\lambda_0)^j \tilde{\Phi}(\varphi_0,\lambda_0)^{n-1-j}E(\varphi_0,\lambda_0) + (4\lambda_0)^n\varphi_0$$

Since $E(-\varphi_0) = -E(\varphi_0) - 4\varphi_0(-\infty)\varphi_{0x}$, we know that $-\varphi_0$ satisfies

$$(-\varphi_0)_t = \tilde{\Phi}^n x(-\varphi_0) - \sum_{j=0}^{n-1}(4\lambda_0)^j [E(-\varphi_0,\lambda_0)] + (4\lambda_0)^n(-\varphi_0)$$

$$- 4\varphi_0(-\infty)\sum_{j=0}^{n-1}(4\lambda_0)^j \tilde{\Phi}^{n-1-j}\varphi_{0x}$$

Now if we define $\overline{u} = \lambda_0 - (-\varphi_0)_x - (-\varphi_0)^2$, then $\overline{u}_t = \lambda_{0t} - (D - 2\varphi_0)(-\varphi_{0t})$, so from lemma 4 we get

$$\overline{u}_t - \Phi(\overline{u})^n(x\overline{u}_x + 2\overline{u})$$
$$= -(D - 2\varphi_0)[(-\varphi_0)_t - \tilde{\Phi}(-\varphi_0,\lambda_0)^n(-x\varphi_{0x})$$
$$+ \sum_{j=0}^{n-1}(4\lambda_0)^j \tilde{\Phi}(-\varphi_0,\lambda_0)^{n-1-j}E(-\varphi_0,\lambda_0) - (4\varphi_0)^n(-\varphi_0)]$$
$$= (D - 2\varphi_0)\sum_{j=0}^{n-1}(4\lambda)^j \tilde{\Phi}(\varphi_0,\lambda_0)^{n-1-j}4\varphi_0(-\infty)\varphi_{0x}$$
$$= \sum_{j=0}^{n-1}(4\lambda)^j \Phi(\overline{u})^{n-1-j}4\varphi_0(-\infty)\overline{u}_x.$$

Now we give the superposition law. The proof is similar to the proof of theorem 2.
Theorem 4. Let u and φ satisfy the assumption of theorem 3, then

$$\overline{\varphi}(x,t,\lambda) = \frac{\lambda - \lambda_0}{\varphi - \varphi_0} - \varphi_0, \quad \lambda \neq \lambda_0$$

satisfies

$$\overline{\varphi}_t = \tilde{\Phi}(\overline{\varphi},\lambda)^n(x\overline{\varphi}_x) - \sum_{j=0}^{n-1}(4\lambda)^j \tilde{\Phi}(\overline{\varphi},\lambda)^{n-1-j}E(\overline{\varphi},\lambda) + (4\lambda)^n\overline{\varphi}$$

$$+ 4\varphi_0(-\infty)\sum_{j=0}^{n-1}(4\lambda_0)^j \tilde{\Phi}(\overline{\varphi},\lambda)^{n-1-j}\overline{\varphi}_x.$$

As the equation (3.1), equation (3.5) is also a non–isospectral n-th order Kdv equation. In fact, equation

$$u_t = \Phi^n(xu_x + 2u) + d\sum_{j=0}^{n-1}(4\lambda_0)^j \Phi^{n-1-j}u_x, \qquad (3.6)$$

has the following Lax pair

$$V_x = \begin{pmatrix} 0 & 1 \\ \lambda - u & 0 \end{pmatrix} V, \quad V_t = \begin{pmatrix} A_n & B_n \\ C_n & -A_n \end{pmatrix}, \qquad (3.7)$$

where $\lambda_t = 2^{2n+1}\lambda^{n+1}, A_n, B_n, C_n$ are defined by

$$B_n = \sum_{j=0}^{n}\overline{b}_{n-j}(4\lambda)^j + d\sum_{j=0}^{n-1}(4\lambda_0)^j \sum_{l=0}^{n-1-j}\overline{\overline{b}}_{n-1-j-l}(4\lambda)^l \qquad (3.8)$$

$$A_n = -\frac{1}{2}B_{n,x}, \quad C_n = -\frac{1}{2}B_{n,xx} + (\lambda - u)B_n \qquad (3.9)$$

Here $\overline{b}_k$ equals the b_k defined by (2.4) and $\overline{\overline{b}}_i$ equals the b_i defined by (3.4).

Define $\varphi = v_2/v_1$, then from the previous procedure we know that the Lax pair (3.7) is equivalent to the following Lax pair:

$$\varphi_x = \lambda - u - \varphi^2 \qquad (3.10a)$$

$$\varphi_t = \tilde{\Phi}^n(x\varphi_x) - \sum_{j=0}^{n-1}(4\lambda)^j \tilde{\Phi}^{n-1-j}E + (4\lambda)^n\varphi + d\sum_{j=0}^{n-1}(4\lambda_0)^j \tilde{\Phi}^{n-1-j}\varphi_x \qquad (3.10b)$$

Now we define

$$\varphi^{(0)} = \varphi \quad \varphi^{(j)} = \frac{\lambda - \lambda_{j-1}}{\varphi^{(j-1)} - \varphi^{(j-1)}_{j-1}} - \varphi^{(j-1)}_{j-1}, \quad \varphi^{(k)}_j = \varphi^{(k)}(x,t,\lambda_j) \qquad (3.11)$$

$$u^{(0)} = u, \quad u^{(j)} = u^{(j-1)} + 2\left(\varphi^{(j-1)}_{j-1}\right)_x \qquad (3.12)$$

From (3.10a),(3.10b) and the proof of theorem 3 we see that if u and φ satisfy the assumption of theorem 3, then $u^{(k)}$ is a solution of the following equation

$$u_t = \Phi^n(xu_x + 2u) + 4\sum_{l=0}^{k-1}\varphi_l^{(l)}(-\infty)\sum_{j=0}^{n-1}(4\lambda_l)^j \Phi^{n-1-j}u_x, \varphi_l^{(l)}(-\infty) = \lim_{x\to-\infty}\varphi_l^{(l)}$$

We have also obtained similar results for MKdv, AKNS and other hierarchies of evolution equations, these results will appear in later papers.

REFERENCES

1. R.M. Miura et al., Lecture Notes in Math., vol.515. Springer–Verlag, 1974.
2. C.Rogers and W.R. Shadwick, Bäcklund transformation. Academic Press, 1982.
3. Li Y.S. , to appear in adv. Math.
4. H.D. Wahlquist and F. Estabrook, J.Math. Phys. 10(1973), 1–7.
5. M.Wadati, H. Sanuki, K. Konno, Prog. Theo. Phys., 53(1975), 417–436.
6. Tian C., Acta Math. Appl. Sinica (English series), 2(1985), 89–94.
7. Gu C.H. and Hu H.S., Lett. Math. Phys., 11(1986), 325.
8. Gu C.H., Lett. Math. Phys., 12(1986), 31
9. Gu C.H. and Zhou Z.X., Lett. Math. Phys., 13(1987), 179.
10. Li Y.S., Gu X.S. and Zou M.R., Acta Math. Sinica, New series, 3(1987), 143.
11. Gu X.S., Ann. Diff. Eqs., 3(1987)13.
12. Li Y.S., Scientia Sinica A, (1982) 385.

Multiple Darboux Transformations and Multiple Pole Solutions for AKNS Hierarchy

Gu Xinshen

Department of Mathematics, University of Science and Technology of China, Hefei 230026, People's Rep. of China

Using a new method—multiple Darboux Transformation method, we gave a set of universal formulas of multiple pole solutions for AKNS hierarchy and its reductions, which include NLS hierarchy and M-KdV hierarchy. In addition, we showed the existance of the bounded multiple pole solutions for NLS hierarchy and M-KdV hierarchy.

§1. Double m-multiple Darboux Transformation (DT)

Let n be a natural number, $\alpha_0, \alpha_1, \ldots, \alpha_n$ be functions of t which are independent of x, A_j, B_j, and $C_j (j = 0, 1, \ldots, n)$ be functions which are defined by following formulas:

$$A_0 = \alpha_0 (\alpha_0 \neq 0), B_0 = C_0 = 0, A_1 = -\frac{\alpha_1}{2A_0},$$

$$B_{j+1} = \frac{i}{2}B_{j,x} + iqA_j, C_{j+1} = -\frac{i}{2}C_{j,x} + irA_j (j = 0, 1, \ldots, n),$$

$$A_j = \frac{1}{2A_0}(-\alpha_j - \sum_{k=1}^{j-1} A_k A_{j-k} - \sum_{k=1}^{j-1} B_k C_{j-k})(if\ n \geq 2, j = 2, \ldots, n).$$

Let λ be a parameter, M and N be 2 × 2 matrices as follows:

$$M = \begin{pmatrix} -i\lambda & q \\ r & i\lambda \end{pmatrix}, N = \sum_{j=0}^{n} \lambda^{n-j} \begin{pmatrix} A_j & B_j \\ C_j & -A_j \end{pmatrix} \quad (1)$$

where q and r are functions of x and t. Then equation

$$M_t - N_x + MN - NM = 0 \quad (2)$$

holds for all λ if and only if equations

$$\begin{cases} q_t = -2iB_{n+1} \\ r_t = 2iC_{n+1} \end{cases} \quad (3)$$

hold. Hierarchy of all systems of equations (3) for all natural numbers n=1,2, ... and all functions $\alpha_0, \alpha_1, \ldots, \alpha_n$ of t is called AKNS hierarchy. Consider 2 × 2 matrix ϕ such that

$$\begin{cases} \phi_x = M\phi \\ \phi_t = N\phi \end{cases} \quad (4)$$

Suppose M and N satisfy (2). then there is a matrix $\phi = (\phi_1, \phi_2) = \begin{pmatrix} \phi_{11} & \phi_{12} \\ \phi_{21} & \phi_{22} \end{pmatrix}$ such that ϕ satisfies (4) and

$$det\phi \neq 0 \qquad (5)$$

We call the above ϕ a fundamental matrix solution of (4). In ref.[1] we have studied the three kinds of DT for AKNS hierarchy. Now we introduce a special multiple DT—double m-multiple DT as follows:

Assume that m is a natural number and $m > 1$, $f_j, g_j (j = 0, 1, \ldots, m, f_0 \neq 0, g_0 \neq 0)$ and $\lambda_1, \lambda_2 (\lambda_1 \neq \lambda_2)$ are complex numbers. Let $\phi(x, t, \lambda)$ be a fundamental matrix solution of (4) in a neighborhood of λ_1 and a neighborhood of λ_2 such that ϕ is analytic on them.let

$$T = \sum_{j=0}^{m} \lambda^{m-j} T_j, T_0 = I = \begin{pmatrix} 1 & 0 \\ 0 & 1 \end{pmatrix}. \qquad (6)$$

Consider the following equations:

$$\begin{cases} \partial_{\lambda_1}^{j-1}(T(\lambda_1)\phi_2(\lambda_1)) = \sum_{s=0}^{j-1} \binom{j-1}{s} f_s \partial_{\lambda_1}^{j-1-s}(T(\lambda_1)\phi_1(\lambda_1)) \\ \partial_{\lambda_2}^{j-1}(T(\lambda_2)\phi_1(\lambda_2)) = \sum_{s=0}^{j-1} \binom{j-1}{s} g_s \partial_{\lambda_2}^{j-1-s}(T(\lambda_2)\phi_2(\lambda_2)) \end{cases} (j=1,\ldots,m) \qquad (7)$$

This is a system of 4m linear algebraic equations for the 4m elements of $T_j (j = 1, \ldots, m)$. Suppose that the coefficient determinant is not equal to zero. Then $T_j (j = 1, \ldots, m)$ is uniquely determined by (7). Regarding T as a transformation matrix, we get a transformation from ϕ to ψ, which depends on parameters $(\lambda_1, f_0, f_1, \ldots, f_{m-1}; \lambda_2, g_0, g_1, \ldots, g_{m-1})$. It is denoted by the following formula:

$$\psi = T\phi = T(\lambda_1, f_0, f_1, \ldots, f_{m-1}; \lambda_2, g_0, g_1, \ldots, g_{m-1})\phi. \qquad (8)$$

The transformation is called a double m-multiple DT. By (7), it can be proved that

$$detT = (\lambda - \lambda_1)^m (\lambda - \lambda_2)^m.$$

Suppose that there are $P^{(m)}$ and $T^{(m-1)}$ which are polynomials of λ

$$P^{(m)} = \sum_{j=0}^{1} \lambda^{1-j} P_j^{(m)}, T^{(m-1)} = \sum_{j=0}^{m-1} \lambda^{m-1-j} T_j^{(m-1)}, P_0^{(m)} = T_0^{(m-1)} = I.$$

such that for $j = 1, \ldots, m-1$,

$$\begin{cases} \partial_{\lambda_1}^{j-1}(T^{(m-1)}(\lambda_1)\phi_2(\lambda_1)) = \sum_{s=0}^{j-1} \binom{j-1}{s} f_s \partial_{\lambda_1}^{j-1-s}(T^{(m-1)}(\lambda_1)\phi_1(\lambda_1)) \\ \partial_{\lambda_2}^{j-1}(T^{(m-1)}(\lambda_2)\phi_1(\lambda_2)) = \sum_{s=0}^{j-1} \binom{j-1}{s} g_s \partial_{\lambda_2}^{j-1-s}(T^{(m-1)}(\lambda_2)\phi_2(\lambda_2)) \end{cases} \qquad (9)$$

and

$$\begin{cases} P^{(m)}(\lambda_1)(T^{(m-1)}(\lambda_1)\phi_2(\lambda_1)) = P^{(m)}(\lambda_1)\sum_{s=0}^{m-1}\binom{m-1}{s}f_s\partial_{\lambda_1}^{m-1-s}(T^{(m-1)}(\lambda_1)\phi_1(\lambda_1)) \\ P^{(m)}(\lambda_2)(T^{(m-1)}(\lambda_2)\phi_1(\lambda_2)) = P^{(m)}(\lambda_2)\sum_{s=0}^{m-1}\binom{m-1}{s}g_s\partial_{\lambda_2}^{m-1-s}(T^{(m-1)}(\lambda_2)\phi_2(\lambda_2)) \end{cases}$$
(10)

It can be checked that $T = P^{(m)}T^{(m-1)}$ is a solution of (7). Inversely, it can be proved that if the coefficient determinant of (7) is not equal to zero, then each of the coefficient determinants of (9) and (10) is also not equal to zero. Thus, the double m-multiple DT has been decomposed into two transformations:

$$\phi^{(m-1)} = T^{(m-1)}\phi, \psi = P^{(m)}\phi^{(m-1)}.$$

Because the first transformation is a double (m-1)-multiple DT, it can be continuously decomposed step by step until

$$\psi = P^{(m)}P^{(m-1)}\ldots P^{(1)}\phi$$

where $P^{(1)} = T^{(1)}$. Obviously, $P^{(1)} = T^{(1)}$ is a DT3. Now we prove that each $P^{(j)}(j = 1,\ldots,m)$ is a DT3. For showing this, suppose that for fixed j, $1 \leq j < m$, $P^{(s)}(1 \leq s \leq j)$ is a DT3, we go to prove that $P^{(j+1)}$ is a DT3. We denote $\phi^{(j)} = P^{(j)}\ldots P^{(1)}\phi$. By the properties of DT3, we have $\phi_x^{(j)} = M^{(j)}\phi^{(j)}, \phi_t^{(j)} = N^{(j)}\phi^{(j)}$, where $M^{(j)}$ and $N^{(j)}$ depend on $q^{(j)}, r^{(j)}$ and λ in the same way as M and N on q,r and λ. Let (noticing $T^{(j)} = P^{(j)}\ldots P^{(1)}$)

$$\tilde{\phi}_2^{(j)} = \partial_{\lambda_1}^j(T^{(j)}(\lambda_1)\phi_2(\lambda_1)) - \sum_{s=0}^{j-1}\binom{j}{s}f_s\partial_{\lambda_1}^{j-s}(T^{(j)}(\lambda_1)\phi_1(\lambda_1)),$$

$$\tilde{\phi}_1^{(j)} = \partial_{\lambda_2}^j(T^{(j)}(\lambda_2)\phi_1(\lambda_2)) - \sum_{s=0}^{j-1}\binom{j}{s}g_s\partial_{\lambda_2}^{j-s}(T^{(j)}(\lambda_2)\phi_2(\lambda_2)),$$

$$\tilde{\phi}^{(j)}(\lambda_1) = (\phi_1^{(j)}(\lambda_1), \tilde{\phi}_2^{(j)}), \tilde{\phi}^{(j)}(\lambda_2) = (\tilde{\phi}_1^{(j)}, \phi_2^{(j)}(\lambda_2)).$$

It can be proved that at λ_1 and λ_2,

$$\tilde{\phi}_x^{(j)} = M^{(j)}\tilde{\phi}^{(j)}, \tilde{\phi}_t^{(j)} = N^{(j)}\tilde{\phi}^{(j)}$$

and

$$det\tilde{\phi}^{(j)}(\lambda_1) = j!(\lambda_1 - \lambda_2)^j det\phi(\lambda_1), det\tilde{\phi}^{(j)}(\lambda_2) = j!(\lambda_2 - \lambda_1)^j det\phi(\lambda_2).$$

So the transformation which matrix is $P^{(j+1)}$ is also a DT3. The proof is completed. Now we suppose that $\psi_x = \bar{M}\psi$. Then we have $T_x + TM = \bar{M}T$. Directly comparing the coefficient of λ^m power in above equality we get

$$\begin{cases} \bar{q} = q + 2iT_{1,12} \\ \bar{r} = r - 2iT_{1,21} \end{cases}$$
(11)

Finally, by the properties of DT3, we get the following theorem:

Theorem. If a matrix T can be uniquely determined by (7) in a neighborhood of (x_0, t_0), then the double m-multiple DT can be decomposed into a product of m DT3 in the neighborhood. Moreover, under the double m-multiple DT:

$$\psi = T\phi$$

ψ satisfies equations

$$\psi_x = \bar{M}\psi, \psi_t = \bar{N}\psi$$

where $\bar{M}$ and $\bar{N}$ depend on $\bar{q}, \bar{r}$ and λ in the same way as M and N on q,r and λ. Thus, in the neighborhood q and r satisfy the same evolution equations (3) as q and r do.

§2. Formulas of m-multiple pole solutions for AKNS hierarchy

Suppose a system of evolution equations (3) is given. Obviously, q=r=0 satisfy (3). We can check that

$$N|_{q=r=0} = \begin{pmatrix} -N_{22}^0 & 0 \\ 0 & N_{22}^0 \end{pmatrix} \qquad (12)$$

where N_{22}^0 depends on λ and t only. we write down ϕ:

$$\phi = \begin{pmatrix} e^{-i\lambda x - \int_0^t N_{22}^0 dt} & 0 \\ 0 & e^{i\lambda x + \int_0^t N_{22}^0 dt} \end{pmatrix} \qquad (13)$$

It is a fundamental matrix solution of (3) when q=r=0. Substitute ϕ into (7), solve $T_{1,12}$ and $T_{1,21}$, by a long complicated calculation, at finally we get, by formulas (11),

$$\begin{cases} \bar{q} = \dfrac{2i}{(m-1)!}(\binom{m-1}{0}g_{m-1}, \binom{m-1}{1}g_{m-2}, \ldots, \binom{m-1}{m-1}g_0)(I - JK)^{-1}\alpha \\ \bar{r} = \dfrac{-2i}{(m-1)!}(\binom{m-1}{0}f_{m-1}, \binom{m-1}{1}f_{m-2}\binom{m-1}{m-1}f_0)(I - KJ)^{-1}\beta \end{cases} \qquad (14)$$

where I is a $m \times m$ unit matrix, K and J are $m \times m$ matrices and α and β are m-dimensional column vectors. Their entries and components are respectively defined as follows:

$$\begin{cases} K_{j,k} = -\partial_{\lambda_2}^{j-1} \sum_{s=0}^{j-k} \dfrac{\binom{m-1-s}{k-1}}{(m-1-s)!} g_{m-k-s}(\lambda_2 - \lambda_1)^{m-1-s} \phi_{11}^2(\lambda_2) \\ J_{j,k} = -\partial_{\lambda_1}^{j-1} \sum_{s=0}^{m-k} \dfrac{\binom{m-1-s}{k-1}}{(m-1-s)!} f_{m-k-s}(\lambda_1 - \lambda_2)^{m-1-s} \phi_{22}^2(\lambda_1) \end{cases} \qquad (15)$$

$$\begin{cases} \alpha_j = \partial_{\lambda_2}^{j-1}[(\lambda_2 - \lambda_1)^m \phi_{22}^2(\lambda_2)] \\ \beta_j = \partial_{\lambda_1}^{j-1}[(\lambda_1 - \lambda_2)^m \phi_{11}^2(\lambda_1)] \end{cases} \qquad (16)$$

where j,k=1,2,...,m. Formulas (14) are just the universal formulas of m-multiple pole solutions of equations (3). They are identical with the formulas in ref.[2]. Because a DT transforms a set of solutions of system (3) in AKNS hierarchy into a set of solutions of the same system. So it is here that we can say that the pair (q,r) given by (14) is exactly a set of solutions of system (3). This is not clear in ref.[2].

Now we write down some special relations in some reduction cases of AKNS hierarchy, some of which have been studied in ref.[3].

(a). Case q=r. M and N satisfy and ϕ should satisfy the following relation:
$$\begin{pmatrix} 0 & 1 \\ 1 & 0 \end{pmatrix} X(\lambda) \begin{pmatrix} 0 & 1 \\ 1 & 0 \end{pmatrix} = X(-\lambda).$$

And $\lambda_2 = -\lambda_1, g_s = (-1)^s f_s, K_{k,j} = (-1)^{k+j+1} J_{k,j}, \beta_j = (-1)^{m+j} \alpha_j$.

(b). Case q=-r. M and N satisfy and ϕ should satisfy the following relation:
$$\begin{pmatrix} 0 & 1 \\ -1 & 0 \end{pmatrix} X(\lambda) \begin{pmatrix} 0 & -1 \\ 1 & 0 \end{pmatrix} = X(-\lambda).$$

And $\lambda_2 = -\lambda_1, g_s = (-1)^{s+1} f_s, K_{k,j} = (-1)^{k+j+1} J_{k,j}, \beta_j = (-1)^{m+j+1} \alpha_j$.

(c). Case $q = r^*$. M and N satisfy and ϕ should satisfy the following relations:
$$\begin{pmatrix} 0 & 1 \\ 1 & 0 \end{pmatrix} X(\lambda) \begin{pmatrix} 0 & 1 \\ 1 & 0 \end{pmatrix} = [X(\lambda^*)]^*.$$

And $\lambda_2 = \lambda_1^*, g_k = f_k^*, J = -K^*, \beta = -\alpha^*$.

(d) Case $r = -q^*$. M and N satisfy and ϕ should satisfy the following relation:
$$\begin{pmatrix} 0 & 1 \\ -1 & 0 \end{pmatrix} X(\lambda) \begin{pmatrix} 0 & -1 \\ 1 & 0 \end{pmatrix} = [X(\lambda^*)]^*.$$

And $\lambda_2 = \lambda_1^*, g_k = -f_k^*, J = -K^*, \beta = \alpha^*$.

In ref.[4], we have given the reductions of DT3 for all above cases. By the reduction formulas of DT3 in the above four cases, we can discuss the boundedness of new pair of potentials gained after DT3 from the boundedness of old potentials. Thus, by theorem , we can get the following conclusions:

(A) The m-multiple pole solutions of each evolution equation in NLS hiarachy are always bounded in the region $\{-\infty < x < +\infty, t \geq 0\}$.

(B) If λ_1 is a pure imaginary number (and not equal to zero) and $f_j (0 \leq j \leq m-1)$ is real, then the m-multiple pole solutions of each evolution equation in M-KdV hierarchy are bounded in the region $\{-\infty < x < +\infty, t \geq 0\}$.

In ref.[3] we have given some examples of m-multiple pole solutions of AKNS hierarchy. By our formulas we get some solutions which are consistent with solutions in ref.[5] and ref.[6].

Acknowlegements

The author would like to thank Professor Abdus Salam, the International Atomic Energy Agency and UNESCO for hospitality at the International Centre for Theoretical Physics, Trieste. Thanks are also due to Professor Li Yishen for his encouragement. This work is supported by the Chinese Science fund.

References

[1] Li Yishen, Gu Xinshen and Zou Maorong, Chinese Journal of Math., Vol.3, No.2, 1987, pp 143-151.

[2] Gu Xinshen, Journal of China University of Science and Technology, Math.Issue 1985, pp 35-49 (in Chinese).

[3] Gu Xinshen, ICTP preprint, Trieste, IC/89/1.

[4] Gu Xinshen, Ann. of Dif. Eqs. (Fuzhou, China), Vol.3, No.1, 1987, pp 13-37.

[5] Miki Wadati and Kenji Ohkuma, J. Phys. Jap., Vol.5, No.6. (1982), pp 2029-2035.

[6] E.Olmedilla, Physica 25D (1987), pp 330-349.

A Lie Algebraic Structure of G.J. and Its Gauge Equivalent Yang Hierarchies

Li Yishen, Cheng Yi, and Zeng Yunbo

Department of Mathematics, University of Science and Technology of China, Hefei 230026, People's Rep. of China

Abstract. We establish in this paper an infinite dimensional Lie algebraic structure of the integrable hamiltonion system associated with.G.J and it gauge equivalent Yang equation. since the recursion operator associated with these two systems are not hereditary, a new approach is needed which make no use the hereditary property.

1. Introduction and notations.

consider the generalized AKNS eigenvalue problem

$$\varphi_x = M\varphi \quad \varphi = \begin{pmatrix} \varphi_1 \\ \varphi_2 \end{pmatrix}, \quad M = \lambda V_0 + V_1, \quad V_0 = \begin{pmatrix} -1 & 0 \\ 0 & 1 \end{pmatrix}, \quad V_1 = \begin{pmatrix} w & u \\ v & -w \end{pmatrix} \quad (1.1)$$

which is proposed by Giachetti and Johnson in [1], and is called G.J. equation for short. its gauge equivalent Yang equation[2], reads

$$\psi_x = \overline{M}\psi \quad \psi = \begin{pmatrix} \psi_1 \\ \psi_2 \end{pmatrix}, \quad \overline{M} = \begin{pmatrix} s & \xi+q+r \\ -\xi & q+r & -s \end{pmatrix} \quad (1.2)$$

They connect by the gauge transformation, $\psi_1 = \varphi_1 + \varphi_2, \ \psi_2 = i\varphi_1 - i\varphi_2$, where λ and ξ are eigenparameters, u, v, w, q, r, s are functions of x, t, they belong to schwatze space, and $\xi = i\lambda, iq = w, r = i(u-v)/2, s = (u+v)/2$; Hierarchies of the nonlinear evolution equations deduced from the isospectral eigenvalue problems of (1.1) and (1.2) respectively are given in [3]

$$p_t = \theta_l \phi^{*^{\alpha+1}} f_0 \quad f_0^T = (0, 0, -2) \quad \alpha = 0, 1, 2... \quad (1.3)$$

$$\bar{p}_t = \bar{\theta}_l \bar{\phi}^{*^{\alpha+1}} \bar{f}_0, \quad \bar{f}_0^T = (-2i, 0, 0), \quad \text{l is parameter} \quad (1.4)$$

where

$$p^T = (u, v, w) \quad \bar{p}^T = (q, r, s) \quad \bar{p} = Tp$$

$$T = \begin{pmatrix} 0 & 0 & -i \\ \frac{i}{2} & -\frac{i}{2} & 0 \\ \frac{1}{2} & \frac{1}{2} & 0 \end{pmatrix}, \quad \theta_l = \begin{pmatrix} 0 & -2 & lu \\ 2 & 0 & -lv \\ -lu & lv & lD \end{pmatrix}, \quad \bar{\theta}_l = \begin{pmatrix} -D & -ls & lr \\ ls & v & -i \\ -lr & i & 0 \end{pmatrix},$$

$$\phi^* = \begin{pmatrix} \frac{1}{2}D+w & 0 & -\frac{v}{2} \\ 0 & -\frac{1}{2}D+w & -\frac{u}{2} \\ D^{-1}u(D+2w) & D^{-1}v(D-2w) & 0 \end{pmatrix}, \quad D^{-1}D = DD^{-1} = I$$

$$\bar{\theta}^* = \begin{pmatrix} 0 & D^{-1}(2sq - rD) & D^{-1}(-2qr - sD) \\ -r & -q & \frac{D}{2} \\ -s & -\frac{D}{2} & -q \end{pmatrix}, \quad D = \frac{\partial}{\partial x}, \quad (1.5)$$

(* denote the conjugate of the operator ϕ or matrix T),

We have pointed out in [3], that (1.3),(1.4) are hamiltonian systems which passess infinite constants of motion in involution, Such that they are integrable hamiltonian systems. we also find that $\theta_t\phi^* \neq \phi\theta_t$ and $\bar{\theta}_t\bar{\phi}^* \neq \bar{\phi}\bar{\theta}_t$. In [4], we prove that operotor $\phi, \bar{\phi}$ are not hereditary and so called $\phi, \bar{\phi}$ the nonhereditary recursion operators.

For given an integrable hamiltonion system associated with an eigenvalue problem, we can construct two sets of infinite number of symmetries and the relevent Lie algebraic structure by using the hereditary property of the recursion operator ϕ. and these two sets of symmetries are connected with isospectral and non-isospectral hierarchy of the eigenvalue problem [5][6].

In this paper, an approach which make no use of hereditary property is used to prove that there are also two sets of infinte number of symmetries which connected with the isospectral and nonisospctral hierarchies of the eigenvalue problem (1.1) and (1.2) for the system (1.3) and (1.4) respectively and the relevent Lie algebraic structure is given .

Throught out this paper , we denote by $[M, N] = MN - NM$, the commutator of 2×2 matrix M, N and $TrM = TrN = 0$, the Lie bracket $\|M, N\| = M'[N] - N'[M]$, where

$$M'[N] = \lim_{\varepsilon \to 0} \frac{\partial}{\partial \varepsilon} M[V_1 + \varepsilon N] \quad (1.6)$$

Given a matrix $M = M_{ij}(i, j = 1, 2)$, we denote by M_D and M_F , its diagonal and off diagonal part respectively, it is easy shown that

$$M = M_0 + M_F, \quad [M_D, N] = [M_D, N_F], \quad [M, N]_D = [M_F, N_F], \quad [M_F, N_F]_F = 0 \quad (1.7)$$

$$\text{If} \quad M_{11} \neq M_{22} \quad \text{then} \quad [M, N] = 0 \quad \text{implies} \quad N_F = 0 \quad (1.8)$$

we start from

$$N_0^0 = V_0, \quad N_0^1 = xV_0 \quad (1.9)$$

and the equations

$$a\delta_{j,v}V_0 - N_{j,x}^a + [V_0, N_{j+1}^a] + [V_1, N_j^a] = 0, \quad a = 0 \text{ or } 1, \quad j = 0, 1, 2... \quad (1.10)$$

To fix the solution, we promise the integral constants of N_D are zero, for excemple

$$N_1^a = x^a V_{1,F}, \quad N_2^0 = \begin{pmatrix} \frac{1}{2}uc & -\frac{1}{2}(u_x - 2wu) \\ \frac{1}{2}(v_x + 2wv) & -frac12uv \end{pmatrix},$$

$$N_2^1 = \begin{pmatrix} \frac{1}{2}(xav) + \frac{1}{2}D^{-1}(uv) & -\frac{1}{2}(u + xu_x) + xwu \\ \frac{1}{2}(v + xv_x) + xwv & -\frac{1}{2}(xuv) - \frac{1}{2}D^{-1}(uv) \end{pmatrix} \quad (1.11)$$

If we introduce the weight

$$\rho(u) + \rho(v) = \rho(D) = -\rho(x) = 1 \quad \rho(constant) = 0 \quad \rho(fg) = \rho(f) + \rho(g)$$

then $\rho(N_j^a) = j - a$, we have $\rho(N_1^0) = 1 \quad \rho(N_1^1) = 0 \quad \rho(N_2^0) = 2 \quad \rho(N_2^1) = 1$

Now, we define the 2×2 matrix F_m^a

$$F_m^a \equiv [V_0, N_{m+1}^a] + l[V_{1F}, N_{m+1,F}^a] - l[V_1, N_{m+1,D}^a] \quad m = 0, 1, 2... \quad (1.12)$$

The system (1.3) [see ref [3] (3.12)] can be written as follows

$$V_{1,t} = F_\alpha^0, \quad \alpha = 0, 1, 2... \quad (1.13)$$

We call the systems (1.13) the isospectral hierarchies of (1.1) and the following systems are called nonisospectral hierarchies of (1.1)

$$V_{1t} = F_\alpha^1 \quad \alpha = 0, 1, 2... \quad (1.14)$$

Two relations which can be proved by the same approach used in [9], will be used

$$\frac{d}{dx} \sum_{s=0}^{j-1} [N_s^a, N_{n+j-s}^b] = a[V_0, N_{n+j}^b] + \delta_{n_1-1} b[N_{j-1}^a, V_0] + [[V_0, N_{n+1}^b], N_j^a]$$

$$+ \sum_{s=0}^{j} [V_0, [N_s^a, N_{n+j+1-s}^b]] + \sum_{s=0}^{j-1} [V_1, [N_s^a, N_{n+j-s}^b]] \quad (n \geq -1, j \geq 1) \quad (1.15)$$

$$\sum_{s=0}^{k} [N_s^a, N_{k-s}^b] = (b-a)(k-1) N_{k-1}^0 \quad (1.16)$$

2. main results.

THEOREM 1.
$$N_0^{a'}[F_n^b] = 0 \quad (2.1)$$

$$N_{m+1}^{a'}[F_n^b] = am N_{n+m}^b + \sum_{s=0}^{m} [N_0^a, N_{n+m+1-s}^b] - l[N_{m+1}^a, N_{n+1,D}^b] \quad (2.2)$$

PROOF:: Since N_0^a is independence with V_1, (2.1) holds travially, We prove (2.2) by induction on m, to prove for $m=0$ is trivial, suppose that (2.2) holds for $k \leq m-1$, we proceed to prove it for m; From (1.10),

$$-N_{m,x}^a + [V_0, N_{m+1}^a] + [V_1, N_m^a] = 0, \quad (N'[M])_x = N_x'[M]$$

$$-(N_m^{a'}[F_n^b])_x + [V_0, N_{m+1}^{a'}[F_n^b]] + [V_1'[F_n^b], N_m^a] + [V_1, N_m^{a'}[F_n^b]] \equiv J_1 + J_2 + J_3 + J_4 = 0 \quad (2.3)$$

$$J_1 = -(N_m^{a'}[F_n^b])_x = -\{a(m-1) N_{n+m-1}^b + \sum_{s=0}^{m-1} [N_s^a, N_{n+m-s}^b]\}_x + l[N_m^a, N_{n+1,D}^b] \equiv I_1 + I_2$$

$$J_3 = [[V_0, N_{n+1}^b], N_m^a] + l[[V_{1F}, N_{n+1,F}^b], N_m^a] - l[[V_1, N_{n+1,D}^b], N_m^a] \equiv J_{31} + J_{32} + J_{33}$$

$$J_4 = [V_1, a(m-1) N_{n+m+1}^b + \sum_{s=0}^{m-1} [N_s^a, N_{n+m-s}^b]] - l[V_1, [N_m^a, N_{n+1,D}^b]] \equiv J_{41} + J_{42}$$

$$I_1 = -[V_0, am N_{n+m}^b + \sum_{s=0}^{m} [N_s^a, N_{n+m+1-s}^b]] - [V_1, a(m-1) N_{n+m-1}^b + \sum_{s=0}^{m-1} [N_s^a, N_{n+m+s}^b]]$$

$$- [[V_0, N_{n+1}^b], N_m^a] \equiv I_{11} + I_{12} + I_{13}$$

$$I_2 = l[[V_0, N^a_{m+1}], N^b_{n+1,D}] + l[[V_1, N^b_m], N^b_{n+1,D}] + l[N^a_m, [V_{1F}, N^b_{n+1,F}]] \equiv I_{21} + I_{22} + I_{23}$$

we note that $J_{41} + I_{12} = 0$, $J_{31} + I_{13} = 0$ $J_{32} + I_{23} = 0$, $J_{33} + J_{42} + I_{22} = 0$, (2.3) is reduced to $J_2 + I_{11} + I_{21} = 0$. i,e

$$[V_0, N^{a'}_{m+1}[F^b_n]] - amN^b_{n+m} + \sum_{s=0}^{m}[N^a_s, N^b_{n+m+1-s}] + l[N^a_{m+1}, N^b_{n+1,D}]] = 0$$

Using (1.8), we conclude that (2.2) holds for offdiagonal part, to complete the proof, we have to prove (2.2) for diagonal part, since each term of it is of the same weight $n + m + 1 - a - b$, we need only to prove

$$(N^{a'}_{m+1}[F^b_n])_{D,x} = am(N^b_{n+m,D})_x + (\sum_{s=0}^{m}[N^a_s, N^b_{n+m+1-s}]_D)_x \tag{2.4}$$

using (1.15). we get

$$(\sum_{s=0}^{m}[N^a_s, N^b_{n+m+1-s}]_D)_x = [[V_0, N^b_{n+1}], N^a_{m+1,F}]_D + \sum_{s=0}^{m}[V_{1,F}, [N^a_s, N^b_{n+m+1-s}]_F]_D$$
$$= [(V'_1[F^b_n])_F, N^a_{m+1,F}]_D + l[[V_{1,F}, N^b_{n+1,D}], N^a_{m+1,F}]_D$$
$$+ [V_{1F}, (N^{a'}_{m+1}[F^b_n])_F]_D - [V_1, amN^b_{n+m,F}]_D + l[V_{1F}, [N^a_{m+1,F}, N^b_{n+1,D}]_F]_D$$
$$= ([V_1, N^a_{m+1,F}]_D)'[F^b_n] - [V_1, amN^b_{n+m,D}]_D = N^{a'}_{m+1,x\ D}[F^b_n] - amN^b_{n+m,D,x}$$

which implies (2.4) The proof is completed.

THEOREM 2. *We have*

$$\|F^0_m, F^0_n\| = 0, \quad \|F^1_m, F^1_n\| = (m-n)F^1_{m+n-1}, \quad \|F^0_m, F^1_n\| = mF^0_{m+n-1} \tag{2.5}$$

proof

$$F^{a'}_m[F^b_n] = [V_0, N^{a'}_{m+1}[F^b_n]] + l[V'_{1F}[F^b_n], N^l_{m+1,F}] + l[V_{1F}, N^{a'}_{m+1,F}[F^b_n]]$$
$$- l[V'_1[F^b_n], N^a_{m+1,D}] - l[V_1, N^{a'}_{m+1,D}[F^b_n]]$$
$$= [V_0, amN^b_{n+m}] + [V_0, \sum_{s=0}^{m}[N^a_s, N^b_{n+m+1-s}]] - l[V_0, [N^a_{m+1}, N^b_{n+1,D}]]$$
$$+ l[V_{1F}, amN^b_{n+m,F}] + l[V_{1F}, \sum_{s=0}^{m}[N^a_s, N^b_{n+m+1-s}]_F] - l^2[V_{1F}, [N^a_{m+1}, N^b_{n+1,D}]_F]$$
$$- l[V_1, amN^b_{n+m,D}] - l[V_1, \sum_{s=0}^{m}[N^a_s, N^b_{n+m+1-s}]_D] + l^2[V_1, [N^a_{m+1}, N^b_{n+1,D}]_D]$$
$$+ l[[V_0, N^b_{n+1}]_F, N^a_{m+1,F}] + l^2[[V_{1F}, N^b_{n+1,F}]_F, N^a_{m+1,F}] - l^2[[V_1, N^b_{n+1,D}]_F, N^a_{m+1,F}]$$
$$- l[[V_0, N^b_{n+1}], N^a_{m+1,D}] - l^2[[V_{1F}, N^b_{n+1,F}], N^a_{m+1,D}] + l^2[[V_1, N^b_{n+1,D}], N^a_{m+1,D}]$$
$$\equiv J_{11} + J_{12} + J_{13} + J_{31} + J_{32} + J_{33} + J_{51} + J_{52} + J_{53} + J_{21} + J_{22} + J_{23}$$
$$+ J_{41} + J_{42} + J_{43}$$

$$F_n^{b'}[F_m^a] = [V_0, bnN_{m+n}^a] + [V_0, \sum_{s=0}^{n}[N_s^b, N_{n+m+1-s}^a]] - l[V_0, [N_{n+1}^b, N_{m+1,D}^a]]$$

$$+ l[V_{1F}, bnN_{m+n,F}^a] + l[V_{1F}, \sum_{s=0}^{n}[N_s^b, N_{n+m+1-s}^a]_F] - l^2[V_{1F}, [N_{n+1}^b, N_{m+1,D}^a]_F]$$

$$- l[V_1, bnN_{n+m,D}^a] - l[V_1, \sum_{s=0}^{n}[N_s^b, N_{n+m+1-s}^a]_D] + l^2[V_1, [N_{n+1}^b, N_{m+1,D}^a]_D]$$

$$+ l[[V_0, N_{m+1}^a]_F, N_{n+1,F}^b] + l^2[[V_{1F}, N_{m+1,F}^a], N_{n+1,F}^b] - l^2[[V_1, N_{m+1,D}^a]_F, N_{n+1,F}^b]$$

$$- l[[V_0, N_{m+1}^a], N_{n+1,D}^b] - l^2[[V_{1F}, N_{m+1,F}^a], N_{n+1,D}^b] + l^2[[V_1, N_{m+1,D}^a], N_{n+1,D}^b]$$

$$\equiv I_{11} + I_{12} + I_{13} + I_{31} + I_{32} + I_{33} + I_{51} + I_{52} + I_{53} + I_{21} + I_{22} + I_{23}$$
$$+ I_{41} + I_{42} + I_{43}$$

Using the relations, $J_{13} = I_{41}$, $J_{33} + J_{23} = I_{42} = 0$, $J_{53} = I_{53} = 0$, $J_{22} = I_{22} = 0$
, $J_{41} = I_{13}$, $J_{12} = I_{33} + I_{23} = 0$, $J_{43} = I_{43}$, $J_{21} = I_{21}$

We get

$$F_m^{a'}[F_n^b] - F_n^{b'}[F_m^a]$$

$$= [V_0, amN_{n+m}^b + \sum_{s=0}^{m}[N_s^a, N_{n+m+1-s}^b]] - [V_0, bnN_{m+n}^a + \sum_{s=0}^{n}[N_s^b, N_{n+m+1-s}^a]]$$

$$+ l[V_{1F}, amN_{n+m}^b + \sum_{s=0}^{m}[N_a^s, N_{n+m+1-s}^b]_F] - l[V_{1F}, bnN_{m+n,F}^a + \sum_{s=0}^{n}[N_s^b, N_{n+m+1-s}^a]_F]$$

$$- l[V_1, amN_{n+m,D}^b + \sum_{s=0}^{m}[N_a^s, N_{n+m+1-s}^b]_D] + l[V_1, bnN_{m+n,D}^a + \sum_{s=0}^{n}[N_s^b, N_{m+n+1-s}^a]_D]$$

Considering the relation (1.16) and

$$\sum_{s=0}^{m}[N_s^a, N_{n+m+1-s}^b] - \sum_{s=0}^{n}[N_s^b, N_{n+m+1-s}^a] = \sum_{s=0}^{n+m+1}[N_s^a, N_{n+m+1-s}^b]$$

it yields that
i) when $a = b = 0$, $F_m^{0'}[F_n^0] - F_n^{0'}[F_m^0] = 0$.
ii) when $a = b = 1$

$$F_m^{1'}[F_n^1] - F_n^{1'}[F_m^1] = (m-n)\{[V_0, N_{n+m}^1] + l[V_{1F}, N_{n+m,F}^1] - l[V_1, N_{n+m,D}^1]\}$$
$$= (m-n)F_{m+n-1}^1.$$

iii) when $a = 0, b = 1$

$$F_m^{0'}[F_n'] - F_n^{1'}[F_m^0] = [V_0, -nN_{m+n}^0] + [V_0, (n+m)N_{n+m}^0] + l[V_{1F}, -nN_{n+m,F}^0]$$
$$+ l[V_{1F}, (n+m)N_{n+m,F}^0] - l[V_1, -nN_{n+m,D}^0] - l[V_1, (n+m)N_{n+m,D}^0]$$
$$= m\{[V_0, N_{n+m}^0] + l[V_{1F}, N_{n+m,F}^0] - l[V_1, N_{n+m,D}^0]\} = mF_{n+m-1}^0.$$

The proof. is completed
Now, we define the vectors

$$k_n(p) = ((F_n^0)_{12}, (F_n^0)_{21}, (F_n^0)_{11})^T, \quad \sigma_n(p) = ((F_n^1)_{12}, (F_n^1)_{21}, (F_n^1)_{11})^T \qquad (2.6)$$

and define the Lie bracket for the vector k,σ, $\|k,\sigma\| = k'[\sigma] - \sigma'[k]$ where, $k'[\sigma] = \lim_{\varepsilon\to 0}\frac{\partial}{\partial\varepsilon}k(p+\varepsilon\sigma)$, Then the relatirns (2.5) can be read

$$\|k_m,k_n\| = 0, \quad \|\sigma_m,\sigma_n\| = (m-n)\sigma_{m+n-1}, \quad \|k_m,\sigma_n\| = mk_{m+n-1} \qquad (2.7)$$

we rewrite the system (1.3) (or 1.13) as follows

$$p_t = k_\alpha(p) \qquad \alpha = 0,1,2... \qquad (2.8)$$

and define, $\tau_n^\alpha = t\alpha k_{n+\alpha-1} + \sigma_n$, by the some approach using in [5], we have the theorem.

THEOREM 3. The G.J. hierarchies (1.3), [or (2.8)], have two sets of symmetries k_m, τ_n^α, They form an infinite dimensional Lie algebra

$$\|k_m,k_n\| = 0, \quad \|k_m,\tau_n^\alpha\| = mk_{m+n-1}, \quad \|\tau_m^\alpha,\tau_n^\alpha\| = (m-n)\tau_{m+n-1}^\alpha \qquad (2.9)$$

we define, the vectors, $\bar{k}_m(\bar{p}) = Tk_n(p)$, $\bar{\tau}_n^\alpha(\bar{p}) = T\tau_n^\alpha(p)$, , The Yang hierarchies (1.4) read

$$\bar{p}_t = \bar{k}_\alpha(\bar{p}) \quad \alpha = 0,1,2... \qquad (2.10)$$

From the relation [8], $\|\bar{k},\bar{\tau}\|_{\bar{p}} = T\|k,\tau\|$, where , $\|\bar{k},\bar{\tau}\|_{\bar{p}} = \bar{k}'_{\bar{p}}[\bar{\tau}] - \bar{\tau}'_{\bar{p}}[\bar{k}]$, $\bar{k}_{\bar{p}}[\bar{\tau}] = \lim_{\varepsilon\to 0}\frac{\partial}{\partial\varepsilon}\bar{k}(\bar{p}+\varepsilon\bar{\tau})$
we have the theorem

THEOREM 4. The Yang hierarchies (1.4) [or (2.10)] have two sets of symmtries, $\bar{k}_m, \bar{\tau}_n^\alpha$, They form an infinite dimensional Lie algebra

$$\|\bar{k}_m,\bar{k}_n\|_{\bar{p}} = 0, \quad \|\bar{k}_m,\bar{\tau}_n^\alpha\|_{\bar{p}} = m\bar{k}_{m+n-1}, \quad \|\bar{\tau}_m^\alpha,\bar{\tau}_n^\alpha\|_{\bar{p}} = (m-n)\bar{\tau}_{m+n-1}^\alpha \qquad (2.11)$$

ACKNOWLEDGMENTS

Many thanks for Zhu Guocheng who read our manuscript and give useful comment, This work was supported by Fok.Ying-Tung Education Fundation and National science Fundation of China .

REFERENCES

1. Giachetti, R; Johnson,R. phys. lett. 102A (1984) 81.
2. Yang.C.N. Commun. math. phys. 112 (1987) 205.
3. Li Yi–shen. The integrable system associatod with G.J.equation and its gange. equivalent Yang equation to appear in Kexue Tongbao (1989).
4. Li Yi–shen. Zeng Yun–bo. on some properties of G.J equation and its gauge equivalent Yang equation.to be published .
5. Li Yi–shen Zhu. G.C. a) J.phys math Gen 19, (1986) 3713, b) scientia sinica vol XXX (1987) 1243.
6. Stramp. W, Olver W. Prog. Theor. Phys. 74(1985) 922.
7. Fuchssteiner. A.S Fokas, Physica 4 D(1981) 47.
8. Li Yi–shen. Comment on some agebraic properties of the gauge equivalent soliton equalions. to be published.
9. Tu Gui-zhang. Lie algebraic structure of $N \times N$ nonisospctral AKNS hierarchy. preprint.

Part II

**Finite Dimensional
Dynamical Systems**

Coupled Nonlinear Oscillators: Symmetries and Integrability

M. Lakshmanan

Department of Physics, Bharathidasan University, Tiruchirapalli 620024, India

It is shown how the Lie's method of invariance analysis involving extended, velocity-dependent vector fields can systematically identify integrable cases of nonlinear dynamical systems. The method is illustrated for the case of coupled nonlinear oscillators involving polynomial potentials with two and three degrees of freedom.

1. INTRODUCTION

It is a fundamental problem in nonlinear dynamics to understand under what conditions a given dynamical system is integrable or not. For a Hamiltonian system with N degrees of freedom, Liouville theorem [1] ensures that the system is integrable whenever it admits N involutive integrals of motion in the Poisson bracket sense. In the search for such involutive integrals of motion, at least three different techniques have been advocated by different authors [2-5] in recent times (apart from the inverse scattering transform technique applicable to lattice soliton systems such as the Toda lattice). They are

1) method of direct search for integrals of motion [6] by starting with assumed form for the integrals, which is a generalization of the method used by Bertrand, Darboux, Chandrasekar and others [7,8];
2) the method of analysing the singularity structure of associated equations of motion [4,9], originally developed by Fuchs, Kovalevskaya, Painlevé and coworkers; and
3) invariance analysis of the equations of motion under appropriate symmetry transformations [10-12], originally proposed by Sophus Lie.

Usually the theory of one-parameter Lie group of transformations is applied to dynamical systems assuming the infinitesimal generators to be functions of the independent and dependent variables alone [10]. However it is more advantageous to consider the invariance under velocity dependent transformations, corresponding to dynamical symmetries. Such an approach was discussed at considerable length by Lutzky [5], who has applied the method to find the symmetries of the one-dimensional harmonic oscillator.

In this article, we are concerned with the invariance analysis under one-parameter continuous group of velocity dependent symmetry transformations and its application to a system of coupled nonlinear pendula given by the Hamiltonian

$$H = \frac{1}{2} \sum_{i=1}^{N} \frac{p_i^2}{2} + V(x_1, x_2, \ldots, x_N), \qquad (1)$$

where V is a polynomial in the coordinate variables. Systems of the type (1) are ubiquitous in physics and need no introduction. They have been studied for their integrability property through combined Painlevé analysis and direct search for integrals of motion by several authors [13,14] recently. It is the aim here to show how the invariance analysis involving Lie's extended vector fields systematically identifies [15] integrable cases for the N=2 and N=3 oscillator cases along with their associated involutive integrals of motion.

The paper is presented as follows. In Sec. 2, we give a brief introduction to Lie's theory of extended vector fields applicable to Lagrangian systems involving two degrees of freedom and derive the conditions under which the equations of motion remain invariant and their connection to the integrals of motion. In Sec. 3, we apply the theory to the two coupled oscillators (N=2) of (1) and obtain the symmetries and integrals of motion by solving just a set of algebraic equations. Sections 3 and 4 then deal with three coupled oscillators (N=3). Finally in Sec. 5, a few critical remarks are made.

2. LIE'S THEORY OF INVARIANCE UNDER EXTENDED VECTOR FIELDS FOR TWO COUPLED OSCILLATORS

Consider the Lagrangian of a system of two coupled pendula

$$L = \frac{1}{2}(\dot{x}^2 + \dot{y}^2) - V(x, y), \qquad (2)$$

where V is the potential which is nonlinear. Then the Euler-Lagrange equations of motion are

$$\ddot{x} = \frac{\partial L}{\partial x} \equiv \alpha_1(x,y), \qquad (3a)$$

$$\ddot{y} = \frac{\partial L}{\partial y} \equiv \alpha_2(x,y). \tag{3b}$$

For example, for the Henon-Heiles system

$$V_{HH} = \frac{1}{2}(Ax^2 + By^2) + (Dx^2y - \frac{C}{3}y^3) \tag{4}$$

so that

$$\alpha_1 = -Ax - 2Dxy, \quad \alpha_2 = -By - Dx^2 + Cy^2. \tag{5}$$

Similarly for the coupled quartic oscillators

$$V_2 = (Ax^2 + By^2) + \alpha x^4 + \beta y^4 + \delta x^2 y^2 \tag{6}$$

and so

$$\alpha_1 = -2Ax - 4\alpha x^3 - 2\delta xy^2, \quad \alpha_2 = -2By - 4\beta y^3 - 2\delta x^2 y. \tag{7}$$

Consider now a transformation of the form

$$\{t,x,y,\dot{x},\dot{y}\} \longrightarrow \{T,X,Y,\dot{X},\dot{Y}\}, \tag{8}$$

where $T = T(t,x,\dot{x},y,\dot{y})$, $X = X(t,x,\dot{x},y,\dot{y})$, $Y = Y(t,x,\dot{x},y,\dot{y})$, and

$$\dot{X} = \frac{dX}{dT} = \frac{\{\frac{\partial X}{\partial t} + \frac{\partial X}{\partial x}\dot{x} + \frac{\partial X}{\partial y}\dot{y} + \frac{\partial X}{\partial \dot{x}}\alpha_1 + \frac{\partial X}{\partial \dot{y}}\alpha_2\}}{\{\frac{\partial T}{\partial t} + \frac{\partial T}{\partial x}\dot{x} + \frac{\partial T}{\partial y}\dot{y} + \frac{\partial T}{\partial \dot{x}}\alpha_1 + \frac{\partial T}{\partial \dot{y}}\alpha_2\}}, \tag{9}$$

and

$$\dot{Y} = \frac{dY}{dT} = \frac{\{\frac{\partial Y}{\partial t} + \frac{\partial Y}{\partial x}\dot{x} + \frac{\partial Y}{\partial y}\dot{y} + \frac{\partial Y}{\partial \dot{x}}\alpha_1 + \frac{\partial Y}{\partial \dot{y}}\alpha_2\}}{\{\frac{\partial T}{\partial t} + \frac{\partial T}{\partial x}\dot{x} + \frac{\partial T}{\partial y}\dot{y} + \frac{\partial T}{\partial \dot{x}}\alpha_1 + \frac{\partial T}{\partial \dot{y}}\alpha_2\}}. \tag{10}$$

Then the transformation (8) maps solution curves into solution curves of the equation of motion

$$X = X(T), \quad Y = Y(T) \tag{11}$$

in the $(T,X,Y,\dot{X},\dot{Y})$ space. Thus if the above transformation is to be a symmetry of the equation of motion, then

$$\ddot{X} = \alpha_1(X,Y), \tag{12}$$
$$\ddot{Y} = \alpha_2(X,Y).$$

Next we need a set of criteria for (8) to be a symmetry of the equations of motion (3). For this purpose, we consider the infinitesimals of a one-parameter group of continuous transformations (8):

$$x \to X = x + \varepsilon \eta_1(t,x,\dot{x},y,\dot{y}) + O(\varepsilon^2), \tag{13a}$$

$$y \to Y = y + \varepsilon \eta_2(t,x,\dot{x},y,\dot{y}) + O(\varepsilon^2), \tag{13b}$$

$$t \to T = t + \varepsilon \xi(t,x,\dot{x},y,\dot{y}) + O(\varepsilon^2), \quad \varepsilon \ll 1 \tag{13c}$$

so that

$$\dot{x} \to \dot{X} = \dot{x} + \varepsilon(\dot{\eta}_1 - \dot{\xi}\dot{x}) + O(\varepsilon^2), \tag{13d}$$

$$\dot{y} \to \dot{Y} = \dot{y} + \varepsilon(\dot{\eta}_2 - \dot{\xi}\dot{y}) + O(\varepsilon^2) \tag{13e}$$

where

$$\dot{\eta}_i = \left(\frac{\partial}{\partial t} + \dot{x}\frac{\partial}{\partial x} + \dot{y}\frac{\partial}{\partial y} + \alpha_1\frac{\partial}{\partial \dot{x}} + \alpha_2\frac{\partial}{\partial \dot{y}}\right)\eta_i, \quad i = 1,2 \tag{13f}$$

and

$$\dot{\xi} = \left(\frac{\partial}{\partial t} + \dot{x}\frac{\partial}{\partial x} + \dot{y}\frac{\partial}{\partial y} + \alpha_1\frac{\partial}{\partial \dot{x}} + \alpha_2\frac{\partial}{\partial \dot{y}}\right)\xi. \tag{13g}$$

Thus the corresponding infinitesimal generator of the group is [5]

$$E = \xi\frac{\partial}{\partial t} + \eta_1\frac{\partial}{\partial x} + \eta_2\frac{\partial}{\partial y} + (\dot{\eta}_1 - \dot{\xi}\dot{x})\frac{\partial}{\partial \dot{x}} + (\dot{\eta}_2 - \dot{\xi}\dot{y})\frac{\partial}{\partial \dot{y}} \tag{14}$$

so that the finite transformations become

$$(X,Y,T,\dot{X},\dot{Y}) = \exp(\varepsilon E).(x,y,t,\dot{x},\dot{y}) \tag{15}$$

Now let us consider any conserved quantity $I(t,x,y,\dot{x},\dot{y})$. Then under the action of the group (15), we have

$$\exp(\varepsilon E)\{I(t,x,y,\dot{x},\dot{y})\} = I\{e^{\varepsilon E}t,\ldots,e^{\varepsilon E}\dot{y}\}$$

$$= I(T,X,Y,\dot{X},\dot{Y})$$

$$\equiv \psi(t,x,y,\dot{x},\dot{y}). \tag{16}$$

If $I(t,x,y,\dot{x},\dot{y}) = C_1 =$ constant for the solution $\{x(t),y(t)\}$, then $I(T,X,Y,\dot{X},\dot{Y}) = C_2$ for $\{X(T),Y(T)\}$, so that $\psi(t,x,y,\dot{x},\dot{y},\varepsilon)$ is also conserved. Now expanding ψ in powers of ε,

$$\psi = I(t,x,y,\dot{x},\dot{y},\varepsilon=0) + \varepsilon E\{I\} + O(\varepsilon^2) \tag{17}$$

so that

$$E\{I\} = \text{const.} \Rightarrow \frac{d}{dt}(E\{I\}) = 0, \tag{18}$$

implying that

$$\ddot{\eta}_1 - \dot{x}\ddot{\xi} - 2\dot{\xi}\alpha_1 = E(\alpha_1), \qquad (19)$$

$$\ddot{\eta}_2 - \dot{y}\ddot{\xi} - 2\dot{\xi}\alpha_2 = E(\alpha_2). \qquad (20)$$

These are the set of underdetermined system of invariance conditions one obtains for the infinitesimals. Solving for ξ, η_1, η_2 consistently, we can then obtain the infinitesimal symmetries of the system.

Finally it can be easily checked as shown by Lutzky [5] that the constant of motion associated with a given set of infinitesimal symmetry ξ, η_i, $i = 1,2$ is

$$\Phi = (\xi\dot{x} - \eta_1)\frac{\partial L}{\partial \dot{x}} + (\xi\dot{y} - \eta_2)\frac{\partial L}{\partial \dot{y}} - \xi L + f, \qquad (21)$$

where $f = f(x,y,t)$ is a function to be determined from the condition

$$E\{L\} + \dot{\xi}L = \dot{f} \qquad (22)$$

4. DETERMINATION OF INTEGRABLE CASES OF THE TWO COUPLED PENDULA

The invariance conditions (19) and (20) form an incomplete system in η_1, η_2 and ξ. So we will have to solve them by assuming specific forms. One specific form we can always assume for time independent potentials $V(x,y)$ in (2) is

$$\eta_1 = \eta_2 = 0, \quad \xi = C = \text{const.}, \qquad (23)$$

corresponding to the time-translation invariance, $t \to T = t + a$. So $\Phi = C[\dot{x}^2 + \dot{y}^2 - L] + f$, $\dot{f} = 0$, which on rescaling gives the integral of motion,

$$\Phi = \frac{1}{2}(\dot{x}^2 + \dot{y}^2) + V(x,y) \equiv H, \qquad (24)$$

which is nothing but the Hamiltonian.

Next we may assume a polynomial form in the velocities,

$$\xi = \sum_{k\ell=0}^{M} a_{k\ell}(x,y,t).(\dot{x})^k(\dot{y})^\ell \quad \text{and} \quad \eta_i = \sum_{k,\ell=0}^{M} b_{k\ell}^i(x,y,t)(\dot{x})^k(\dot{y})^\ell,$$

$i = 1,2$. Substituting these forms in (19) and (20), we can consistently determine the form of the various unknown coefficients a's and b's.

To begin with we may start with a linear form

$$\xi = a_1 + a_2\dot{x} + a_3\dot{y}, \tag{25a}$$

$$\eta_1 = b_1 + b_2\dot{x} + b_3\dot{y}, \tag{25b}$$

$$\eta_2 = c_1 + c_2\dot{x} + c_3\dot{y}, \tag{25c}$$

where $a_i = a_i(x,y,t)$, $b_i = b_i(x,y,t)$, $c_i = c_i(x,y,t)$, $i = 1,2,3$.

Substituting (25) in the invaraince conditions, regrouping and equating the various coefficients of $\dot{x}^m \dot{y}^n$, $m,n = 0,1,2,3,4$ an overdetermined system of 24 linear pdfs arise [15]. The first few of them are as follows.

$$a_{2xx} = 0, \quad 2a_{2xy} + a_{3xx} = 0, \quad a_{2yy} + 2a_{3xy} = 0, \quad a_{3yy} = 0, \tag{26}$$

$$b_{2yy} + 2b_{3xy} - a_{1yy} - 2a_{3yt} = 0, \quad b_{3yy} = 0 \tag{27}$$

and so on. These equations can be solved consistently after considerable labour [15].

However an analysis of the resulting symmetries suggests that it is adequate to consider time independent symmetries alone:

$$\xi = 0, \quad \eta_1, \eta_2 : \text{independent of } t. \tag{28}$$

Then the resultant determining equations are

$$a_i = 0, \quad i = 1,2,3 \tag{29a}$$

$$b_{2xx} = 0, \quad b_{3xx} + 2b_{2xy} = 0, \quad b_{2yy} + 2b_{3xy} = 0,$$

$$b_{1xx} = 0, \quad b_{1xy} = 0, \quad b_{1yy} = 0, \quad b_{3yy} = 0. \tag{29b}$$

$$c_{2xx} = 0, \quad c_{3xx} + 2c_{2xy} = 0, \quad c_{2yy} + 2c_{3xy} = 0,$$

$$c_{1xx} = 0, \quad c_{1xy} = 0, \quad c_{1yy} = 0, \quad c_{3yy} = 0. \tag{29c}$$

$$3\alpha_1 b_{2x} + b_3 \alpha_{2x} + 2\alpha_2 b_{3x} + \alpha_2 b_{2y} - c_2 \alpha_{1y} = 0,$$

$$b_2 \alpha_{1y} + 2\alpha_1 b_{2y} + b_3 \alpha_{2y} + 3\alpha_2 b_{3y} + \alpha_1 b_{3y} - b_3 \alpha_{1x} - c_3 \alpha_{1y} = 0,$$

$$\alpha_1 b_{1x} + \alpha_2 b_{1y} - b_1 \alpha_{1x} - c_1 \alpha_{1y} = 0. \tag{29d}$$

$$c_2 \alpha_{1x} + 3\alpha_1 c_{2x} + (c_3 - b_2)\alpha_{2x} + \alpha_2(2c_{3x} + c_{2y}) - c_2 \alpha_{2y} = 0,$$

$$c_2 \alpha_{1y} + \alpha_1(2c_{2y} + c_{3x}) + 3\alpha_2 c_{3y} - b_3 \alpha_{2x} = 0,$$

$$\alpha_1 c_{1x} + \alpha_2 c_{1y} - b_1 \alpha_{2x} - c_1 \alpha_{2y} = 0. \tag{29e}$$

Eqs. (29b) and (29c) can be very easily solved:

$$b_1 = b_{10}x + b_1 y + b_{12}$$

$$b_2 = b_{20}y^2 + b_{21}xy + b_{22}x + b_{23}y + b_{24}$$

$$b_3 = -b_{21}x^2 - b_{20}xy + b_{31}x + b_{32}y + b_{33} \tag{30a}$$

and

$$c_1 = c_{10}x + c_{11}y + c_{12},$$

$$c_2 = c_{20}y^2 + c_{21}xy + c_{22}x + c_{23}y + c_{24},$$

$$c_3 = -c_{21}x^2 - c_{20}xy + c_{31}x + c_{32}y + c_{34}, \tag{30b}$$

where the b_{ij}'s and c_{ij}'s are constants. Thus we have 22 unknown constants in eqs. (30) which are to be determined from eqs.(29d) and (29e). Substituting the forms (30) for b_i's and c_i's in eqs. (29d,e), we ultimately end up with purely algebraic equations. Equating various powers of $x^m y^n$ to zero, then we can find the parameteric values at which the system is integrable. The same procedure can then be extended straightforwardly to cubic polynomials in velocities also.

Using this procedure, we have identified three parametric choices for the Henon-Heiles system and four parametric choices for the coupled quartic anharmonic oscillator. The associated symmetries and integrals of motion are given in Table I. They agree with the known results fully [13].

5. THREE COUPLED OSCILLATORS: SYMMETRIES AND INTEGRABILITY

The symmetry analysis for two coupled oscillators can be extended to three coupled oscillators in a systematic way. We now consider the Lagrangian

$$L = \tfrac{1}{2}(\dot{x}^2 + \dot{y}^2 + \dot{z}^2) - V(x,y,z) \tag{31}$$

Table I: Infinitesimal Symmetries and Integrals of Motion of Two-coupled Anharmonic Oscillator (AHO) Systems

System	Parametric restrictions	ξ	Infinitesimal symmetries n_1	n_2	Integral of motion
Henon-Heiles	$A=B$, $\alpha=-\beta$	0	$k\dot{y}$	$k\dot{x}$	$\dot{x}\dot{y}+Axy+\frac{1}{3}\alpha x^3+\alpha xy^2$
	A,B arbitrary, $6\alpha=-\beta$	0	$4\alpha(x\dot{y}-2\dot{x}y)+2(4A-B)\dot{x}$	$4\alpha x\dot{x}$	$4L_{xy}\dot{x}+(4Ay+x^2+4y^2)\dot{x}^2+(4A-B)(\dot{x}^2+Ax^2)$
	$16A=B$, $16\alpha=-\beta$	0	$4\dot{x}^3+4(A+2\alpha y)x^2\dot{x} - \frac{4}{3}x^3\dot{y}$	$-\frac{4}{3}\alpha x^3\dot{x}$	$\dot{x}^4+2(A+2\alpha y)x^2\dot{x}^2 - \frac{4}{3}x^3\dot{x}\dot{y} + A^2x^4 - \frac{4}{3}\alpha(A+\alpha y)x^4y - \frac{2}{9}\alpha^2x^6$
Quartic AHO	A,B arbitrary, $\alpha=\beta$, $\delta=2\alpha$	0	$-2L_{yx}y+2/\alpha(B-A)\dot{x}$	$2L_{xy}\dot{x}$	$L^2_{xy}+2/\alpha(B-A)(\frac{1}{2}\dot{x}^2+Ax^2+\alpha r^2x^2)$
	$A=B$, $\alpha=\beta$, $\delta=6\beta$	0	$k\dot{y}$	$k\dot{x}$	$\dot{x}\dot{y}+2(A+3\alpha r^4+4\alpha x^2y^2)xy$
	$A=4B$, $\alpha=16\beta$, $\delta=12\beta$	0	$y\ddot{y}$	$y\dot{x}-2x\dot{y}$	$-L_{xy}y+2(B+2\beta r^2+2\beta x^2)xy^2$
	$A=4B$, $\alpha=8\beta$, $\delta=6\beta$	0	$8\beta(y\dot{x}-2x\dot{y})y^3 + 5\beta x^2)y^2\dot{y}$	$4\dot{y}^3+8(B+\beta r^2+5\beta x^2)y\dot{y}^2-16\beta xy^3\dot{x}\dot{y}+$	$\dot{y}^4+4(B+\beta r^2+5\beta x^2)y^2\dot{y}^2-16\beta xy^3\dot{x}\dot{y}+4\beta y^4x^2+4\beta(B+2\beta r^2+2\beta x^2)+4\beta^2(r^2+x^2)^2y^4$

where $L_{xy} = x\dot{y}-y\dot{x}$, $r^2 = x^2+y^2$

so that the equations of motion become

$$\ddot{x} = \frac{\partial L}{\partial x} = \alpha_1(x,y,z), \quad (32a)$$

$$\ddot{y} = \frac{\partial L}{\partial y} = \alpha_2(x,y,z), \quad (32b)$$

$$\ddot{z} = \frac{\partial L}{\partial z} = \alpha_3(x,y,z) \quad (32c)$$

Considering now the group of transformations

$$\{t,x,y,z,\dot{x},\dot{y},\dot{z}\} \rightarrow \{T,X,Y,Z,\dot{X},\dot{Y},\dot{Z}\}, \quad (33)$$

where the primed quantities are obvious generalizations of (8)-(10), which maps the solution curves of (32) into solution curves, we can derive the invariance criteria as in the previous section. Considering the infinitesimal transformations

$$x \rightarrow X = x + \varepsilon \eta_1(t,x,y,z,\dot{x},\dot{y},\dot{z}) + O(\varepsilon^2) \quad (34a)$$

$$y \rightarrow Y = y + \varepsilon \eta_2(t,x,y,z,\dot{x},\dot{y},\dot{z}) + O(\varepsilon^2) \quad (34b)$$

$$z \rightarrow Z = z + \varepsilon \eta_3(t,x,y,z,\dot{x},\dot{y},\dot{z}) + O(\varepsilon^2) \quad (34c)$$

$$t \rightarrow T = t + \varepsilon \xi(t,x,y,z,\dot{x},\dot{y},\dot{z}) + O(\varepsilon^2), \quad \varepsilon << 1 \quad (34d)$$

the associated infinitesimal generator becomes

$$E = \xi \frac{\partial}{\partial t} + \eta_1 \frac{\partial}{\partial x} + \eta_2 \frac{\partial}{\partial y} + \eta_3 \frac{\partial}{\partial z} + (\dot{\eta}_1 - \xi\dot{x}) \frac{\partial}{\partial \dot{x}}$$
$$+ (\dot{\eta}_2 - \xi\dot{y}) \frac{\partial}{\partial \dot{y}} + (\dot{\eta}_3 - \xi\dot{z}) \frac{\partial}{\partial \dot{z}}. \quad (35)$$

Then we are led to the invariance conditions

$$\ddot{\eta}_1 - \dot{x}\dot{\xi} - 2\dot{\xi}\alpha_1 = E(\alpha_1) \quad (36a)$$

$$\ddot{\eta}_2 - \dot{y}\dot{\xi} - 2\dot{\xi}\alpha_2 = E(\alpha_2) \quad (36b)$$

$$\ddot{\eta}_3 - \dot{z}\dot{\xi} - 2\dot{\xi}\alpha_3 = E(\alpha_3) \quad (36c)$$

The associated constants of motion are found to be

$$\Phi = (\xi\dot{x} - \eta_1)\frac{\partial L}{\partial \dot{x}} + (\xi\dot{y} - \eta_2)\frac{\partial L}{\partial \dot{y}} + (\xi\dot{z} - \eta_3)\frac{\partial L}{\partial \dot{z}} - \xi L + f,$$

$$E\{L\} + \dot{\xi}L = \dot{f}. \quad (37)$$

As in the two coupled oscillator case, we assume polynomial forms in velocities for the infinitesimals ξ, η_i, $i = 1, 2, 3$.

i) $\eta_1 = \eta_2 = \eta_3 = 0$, $\xi = \text{const.} \rightarrow \Phi = H = $ Hamiltonian (38)

ii) $\xi = a_1 + a_2\dot{x} + a_3\dot{y} + a_4\dot{z}$, $\eta_1 = b_1 + b_2\dot{x} + b_3\dot{y} + b_4\dot{z}$ (39a)

$\eta_2 = c_1 + c_2\dot{x} + c_3\dot{y} + c_4\dot{z}$, $\eta_3 = d_1 + d_2\dot{x} + d_3\dot{y} + d_4\dot{z}$ (39b)

Using these forms in the invariance conditions, we again obtain a system of 60 linear partial differential equations which can then be solved consistently. Or more simply, we may assume a time independent form for the infinitesimals which simplifies the analysis more drastically.

iii) $\xi = \sum_{k,\ell,m=0}^{M} a_{k\ell m}(\dot{x})^k(\dot{y})^\ell(\dot{z})^m$, $\eta_i = \sum_{k,\ell,m=0}^{M} b_{k\ell m}^i(\dot{x})^k(\dot{y})^\ell(\dot{z})^m$.

Again the analysis proceeds as in the two dimensional system.

In Table II we give the results of our analysis for a system of three quartically coupled oscillators and cubically coupled oscillators.

6. CONCLUSIONS

In this work we have shown that Lie's invariance method of extended vector fields provides an effective and systematic way of identifying integrable dynamical systems for potentials of polynomial type. The only lacuna is that the method requires the assumption of a specific form the integrals of motion in the velocities. We have illustrated the procedure for specific examples of cubic and quartically coupled oscillators with two and three degrees of freedom. The method can be extended in a straightforward way to the case of arbitrary number of coupled oscillators also.

Acknowledgements. I thank Dr R. Sahadevan for his collaboration in this work. The work reported here forms part of a project supported by the Department of Science and Technology, Government of India.

Table 11a Generalised Lie Symmetries and Integrals of Motion of 3-coupled Quartic Nonlinear Oscillators

Cases	ξ	Infinitesimal Symmetries η_1	η_2	η_3	Integrals of Motion
(i)	0	$2/\alpha\ (B-A)\dot{x}+2yL_{yx}+2zL_{zx}$	$2xL_{xy}+2\dot{z}L_{zy}$	$2xL_{xz}+2\dot{y}L_{yz}$	$\vec{L}^2 + 2/\alpha\ [(A-B)P_y^2 + (A-C)P_z^2$
	0	$2/\alpha\ (B-A)(B-C)\dot{x}+2(B-C)yL_{yx}$	$2(B-A)\dot{z}L_{zy} + 2(B-C)xL_{xy}$	$2/\alpha\ (B-A)(B-C)\dot{z}+2(B-A)yL_{yz}$	$2/\alpha\ (B-A)(B-C)(P_y^2 + P_z^2)$
					$+ (B-A)L_{yz}^2 + (B-C)L_{xy}^2$
(ii)	0	$2zL_{zx}$	0	$2xL_{xz}$	L_{xz}^2
	0	$2[\dot{x}\dot{y} + 2(A + 2\alpha\rho^2)xy]\dot{y} + 16\alpha L_{xz}y^2z$	$2(\dot{x}^2 + \dot{z}^2)\dot{y} + 4(A+2\alpha\rho^2)y(x\dot{x}+z\dot{z})$	$2[\dot{y}\dot{z} + 2(A + 2\alpha\rho^2)yz]\dot{y} + 16\alpha L_{xz}xy^2$	$[(x\dot{x}+z\dot{z})\dot{y} +2(A+2\alpha\rho^2)y\hat{y}^2]^2 \hat{y}^2$
					$+ L_{xz}^2[\hat{y}^2\dot{\hat{y}}^2 + 8\alpha y^2]$
(iii)	0	$2zL_{zx}$	0	$2xL_{xz}$	L_{xz}^2
	0	$2y(L_{yx}\dot{y} + 2\alpha xy^2)\dot{y} + 16\beta L_{zx}y^4z$	$2y(-\dot{y}^2+2\alpha y^2)(x\dot{x}+z\dot{z}) +2y^2(\dot{x}^2+\dot{z}^2)\dot{y} - 4(L_{xy}x\dot{y}+L_{yz}z\dot{y}+2\alpha\hat{y}^2y^2)$	$2y(L_{yz}\dot{y} + 2\alpha y^2z)\dot{y} + 16\beta L_{xz}xy^4$	$[L_{xy}x\dot{y} + L_{yz}\dot{y}z + 2\alpha\hat{y}^2 y^2]^2 \hat{y}^2$
					$+ L_{xz}^2[\hat{y}^2 \dot{\hat{y}}^2 + 8\beta y^2]^2$
					$\Omega = B + 2\rho^2 + 2\hat{y}^2$
(iv)	0	$2zL_{zx}$	0	$2xL_{xz}$	L_{xz}^2
	0	$8\beta\dot{y}^3(L_{yx} - x\dot{y})$	$4\dot{y}^3+8y^2(B+5\beta\hat{y}+\beta\rho^2)\dot{y} -16\beta(x\dot{x}+z\dot{z})y^3$	$8\beta\dot{y}3(L_{yz}-z\dot{y})$	$\dot{y}^4+4y^2[B+5\beta y+\beta\ \rho^2]\dot{y}^2-16\beta\hat{y}^3(x\dot{x}+z\dot{z})\dot{y}$
					$+ 4[B(\dot{x}\dot{z}+\dot{z}^2) +B(B+2\beta\hat{y}+2\beta\ \rho^2)$
					$+ \beta^2(\rho^2 + \hat{y}^2)^2]y^4$
(v)	0	0	$2zL_{zy}$	$2yL_{zy}$	L_{yz}^2
	0	$y\dot{y} + z\dot{z}$	$L_{yx} - x\dot{y}$	$L_{zx} - z\dot{x}$	$L_{yx}\dot{y} + L_{zx}\dot{z} +2[B+2\beta x^2+2B\rho^2]x\hat{x}^2$
(vi)	0	0	$2zL_{zy}$	$2yL_{zy}$	L_{yz}^2
	0	$8\beta[yL_{yx} +zL_{zx}- (y\dot{y} +z\dot{z})\dot{x}]\hat{x}^2$	$4(\dot{y}^2+\dot{z}^2)\dot{y}+8\beta\hat{x}^2 (B+\beta x^2+B\rho^2)\dot{y}+32\beta$	$4(\dot{y}^2+\dot{z}^2)\dot{z}+8\beta x^2(B+\beta x^2+\beta\rho^2)\dot{z}$	$(\dot{y}^2+\dot{z}^2)^2+4\{\hat{x}^2(B+\beta x^2+\beta\rho^2)(\dot{y}^2+\dot{z}^2)$
			$(B+\beta x+B\ \rho^2)\dot{y}+32\beta$ $(y\dot{y}+z\dot{z})x^2y-16\beta\hat{x}^2xy\dot{x}$	$+32\beta(y\dot{y}+z\dot{z})x\cdot z - 16\beta\hat{x}^2xz\dot{x}$	$+16\beta x^2(y\dot{y}+z\dot{z})^2 -16\beta\hat{x}^2x(y\dot{y}+z\dot{z})\dot{x}$
					$+4[\dot{x}^2+B^2+2B\beta(x^2+\rho^2)+B^2(\rho^2+x^2)^2]\hat{x}^4$

		$L_{yx} - x\dot{y}$	$L_{xz} - z\dot{x}$	$L_{xz}\dot{z} + L_{yx}\dot{y} + 2(B+2(\rho^2+2x^2)x(y^2-z^2)$
(vii)	0	$y\dot{y} - z\dot{z}$	$2\dot{y}\dot{z}+16\beta Lxzxy^2 +$	$\dot{y}^2 y + 8B(x^2 y^2 \dot{z}^2 + y^2 z^2 \dot{x}^2 + x^2 z^2 \dot{y}^2) +$
	0	$16\beta yz(yz\dot{x} - xz\dot{y} - xy\dot{z})$	$4yz(B+6\beta x^2+2\beta\rho^2)\dot{z}$	$8\beta yz[(3x^2+\rho^2)\dot{y}\dot{z}-2x(\dot{y}z+y\dot{z})\dot{x}]+$
				$4Byz(y\dot{z}+y\dot{z})+16\beta(Bx^2\Gamma^2+\beta\Gamma^4.z^2)y^2$
(viii)	0	$y\dot{z} + z\dot{y}$	$L_{yx} - x\dot{y}$	$(\dot{y}z+y\dot{z})z-2xy\dot{y}\dot{z}+4\beta(B+\beta x^2+\beta(\rho^2)xyz$
	0	$8\beta(y^2-z^2)^2\dot{x}-16\beta x(y^2-z^2)$	$-4(\dot{y}^2-\dot{z}^2)\dot{z}-8(y^2-z^2)$	$(\dot{y}^2-\dot{z}^2)^2+4\beta(y^2-z^2)(B+(\rho^2+x^2)(\dot{y}^2-\dot{z}^2)$
		$(y\dot{y}-z\dot{z})$	$(B+\rho^2+x^2)\dot{y}+$	$+16\beta x[x(y\dot{y}-z\dot{z})-(y^2-z^2)\dot{x}](y\dot{y}-z\dot{z})$
			$+32\beta x^2 y(y\dot{y}-z\dot{z})$	
			$-16\beta xy(y^2-z^2)\dot{x}\cdots+16\beta xz(y^2-z^2)\dot{x}\cdots$	$+4[\beta x^2+(B+2(\rho^2+2x^2)+\beta^2(x^2+\rho^2)](y^2-z^2)$.

where $\rho^2 = x^2+y^2+z^2$; $\vec{L} = L_{xv}+L_{vz}+L_{zx}$; $\hat{x} = y^2+z^2$, $\hat{y} = x^2+z^2$; $L_{xv} = (x\dot{y}-y\dot{x})$,...; $P_x = \frac{1}{2}\dot{x}^2+Ax^2+2\alpha x^2\rho^2$, $P_z = \frac{1}{2}\dot{z}^2+Cz^2+2\alpha z^2\rho^2$, $\Gamma^2 = (x^2+\rho^2)$.

Table IIb : Generalised Lie Symmetries and Integrals of Motion of 3-coupled Cubic Nonlinear Oscillators

$(V = \frac{1}{2}(Ax^2 + By^2 + Cz^2) + \alpha x^2 y + \gamma yz^2 + \frac{1}{3}\beta y^3)$

Cases	Parametric Restrictions	ξ	Infinitesimal Symmetries η_1	η_2	η_3	Integrals of Motion
(i)	$A = B = C, \alpha = \beta = -\gamma$	0	$2zL_{xz}$	0	$2xL_{xz}$	L^2_{xz}
		0	$2(\dot{x}\dot{y}+Axy+\alpha xy^2+ \alpha/3\hat{y}^2)y + 8/3 L_{xz}yz$	$2(\dot{x}^2+\dot{z}^2)+2(Ay+\alpha y^2 +\alpha/3y^2)(x\dot{x}+z\dot{z})$	$2(\dot{y}\dot{z}+Ayz+\alpha y^2 z + \alpha/3\hat{y}^2 z)y + 8/3 L_{xz}xy$	$[(x\dot{x}+z\dot{z})y + A\hat{y}\dot{y}+\alpha\hat{y}^2y + \alpha/3\hat{y}^4]z\hat{y}^2$ $+L^2_{xz}(\dot{y}^2\hat{y}^2+4/3 \alpha y)$
(ii)	$A = C, B$ = arbit.,	0	$2zL_{zx}$	0	$2xL_{xz}$	L^2_{xz}
	$\alpha = \beta, \gamma = -6\alpha$	0	$4(L_{xy} - y\dot{x})+$ $2/\alpha(4A-B)\dot{x}$	$4(x\dot{x} + z\dot{z})$	$4(L_{yz} - y\dot{z}) +$ $2/\alpha(4A-B)\dot{z}$	$4(\dot{z}L_{yz}-\dot{x}L_{xy})+(4A-B)/\alpha(\dot{x}^2+\dot{z}^2+A\hat{y}^2)$ $+\alpha\hat{y}^4+4\alpha\hat{y}^2y^2+4Ay\hat{y}^2$
(iii)	$A = C, B = 16A,$	0	$2zL_{zx}$	0	$2xL_{xz}$	L^2_{xz}
	$\alpha = \beta, \gamma = -16\alpha$	0	$4(\dot{x}^2+\dot{z}^2)\dot{x}+4(A+\alpha y)$ $x\hat{y}^2+4/3\alpha(\hat{y}^2x\dot{y}+4L_{xy}yz)$	$-4/3 \alpha(x\dot{x}+z\dot{z})\hat{y}^2$	$4(\dot{x}^2+\dot{z}^2)\dot{z}+4(A+2\alpha y)\dot{z}\hat{y}^2$ $-4/3\alpha(\hat{y}^2z\dot{y}+4L_{xz}xy)$	$(\dot{x}^2+\dot{z}^2)^2 + 2(A+2\alpha y)(\dot{x}^2+\dot{z}^2)\hat{y}^2 -$ $-4/3\alpha(\hat{y}^2(x\dot{x}+z\dot{z})\dot{y}+3/4A^2\hat{y}^4-(A+\alpha y)$ $\hat{y}^4\dot{y}^2+2/3\alpha\hat{y}^6+2L^2_{xz}y]$

where $\hat{y}^2 = x^2 + z^2$.

REFERENCES

[1] V. I. Arnold, **Dynamical Systems**, Vol. III (Springer-Verlag, New York, 1988).
[2] V. V. Kozlev, Russ. Math. Surveys **38** (1983) 1.
[3] B. Dorizzi, B. Grammaticos and A. Ramani, J. Math. Phys. **25** (1984) 481.
[4] T. Bountis, H. Segur and F. Vivaldi, Phys. Rev. **A25** (1982) 1257.
[5] M. Lutzky, J. Phys. **11A** (1978) 249; **12A** (1979) 973.
[6] L. S. Hall, Physica **8D** (1983) 90.
[7] E. T. Whittker, **A Treatise on the Analytical Dynamics of Particles and Rigid Bodies** (Cambridge University Press, London, 1937).
[8] S. Chandrasekhar, **Principles of Stellar Dynamics** (Dover, New York, 1942), Ch. III.
[9] M. J. Ablowitz, A. Ramani and H. Segur, J. Math. Phys. **21** (1980) 715.
[10] G. W. Bluman and J. D. Cole, **Similarity Methods for Differential Equations** (Springer, New York, 1974).
[11] L. V. Ovsiannikov, **Group Analysis of Differential Equations** (Academic Press, New York, 1982).
[12] P. J. Olver, **Applications of Lie Groups to Differential Equations** (Springer, New York, 1986).
[13] M. Lakshmanan and R. Sahadevan, Phys. Rev. **A31** (1985) 861.
[14] B. Grammaticos, B. Dorrizzi and A. Ramani, J. Math. Phys. **24** (1983) 2289.
[15] R. Sahadevan and M. Lakshmanan, J. Phys. **19A** (1986) L949.

Classical Integrable Systems Generated Through Nonlinearization of Eigenvalue Problems

Cao Cewen and Geng Xianguo

Department of Mathematics, Zhengzhou University,
Zhengzhou, People's Rep. of China

1. Introduction

   It is a challenge for us to look for new finite-dimensional completely integrable systems. H. Flaschka [1] pointed out an important principle to obtain finite-dimensional integrable systems by constraining infinite-dimensional integrable systems on finite-dimensional invariant subset.

   An approach is given in the present paper in realizing the above principle, through which quite a few finite-dimensional integrable systems are effectively obtained.

   Consider the eigenvalue problem:

$$L(u,\lambda)y = 0 \qquad (1)$$

where the matrix differential operator L is a rational function of the eigenparameter λ. We have found that there frequently exists a constraint relation between the potential $u(x)$ and a set of eigenfunctions $q(x) = (q_1(x),\ldots,q_N(x))$, $u = f(q)$, under which (1) is nonlinearized to be a completely integrable system in Liouville sense:

$$L(f(q),\Lambda)q = 0, \quad \Lambda = \mathrm{diag}(\lambda_1,\ldots,\lambda_N). \qquad (2)$$

Sometimes a geometrical condition is attached to the phase space:

$$L(f(q),\Lambda)q = 0, \quad g(p,q) = 0, \qquad (3)$$

which is integrable on the manifold given by the condition $g = 0$.

   The constraint relations f and g have a profound background. (2) is called the Bargmann system, while (3) is called the Neumann system, following the terminology associated with the Schrödinger-KdV system, the most typical one in the soliton analysis.

2. Schrödinger-KdV system

   The Schrödinger-KdV system is a motive and a starting point for our investigation. Let $u(x)$ be the reflectionless (or Bargmann) potential for the Schrödinger equation. In every bound state, the eigenvalue λ_j

$= -\frac{1}{4}\alpha_j$ and eigenfunction $q_j(x)$ satisfy:

$$-q_j'' + uq_j = -\frac{1}{4}\alpha_j q_j, \quad j = 1,\ldots,N.$$

or

$$-q'' + uq = -\frac{1}{4}Aq, \quad A = \text{diag}(\alpha_1,\ldots,\alpha_N) \tag{4}$$

The Bargmann potential can be represented as the squared sum of eigenfunctions (with suitable normed conditions: $\|q_j\|^2 = \sqrt{\alpha_j}$) [2] :

$$u(x) = -2\langle q,q \rangle = f(q) \tag{5}$$

$\langle \xi,\eta \rangle = \sum_{j=1}^{N} \xi_j \eta_j$ being the standard inner-product in R^{2N}. (4) is nonlinearized by (5) into :

$$-q'' - 2\langle q,q \rangle q = -\frac{1}{4}Aq \tag{6}$$

which has been verified recently [3] to be a completely integrable system in the symplectic space $(R^{2N}, dp \wedge dq), (p = q')$, with the Hamiltonian $H = F_0$ and the independent, the involutive conserved integrals $\{F_m\}$:

$$H = F_0 = \frac{1}{2}\langle p,p \rangle + \frac{1}{2}\langle q,q \rangle^2 - \frac{1}{8}\langle Aq,q \rangle \tag{7}$$

$$F_m = \frac{1}{2}\langle A^m p,p \rangle + \frac{1}{2}\langle A^m q,q \rangle \langle q,q \rangle - \frac{1}{8}\langle A^{m+1}q,q \rangle$$

$$+2 \sum_{i+j=m-1} \begin{vmatrix} \langle A^i q,q \rangle & \langle A^i q,p \rangle \\ \langle A^j p,q \rangle & \langle A^j p,p \rangle \end{vmatrix} \tag{8}$$

In the case when $u(x)$ is an N-gap potential for the Schrödinger-Hill equation, the eigenvalue $\lambda = \alpha_j$ given by the right-hand side end point of the spectral gap and the corresponding eigenfunction $q_j(x)$ satisfy:

$$-q_j'' + uq_j = \alpha_j q_j, \quad (j = 0,1,\ldots,N)$$

or

$$-q'' + uq = Aq, \quad A = \text{diag}(\alpha_0, \alpha_1,\ldots,\alpha_N). \tag{9}$$

We have the well-known Mckean-Trubowitz identity [2]

$$\langle q,q \rangle = 1. \tag{10}$$

By differentiating it and using (9), we obtain

$$u(x) = \langle Aq,q \rangle - \langle p,p \rangle, \quad (p = q') \tag{11}$$

(9),(10) and (11) constitute the famous C. Neumann system, which is completely integrable on the tangent bundle of sphere TS^N, whose integrability is given by the Moser constraint [4] of an integrable system $(R^{2N+2}, dp \wedge dq, H = \frac{1}{2} F_1)$ restricted on TS^N. The involutive conserved integrals are

$$F_0 = \langle q,q \rangle, \quad F_m = \langle A^m q,q \rangle + \sum_{i+j=m-1} \begin{vmatrix} \langle A^i q,q \rangle & \langle A^i q,p \rangle \\ \langle A^j p,q \rangle & \langle A^j p,p \rangle \end{vmatrix} \quad (12)$$

Denote the tangent bundle of sphere by:

$$TS^N = \left\{ (p,q) \in R^{2N+2} \mid F = \frac{1}{2}(\|q\|^2 - 1) = 0, G = \langle p,q \rangle = 0 \right\},$$

then the Hamiltonian of the Neumann system is $H^* = H - \mu F$, with the involutive conserved integrals $F_m^* = F_m - \mu_m F$, where the Lagrangian multipliers are:

$$\mu = \frac{(H,G)}{(F,G)} \bigg|_{TS^N} = \langle Aq,q \rangle$$

$$\mu_m = \frac{(F_m,G)}{(F,G)} \bigg|_{TS^N} = (F_m,G) \bigg|_{TS^N}$$

$(.,.)$ is the Poisson bracket in the symplectic space $(R^{2N+2}, dp \wedge dq)$.

3. Bargmann and Neumann constraint

The nonlineatized conditions

I. $u = -2 \langle q,q \rangle$ (Bargmann potential)

II. $\begin{cases} \langle q,q \rangle = 1 & \text{(Mckean-Trubowitz identity)} \\ u = \langle Aq,q \rangle - \langle p,p \rangle & \text{(Neumann potential)} \end{cases}$

play a central role in obtaining the completely integrable Bargmann system (6) and C. Neumann system (9),(10),(11). Next we shall write them in general form, resorting to the functional gradients of eigenvalues and the so-called Lenard's gradients.

The functional gradient $\delta\lambda/\delta u$ of (1)'s eigenvalue λ with regard to the potential $u(x)$ can be calculated under a certain boundary conditions, i.e. periodic conditions or decaying at ∞ conditions. It satisfies a linear differential equation in quite a few cases:

$$K\left(\frac{\delta\lambda_j}{\delta u}\right) = c\lambda_j J\left(\frac{\delta\lambda_j}{\delta u}\right) \quad (13)$$

K, J are called Lenard pair, with which the Lenard gradient series G_j can be determined recursively :

$$KG_{j-1} = JG_j \quad , \quad j = 0,1,2,\ldots$$

Sometimes the element from kerJ is taken as the starting point G_{-1} : $JG_{-1} = 0$. We shall have k series when kerJ has the dimension k. We consider only the one-dimensional case for simplicity. $X_j = JG_j$ is the isospectral vector field of the eigenvalue problem (1), and $u_t = X_j(u)$ is the usual soliton equation.

Consider the following two kinds of constraints :

Condition I. $\quad G_0 = \sum_{j=1}^{N} \gamma_j \dfrac{\delta \lambda_j}{\delta u}$ (Bargmann constraint) $\qquad$ (14)

Condition II. $\quad G_{-1} = \sum_{j=0}^{N} \gamma_j \dfrac{\delta \lambda_j}{\delta u}$ (Neumann constraint) $\qquad$ (15)

As a special case, for the Schrödinger equation, $L = -\partial^2 + u - \lambda$, ($\partial = \partial/\partial x$); $\delta\lambda_j/\delta u = q_j^2(x)$; $K = \partial^3 - 2(u\partial + \partial u)$, $J = \partial$. The Lenard gradients are :

$$G_{-1} = -\tfrac{1}{2} , \quad G_0 = u , \quad G_1 = u'' - 3u^2 , \quad \ldots$$

$X_j = JG_j$ is the KdV vector field. I, II become the GGKM formula $u = -2\langle q,q \rangle$ and MT identity $\langle q,q \rangle = 1$ by taking $\gamma_j = -2$ and $\gamma_j = -\tfrac{1}{2}$ respectively.

(14) is usually equivalent to

$$\text{I.} \quad u = f(q) , \qquad (16)$$

while (15) is equivalent to $g(q) = 0$. The latter yields another $u = f(q)$, resorting to the equation (1). Thus :

$$\text{II.} \quad u = f(q), \; g(q) = 0 . \qquad (17)$$

The eigenvalue problem (1) is nonlinearized into :

$$\text{I.} \quad L(f(q), \Lambda)q = 0 . \qquad (18)$$

$$\text{II.} \quad L(f(q), \Lambda)q = 0 , \; g(q) = 0 . \qquad (19)$$

It is a surprising fact that they are finite-dimensional completely integrable system in Liouville sense for quite a few systems. We shall show this fact with a series of examples.

Moreover, the solution q of (18) or (19) is mapped by f into a solution u = f(q) of the stationary soliton equation :

$$X_N + c_1 X_{N-1} + \ldots + c_N X_0 = 0 .$$

The solution is well-known to be the finite-band potential of the eigenvalue problem.

The heart of the Liouville integrability is the existence of involutive system of conserved integrals. In some cases they can be obtained in the following way. Expressing the soliton equation $u_{t_m} = X_m$ in the Lax form :

$$\frac{\partial}{\partial t_m} L = [V_m, L]$$

then the time part of the Lax pair $\frac{\partial q}{\partial t_m} = V_m q$ is reduced to a finite -dimensional Hamiltonian system under the above constraint, whose Hamiltonian F_m constitute the involutive system of the Bargmann system (18) , or that of the Neumann system (19) through the Moser constraining procedure. But the concrete constructing process is not always simple, it should undergo some modifications. The results are sometimes guessed , especially for the Neumann system.

4. Applications

1). Jaulent-Miodek eigenvalue problem [5]

$$L(u,v,\lambda)y = y'' + (\lambda^2 - \lambda v(x) - u(x))y = 0 . \quad (20)$$

$$\nabla \lambda_j \triangleq \begin{bmatrix} \delta\lambda_j/\delta u \\ \delta\lambda_j/\delta v \end{bmatrix} = \begin{bmatrix} q_j^2(x) \\ \lambda_j q_j^2(x) \end{bmatrix} , \quad \int_\Omega (2\lambda_j - v) q_j^2 dx = 1$$

Ω is the underlying interval. The Lenard pair is

$$K = \begin{bmatrix} -\frac{1}{4}\partial^3 + \frac{1}{2}(u\partial + \partial u) & 0 \\ 0 & \partial \end{bmatrix} , \quad J = \begin{bmatrix} -\frac{1}{2}(v\partial + \partial v) & \partial \\ \partial & 0 \end{bmatrix}$$

kerJ is spanned by the linearly independent solutions G_{-2}, G_{-1} of the equation JG = 0 . The first few Lenard gradients are :

$$G_{-2} = \begin{bmatrix} 0 \\ 1 \end{bmatrix} , \quad G_{-1} = \begin{bmatrix} 1 \\ \frac{1}{2}v \end{bmatrix} , \quad G_0 = \begin{bmatrix} \frac{1}{2}v \\ \frac{1}{2}u + \frac{3}{8}v^2 \end{bmatrix} .$$

I. Bargmann constraint is given by $G_0 = \sum_{j=1}^{N} \frac{1}{2} \nabla \lambda_j$, which is equivalent to

$$u = \langle \Lambda q, q \rangle - \frac{3}{4} \langle q, q \rangle^2 \;,\; v = \langle q, q \rangle \qquad (21)$$

II. Neumann constraint is given by $G_{-1} = \sum_{j=1}^{N} \nabla \lambda_j$, which is equivalent to

$$\langle q, q \rangle = 1 \;,\; v = 2\langle \Lambda q, q \rangle \qquad (22)$$

By differentiating (22) and using (20) we have ($p = q'$):

$$u = -\langle p, p \rangle + \langle \Lambda^2 q, q \rangle - 2\langle \Lambda q, q \rangle^2 \qquad (23)$$

The nonlinearization of (20) under (21) yields the Bargmann system:

$$\{q'' + \Lambda^2 q - v\Lambda q - uq = 0 \text{ , } u,v \text{ given by (21)}\} \qquad (24)$$

which is an integrable Hamiltonian system $\{R^{2N}, dp \wedge dq, H = F_0\}$ with the involutive system of conserved integrals:

$$F_0 = \frac{1}{2}\langle p,p \rangle + \frac{1}{2}\langle \Lambda^2 q, q \rangle - \frac{1}{2}\langle \Lambda q, q \rangle \langle q, q \rangle + \frac{1}{8}\langle q, q \rangle^3$$

$$F_m = \frac{1}{2}\langle \Lambda^m p, p \rangle + \frac{1}{2}\langle \Lambda^{m+2} q, q \rangle - \frac{1}{4}\langle \Lambda q, q \rangle \langle \Lambda^m q, q \rangle - \frac{1}{4}\langle q, q \rangle \langle \Lambda^{m+1} q, q \rangle$$

$$+ \frac{1}{8}\langle q, q \rangle^2 \langle \Lambda^m q, q \rangle + \frac{1}{4} \sum_{i+j=m-1} \begin{vmatrix} \langle \Lambda^i q, q \rangle & \langle \Lambda^i q, p \rangle \\ \langle \Lambda^j p, q \rangle & \langle \Lambda^j p, p \rangle \end{vmatrix}$$

This result is proved by Mou Weihua.

The nonlinearization of (20) under (22), (23) gives the Neumann system:

$$\{q'' + \Lambda^2 q - v\Lambda q - uq = 0, u,v \text{ given by (22),(23); } \langle q, q \rangle = 1\} \qquad (25)$$

Consider the involutive system in $(R^{2N}, dp \wedge dq)$:

$$F_m = \frac{1}{2}\langle q, q \rangle \langle \Lambda^{m+1} q, q \rangle - \frac{1}{2}\langle \Lambda q, q \rangle \langle \Lambda^m q, q \rangle$$

$$+ \frac{1}{2} \sum_{i+j=m-1} \begin{vmatrix} \langle \Lambda^i q, q \rangle & \langle \Lambda^i q, p \rangle \\ \langle \Lambda^j p, q \rangle & \langle \Lambda^j p, p \rangle \end{vmatrix} , m = 1,2,3,\ldots$$

whose Moser constraint to TS^{N-1}:

$$H^* = F_1 - \mu_1 F \;,\; \mu_1 = 2\langle \Lambda^2 q, q \rangle - 2\langle \Lambda q, q \rangle^2$$

$$F_m^* = F_m - \mu_m F \;,\; \mu_m = (F_m, G) / (F, G)\big|_{TS^N}$$

with $F = \frac{1}{2}(<q,q> - 1)$, $G = <p,q>$ determines an integrable system $(TS^{N-1}, dp \wedge dq|_{TS^{N-1}}, H^*|_{TS^{N-1}})$, which is equivalent to the Neumann system (25).

2) Tu Guizhang's eigenvalue problem [6]

$$L(u,v,\lambda)y = y'' + (\lambda - u - \lambda^{-1}v)y = 0 .\qquad(26)$$

$$\nabla \lambda_j = \begin{bmatrix} \delta\lambda_j/\delta u \\ \delta\lambda_j/\delta v \end{bmatrix} = \begin{bmatrix} q_j^2(x) \\ \lambda_j^{-1} q_j^2(x) \end{bmatrix} , \quad K = \begin{bmatrix} 2\partial & 0 \\ 0 & v\partial + \partial v \end{bmatrix},$$

$$J = \begin{bmatrix} 0 & 2\partial \\ 2\partial & \frac{1}{2}\partial^3 - (u\partial + \partial u) \end{bmatrix}, \quad G_{-2} = \begin{bmatrix} 1 \\ 0 \end{bmatrix}, \quad G_{-1} = \begin{bmatrix} \frac{1}{2}u \\ 1 \end{bmatrix} \text{ span ker} J .$$

The Bargmann constraint seems to be an implicit one. The Neumann constraint $G_{-1} = \sum_{j=1}^{N} \nabla\lambda_j$ yields:

$$<\Lambda^{-1}q,q> = 1, \quad u = 2<q,q>, \quad v = -\frac{<q,q> + <\Lambda^{-1}p,p>}{<\Lambda^{-2}q,q>}, (p = q') \qquad (27)$$

The background involutive system in R^{2N} is:

$$H = F_0 = \frac{1}{2}<p,p> + \frac{1}{2}<\Lambda q,q> - \frac{1}{2}<q,q>^2$$

$$F_m = \frac{1}{2}<\Lambda^m p,p> + <\Lambda^{m+1} q,q> - \frac{1}{2}<q,q><\Lambda^m q,q>$$

$$+ \frac{1}{2}\sum_{i+j=m-1} \begin{vmatrix} <\Lambda^i q,q> & <\Lambda^i q,p> \\ <\Lambda^j p,q> & <\Lambda^j p,p> \end{vmatrix}, \quad m = 1,2,3,\ldots$$

The Tu's eigenvalue problem is nonlinearized under (27) to be an integrable system on the tangent bundle of ellipsoid

$$TQ^{N-1} = \left\{ (p,q) \in R^{2N} \mid F = \frac{1}{2}(<\Lambda^{-1}q,q> - 1) = 0, G = <\Lambda^{-1}p,q> = 0 \right\}$$

with Hamiltonian $H^* = H - \mu F|_{TQ^{N-1}}$, the Lagrange multiplier μ being

$$\mu = \frac{(H,G)}{(F,G)}\bigg|_{TQ^{N-1}} = -\frac{<q,q> + <\Lambda^{-1}p,p>}{<\Lambda^{-2}q,q>}$$

The involutive conserved integrals are $F_m^* = F_m - \mu_m F$ with

$$\mu_m = (F_m,G)/(F,G)\big|_{TQ^{N-1}}$$

3) The coupled Harry Dym hierarchy [7]

$$L(u,v,\lambda)y = y'' + (\lambda^2 v + \lambda u - \alpha)y = 0 .\tag{28}$$

$$\nabla\lambda_j = \begin{bmatrix} \delta\lambda_j/\delta u \\ \delta\lambda_j/\delta v \end{bmatrix} = \begin{bmatrix} \lambda_j q_j^2(x) \\ \lambda_j^2 q_j^2(x) \end{bmatrix} , \quad K = \begin{bmatrix} 0 & \frac{1}{2}\partial^3 - 2\alpha\partial \\ \frac{1}{2}\partial^3 - 2\alpha\partial & u\partial + \partial u \end{bmatrix}$$

$$J = \begin{bmatrix} \frac{1}{2}\partial^3 - 2\alpha\partial & 0 \\ 0 & -(v\partial + \partial v) \end{bmatrix} , \quad G_{-1} = \begin{bmatrix} 1 \\ 1/\sqrt{v} \end{bmatrix} , \quad G_0 = \begin{bmatrix} 1/\sqrt{v} \\ -u/v\sqrt{v} \end{bmatrix} .$$

Consider the Bargmann constraint $G_0 = \sum_{j=1}^{N} \nabla\lambda_j$, which yields

$$u = -2 <\Lambda^2 q,q><\Lambda q,q>^{-3} , \quad v = <\Lambda q,q>^{-2} .\tag{29}$$

The nonlinearized (28) under (29)

$$q' = p , \quad p' = 2<\Lambda^2 q,q><\Lambda q,q>^{-3}\Lambda q - <\Lambda q,q>^{-2}\Lambda^2 q + \alpha q$$

is an integrable Hamiltonian system (R^{2N}, $dp \wedge dq$, H) with

$$H = \frac{1}{2}<\Lambda^2 q,q><\Lambda q,q>^{-2} + \frac{1}{2}<p,p> - \frac{1}{2}\alpha<q,q>$$

whose involutive system of conserved integrals is

$$F_m = <q,p><\Lambda^{m-1}q,p> - \frac{1}{2}<p,p><\Lambda^{m-1}q,q> - \frac{1}{2}<q,q><\Lambda^{m-1}p,p>$$

$$- \frac{<\Lambda^2 q,q><\Lambda^{m-1}q,q>}{2<\Lambda q,q>^2} + \frac{<\Lambda^m q,q>}{2<\Lambda q,q>} + \frac{1}{2}\sum_{i+j=m-1}\begin{vmatrix} <\Lambda^i q,q> & <\Lambda^i q,p> \\ <\Lambda^j p,q> & <\Lambda^j p,p> \end{vmatrix} ,$$

$$m = 1, 2, \ldots$$

4) Kaup-Newell hierarchy [8]

$$L(u,v)y = y_x - \begin{bmatrix} -\lambda & \lambda u(x) \\ v(x) & \lambda \end{bmatrix} y = 0 \tag{30}$$

$$\nabla\lambda_j = (\delta\lambda_j/\delta u , \delta\lambda_j/\delta v)^T = (\lambda_j p_j^2(x), -q_j^2(x))^T ,$$

here $y = (q_j(x), p_j(x))^T$ is the eigenfunction corresponding to the eigenvalue λ_j. $G_0 = (v,u)^T$, hence the Bargmann constraint $G_0 = \sum_{j=1}^{N} \nabla\lambda_j$ gives

$$u = -<q,q> , \quad v = <\Lambda p,p> .\tag{31}$$

The nonlinearized (30) under (31)

$$q' = -\Lambda q - \langle q,q\rangle \Lambda p$$
$$p' = \Lambda p + \langle \Lambda p,p\rangle q$$

is an integrable Hamiltonian $(R^{2N}, dp\wedge dq, H)$ with

$$H = -\langle \Lambda q,p\rangle - \frac{1}{2}\langle q,q\rangle\langle \Lambda p,p\rangle$$

whose involutive system of conserved integrals is

$$F_m = \frac{1}{2}\langle p,p\rangle\langle \Lambda^m q,q\rangle - \langle \Lambda^m q,p\rangle - \langle q,p\rangle\langle \Lambda^m q,p\rangle$$

$$-\frac{1}{2}\sum_{j=0}^{m}\begin{vmatrix}\langle \Lambda^j q,q\rangle & \langle \Lambda^j q,p\rangle \\ \langle \Lambda^{m-j}p,q\rangle & \langle \Lambda^{m-j}p,p\rangle\end{vmatrix}, \quad m=0,1,2,\ldots$$

5) TD hierarchy [9]

$$L(u,v,\lambda)y = y_x - \begin{bmatrix}-\lambda + \frac{1}{2}u & v \\ v & \lambda - \frac{1}{2}u\end{bmatrix}y = 0. \quad (32)$$

$$\nabla \lambda_j = (\delta\lambda_j/\delta u, \delta\lambda_j/\delta v)^T = (q_j(x)p_j(x), p_j^2(x) - q_j^2(x))^T,$$

$$G_{-2} = (1,0)^T, \quad G_{-1} = (0,v)^T, \quad G_0 = (\frac{1}{4}v^2, \frac{1}{2}uv)^T.$$

Consider the Bargmann constraint $G_0 = \sum_{j=1}^{N}\nabla\lambda_j$, which is equivalent to

$$u = (\langle p,p\rangle - \langle q,q\rangle)/\sqrt{\langle q,p\rangle}, \quad v = 2\sqrt{\langle q,p\rangle}. \quad (33)$$

The Bargmann system {(32),(33)}

$$q' = -\Lambda q + \frac{\langle p,p\rangle - \langle q,q\rangle}{2\sqrt{\langle q,p\rangle}}q + 2\sqrt{\langle q,p\rangle}\,p$$

$$p = \Lambda p - \frac{\langle p,p\rangle - \langle q,q\rangle}{2\sqrt{\langle q,p\rangle}}p + 2\sqrt{\langle q,p\rangle}\,q$$

has the Hamiltonian $H = F_0$ and the conserved integrals involutive in pairs F_m:

$$F_0 = -\langle \Lambda q,p\rangle + \sqrt{\langle p,q\rangle}(\langle p,p\rangle - \langle q,q\rangle)$$

$$F_m = -\langle \Lambda^{m+1}q,p\rangle + \sqrt{\langle p,q\rangle}(\langle \Lambda^m p,p\rangle - \langle \Lambda^m q,q\rangle)$$

$$-\sum_{i+j=m-1}\begin{vmatrix}\langle \Lambda^i q,q\rangle & \langle \Lambda^i q,p\rangle \\ \langle \Lambda^j p,q\rangle & \langle \Lambda^j p,p\rangle\end{vmatrix}, \quad m=0,1,2,\ldots$$

6) C-KdV hierarchy [10]

$$L(u,v,\lambda)y = y_x - \begin{bmatrix} -\frac{1}{2}\lambda + \frac{1}{2}u & -v \\ 1 & \frac{1}{2}\lambda - \frac{1}{2}u \end{bmatrix} y = 0 .$$ (34)

$$\nabla \lambda_j = (\delta\lambda_j/\delta u , \delta\lambda_j/\delta v)^T = (p_j(x)q_j(x), -p_j^2(x))^T ,$$

$$G_{-2} = (1,0)^T, G_{-1} = (0,1)^T, G_0 = (v,u)^T.$$

The Bargmann constraint $G_0 = \sum_{j=1}^{N} \nabla \lambda_j$ is

$$u = -<p,p> , v = <q,p> .$$ (35)

The nonlinearized (34) under (35)

$$q' = -\frac{1}{2}\Lambda q - \frac{1}{2}<p,p> q - <q,p> p = \partial H/\partial p$$

$$p' = \frac{1}{2}\Lambda p + q + \frac{1}{2}<p,p> p = -\partial H/\partial q$$

has the Hamiltonian $H = F_0$ and the involutive system $\{F_m\}$:

$$H = F_0 = -\frac{1}{2}<q,p><p,p> - \frac{1}{2}<q,q> - \frac{1}{2}<\Lambda q,p>$$

$$F_m = -\frac{1}{2}<q,p><\Lambda^m p,p> - \frac{1}{2}<\Lambda^m q,q> - \frac{1}{2}<\Lambda^{m+1}q,p>$$

$$+ \frac{1}{2}\sum_{i+j=m-1} \begin{vmatrix} <\Lambda^i q,q> & <\Lambda^i q,p> \\ <\Lambda^j p,q> & <\Lambda^j p,p> \end{vmatrix} , m=0,1,2,\ldots$$

7) An eigenvalue problem studied by Hu Xingbiao [11]

$$L(u,v,\lambda)y = y_x - \begin{bmatrix} \lambda + v & u \\ \lambda u & -\lambda - v \end{bmatrix} y = 0 .$$ (36)

$$\nabla \lambda_j = (\delta\lambda_j/\delta u, \delta\lambda_j/\delta v)^T = (p_j^2(x) - \lambda_j q_j^2(x), 2p_j(x)q_j(x))^T$$

$$G_{-2} = (0,1)^T, G_{-1} = (-u,0)^T, G_0 = (\frac{1}{2}u^3 + uv, \frac{1}{2}u^2)^T .$$

The Bargmann constraint $G_0 = \sum_{j=1}^{N} \nabla \lambda_j$ is

$$u = 2\sqrt{<q,p>} , v = (<p,p> - <\Lambda q,q>)/2\sqrt{<q,p>} - 2<q,p> .$$ (37)

The nonlinearized (36) under (37)

$$q' = \Lambda q + \left(\frac{<p,p> - <\Lambda q,q>}{2\sqrt{<q,p>}} - 2<q,p> \right)q + 2\sqrt{<q,p>}\, p = \frac{\partial H}{\partial p}$$

$$p' = -\Lambda p - \left(\frac{<p,p> - <\Lambda q,q>}{2\sqrt{<q,p>}} - 2<q,p> \right)p + 2\sqrt{<q,p>}\,\Lambda q = -\frac{\partial H}{\partial q}$$

has the Hamiltonian $H = -F_1$ and the involutive system of conserved integrals $\{F_m\}$:

$$F_0 = \sqrt{<q,p>}\,(<q,q> - <\bar{\Lambda}^1 p,p>) - <q,p> + <\bar{\Lambda}^1 p,p><q,q>$$

$$F_m = \sqrt{<q,p>}\,(<\Lambda^m q,q> - <\Lambda^{m-1}p,p>) + <\Lambda^{m-1}p,p><q,q> - <\Lambda^m q,p>$$

$$- \sum_{i+j=m-1} \begin{vmatrix} <\Lambda^i q,q> & <\Lambda^i q,p> \\ <\Lambda^j p,q> & <\Lambda^j p,p> \end{vmatrix}$$

The project supported by National Natural Science Foundation of China

References

1 Flaschka,H.,Relations between infinite-dimensional and finite-dimensional isospectral equations,Proc. RIMS Symp. on Nonlinear Integrable Systems,Kyoto,Japan,World Sci. Pub.,Singapore,1983,219-239 .
2 Ablowitz,M.J.,and Segur,H.,Solitons and the Inverse Scattering Transform,SIAM Studies in Appl. Math.,Philadelphia,1981.
3 Cao Cewen,A classical integrable system and the involutive representation of solutions of the KdV equation.(preprint)
4 Moser,J.,Integrable Hamiltonian system and spectral theory,Proc.1983 Beijing Symp. on Diff. Geom.and Diff.Equ's,Science Press Beijing , 1986,157-229.
5 Jaulent,M.,Miodek,I.,Lett. in Math. Phys.,1(1976)243-250.
6 Tu.G.Z ,Nuovo Cimento 73B,15(1983)15-16.
7 Marek Antonowicz and Allan P. Fordy,J. Phys.A:Math. Gen. 21(1988) L269-275.
8 Li yishen and Zhuang Dawei,Scientia Sinica A2(1983)107-118.
9 Tu Guizharg and Meng Dazhi,Acta Mathematicae Applicatae Sinica, English Series,5,1(1989)89-96.
10 Levi,D.,Sym,A.,Wojciechowsk,S.,J. Phys. A:Math. Gen.,16(1983)2423 -2432.
11 Hu Xingbiao,Three kinds of nonlinear evolution equations and their Hamiltonian structures,(preprint).

The Confocal Involutive System and the Integrability of the Nonlinearized Lax Systems of AKNS Hierarchy

Ma Wenxiu

Computing Centre of Academia Sinica, Beijing 100080, People's Rep. of China

§1. INTRODUCTION

Recently Cao[1] presented an important idea, the nonlinearization of the Lax system, and successfully applied[1,2,3] this idea to AKNS, KdV and Harry-Dym equations. Zeng and Li[4,5] considered carefully the corresponding problems for the whole KdV and classical Boussinesq hierarchies and proposed a general method based on the Cao's idea.

After nonlinearizing the Lax systems of a hierarchy of soliton equations, we can obtain a finite-dimensional Hamiltonian system and a series of time evolution equations which are naturally consistent. In general, this finite-dimensional system should be completely integrable(see Ref. [6]). But its integrability is not easily deduced because we haven't yet known how to apply the isospectral technique of Lax[7] to it. Very recently this work has made a little progress. Zeng and Li[8] proved that the nonlinearized eigenvalue problem of the KdV hierarchy is completely integrable.

In this paper, we discuss the integrability of the nonlinearized Lax systems of AKNS hierarchy. We find that these nonlinearized Lax systems are all closely related to the confocal involutive system. The next section proposes a sort of completely integrable Hamiltonian systems generated by the general confocal involutive system, and then Section 3 proves that the nonlinearized eigenvalue problem of AKNS hierarchy is such an integrable system. Finally, we show that the time evolution equations for $n \leq 3$ obtained by nonlinearizing the time parts of Lax systems of AKNS hierarchy are Liouville integrable under the constraint of the spatial part in Section 4.

§2. THE CONFOCAL INVOLUTIVE SYSTEM

Let us consider the confocal involutive system and its completely integrable Hamiltonian systems. Set $\lambda_1, \lambda_2, \cdots, \lambda_N$ are N distinct real numbers. We consider

$$G_i = \sum_{j=1, j \neq i}^{N} \frac{(\phi_{1i}\phi_{2j} - \phi_{1j}\phi_{2i})^2}{\lambda_i - \lambda_j}, \qquad i = 1, 2, \cdots, N \qquad (2.1)$$

which is an involutive system [9,10] with respect to the Poisson bracket

$$\{f, g\} = \sum_{i=1}^{N} \left(\frac{\partial f}{\partial \phi_{1i}} \frac{\partial g}{\partial \phi_{2i}} - \frac{\partial f}{\partial \phi_{2i}} \frac{\partial g}{\partial \phi_{1i}} \right) = < \frac{\partial f}{\partial \phi_1}, \frac{\partial g}{\partial \phi_2} > - < \frac{\partial f}{\partial \phi_2}, \frac{\partial g}{\partial \phi_1} > \qquad (2.2)$$

where $\phi_i = (\phi_{i1}, \phi_{i2}, \cdots, \phi_{iN})^T$, $i = 1, 2$, and $< \cdot, \cdot >$ denotes the inner product of R^N. We can easily prove the following equalities through direct computation

$$\sum_{k=1}^{N} G_k = 0 \quad , \tag{2.3a}$$

$$\sum_{k=1}^{N} \lambda_k^s G_k = \sum_{\substack{i+j=s-1 \\ i,j \geq 0}} \begin{vmatrix} <A^i\phi_1,\phi_1> & <A^i\phi_1,\phi_2> \\ <A^j\phi_2,\phi_1> & <A^j\phi_2,\phi_2> \end{vmatrix} \quad , \quad s = 1,2,\cdots, \tag{2.3b}$$

where $A = \text{diag}(\lambda_1, \lambda_2, \cdots, \lambda_N)$.

Given three real numbers satisfying $a^2 + b^2 + c^2 \neq 0$, we define

$$F_s = F_s(a,b,c) = \sum_{i=1}^{N} \lambda_i^s E_i \quad , \quad s = 0,1,2,\cdots, \tag{2.4}$$

where $\{E_i\}_{i=1}^{N}$ is the confocal involutive system[10], i.e.,

$$E_i = a\phi_{1i}^2 + 2b\phi_{1i}\phi_{2i} + c\phi_{2i}^2 + G_i \quad , \quad i = 1,2,\cdots,N. \tag{2.5}$$

It follows from the involution of $\{E_i\}_{i=1}^{N}$ that $\{F_s\}_{s=0}^{\infty}$ is an involutive system, and by (2.3), it is easy to see that

$$F_0 = a<\phi_1,\phi_1> + 2b<\phi_1,\phi_2> + c<\phi_2,\phi_2> \quad , \tag{2.6a}$$

$$F_s = a<A^s\phi_1,\phi_1> + 2b<A^s\phi_1,\phi_2> + c<A^s\phi_2,\phi_2>$$
$$+ \sum_{\substack{i+j=s-1 \\ i,j \geq 0}} \begin{vmatrix} <A^i\phi_1,\phi_1> & <A^i\phi_1,\phi_2> \\ <A^j\phi_2,\phi_1> & <A^j\phi_2,\phi_2> \end{vmatrix} \quad , \quad s \geq 1. \tag{2.6b}$$

We choose a region $\Omega \sqsubseteq R^{2N}$ on which the N 1-forms $dE_1, dE_2, \cdots, dE_N$ are everywhere linearly independent. Then we can see by (2.4) that $F_0, F_1, \cdots, F_{N-1}$ are also independent over Ω. Thus we have

THEOREM 1. *Let r be a positive integer. If $F : R^r \to R$ is differentiable, then the Hamiltonian equation*

$$\phi_{1t} = -\frac{\partial H}{\partial \phi_2} \quad , \quad \phi_{2t} = \frac{\partial H}{\partial \phi_1} \quad \text{with} \quad H = F(F_0, F_1, \cdots, F_{r-1})$$

is a completely integrable system over Ω. In fact, it possesses the involutive integrals of motion $\{F_s\}_{s=0}^{\infty}$ of which $F_0, F_1, \cdots, F_{N-1}$ are independent over Ω.

We choose $H = F_1$. When $(a,b,c) = (1,0,0)$ and $(0,0,1)$, we can deduce from Theorem 1 the geodesics on an ellipsoid[9,11] and the Neumann system[9,11], respectively.

Remark: When $\det A \neq 0$, i.e., $\lambda_i \neq 0 \; \forall \; 1 \leq i \leq N$, for arbitrary $0 \leq i_1 < i_2 < \cdots < i_N$, $F_{i_1}, F_{i_2}, \cdots, F_{i_N}$ are also independent over Ω.

§3. THE INTEGRABILITY OF THE NONLINEARIZED EIGENVALUE PROBLEM OF AKNS HIERARCHY

For the Lax systems of AKNS hierarchy, we introduce the distinct eigenvalues $\lambda_1, \lambda_2, \cdots, \lambda_N$, and denote by $(\phi_{1j}, \phi_{2j})^T$, $j = 1, 2, \cdots, N$, the corresponding eigenvalue functions of Lax systems. Through making the nonlinear transformation[1,5]

$$q = -<\phi_1,\phi_1> = -\sum_{j=1}^{N}\phi_{1j}^2 \quad , \quad r = <\phi_2,\phi_2> = \sum_{j=1}^{N}\phi_{2j}^2 \quad , \tag{3.1}$$

we obtain from the Lax systems of AKNS hierarchy

$$\begin{cases} \phi_{1jx} = -\lambda_j \phi_{1j} - <\phi_1,\phi_1> \phi_{2j} \\ \phi_{2jx} = <\phi_2,\phi_2> \phi_{1j} + \lambda_j \phi_{2j} \end{cases}, \quad j=1,2,\cdots,N, \quad (3.2)$$

and

$$\begin{bmatrix} \phi_{1j} \\ \phi_{2j} \end{bmatrix}_{t_n} = V^{(n)}(\lambda_j)|_B \begin{bmatrix} \phi_{1j} \\ \phi_{2j} \end{bmatrix}, \quad j=1,2,\cdots,N, \quad (3.3)$$

where the subscript B means to substitute (3.1) into the expression, and (see Ref. [12])

$$V^{(n)}(\lambda) = \sum_{m=0}^{n} V_m \lambda^{n-m} = \sum_{m=0}^{n} \begin{bmatrix} a_m & b_m \\ c_m & -a_m \end{bmatrix} \lambda^{n-m}, \quad n \geq 0$$

with $a_0 = -2$, $b_0 = c_0 = 0$ and

$$\begin{cases} a_{mx} = qc_m - rb_m \\ b_{mx} = -2b_{m+1} - 2qa_m \\ c_{mx} = 2c_{m+1} + 2ra_m \end{cases}, \quad m=0,1,2,\cdots.$$

Some properties of (3.2) and (3.3) have already been discussed in References [1],[5].
For the sake of conveninence, we set

$$K_s = F_s(0,1,0), \quad s = 0,1,2,\cdots. \quad (3.4)$$

The integrability of (3.2) is showed in the following theorem which can be verified by computation.

THEOREM 2. *The nonlinearized eigenvalue problem (3.2) is a completely integrable Hamiltonian system over Ω*

$$\phi_{1x} = -\frac{\partial H}{\partial \phi_2}, \quad \phi_{2x} = \frac{\partial H}{\partial \phi_1} \quad \text{with} \quad H = \frac{1}{2}K_1 + \frac{1}{4}K_0^2,$$

and possesses an infinite number of involutive integrals of motion $\{K_s\}_{s=0}^{\infty}$ of which $K_0, K_1, \cdots, K_{N-1}$ are independent over Ω.

§4. THE INTEGRABILITY OF THE NONLINEARIZED TIME PARTS OF LAX SYSTEMS OF AKNS HIERARCHY

We can obtain from the definition of V_m in Section 3

$$V_0 = \begin{bmatrix} -2 & 0 \\ 0 & 2 \end{bmatrix}, \quad V_1 = \begin{bmatrix} 0 & 2q \\ 2r & 0 \end{bmatrix}, \quad V_2 = \begin{bmatrix} qr & -q_x \\ r_x & -qr \end{bmatrix},$$

$$V_3 = \begin{bmatrix} \frac{1}{2}(qr_x - rq_x) & \frac{1}{2}q_{xx} - q^2 r \\ \frac{1}{2}r_{xx} - qr^2 & -\frac{1}{2}(qr_x - rq_x) \end{bmatrix}.$$

Consider the integrability of the time evolution equations (3.3) for $n \leq 3$.
When $n = 0$ or 1, (3.3) is obviously integrable without preconditions in the sense of Liouville.
Now let $n = 2$. Since

$$V^{(2)}(\lambda) = V_2 + V_1 \lambda + V_0 \lambda^2 = \begin{bmatrix} -2\lambda^2 + qr & 2q\lambda - q_x \\ 2r\lambda + r_x & 2\lambda^2 - qr \end{bmatrix},$$

the evolution equation (3.3) becomes

$$\begin{bmatrix} \phi_1 \\ \phi_2 \end{bmatrix}_{t_2} = W^{(2)}\bigg|_B \begin{bmatrix} \phi_1 \\ \phi_2 \end{bmatrix} = \begin{bmatrix} -2A^2 + qrE_{N \times N} & 2qA - q_xE_{N \times N} \\ 2rA + r_xE_{N \times N} & 2A^2 - qrE_{N \times N} \end{bmatrix}\bigg|_B \begin{bmatrix} \phi_1 \\ \phi_2 \end{bmatrix}$$

(Here $E_{N \times N}$ is the identity matrix of order N.), namely,

$$\begin{cases} \phi_{1t_2} = -2A^2\phi_1 - <\phi_1, \phi_1><\phi_2, \phi_2>\phi_1 - 2<\phi_1, \phi_1>A\phi_2 + 2<\phi_1, \phi_{1x}>\phi_2 \\ \phi_{2t_2} = 2<\phi_2, \phi_2>A\phi_1 + 2<\phi_2, \phi_{2x}>\phi_1 + 2A^2\phi_2 + <\phi_1, \phi_1><\phi_2, \phi_2>\phi_2 \end{cases}$$

(4.1)

THEOREM 3. *If (ϕ_1, ϕ_2) satisfies (3.2), then the evolution equation (4.1) can be transformed into the following Hamiltonian system*

$$\begin{bmatrix} \phi_1 \\ \phi_2 \end{bmatrix}_{t_2} = J \begin{bmatrix} \frac{\delta H_2}{\delta \phi_1} \\ \frac{\delta H_2}{\delta \phi_2} \end{bmatrix} = \begin{bmatrix} 0 & -E_{N \times N} \\ E_{N \times N} & 0 \end{bmatrix} \begin{bmatrix} \frac{\delta H_2}{\delta \phi_1} \\ \frac{\delta H_2}{\delta \phi_2} \end{bmatrix}$$ (4.2)

where $H_2 = K_2 + \frac{1}{2}K_0 K_1 + \frac{1}{8}K_0^3$, and it possesses a series of involutive integrals of motion $\{K_s\}_{s=0}^{\infty}$, therefore the system (4.1) is integrable in the sense of Liouville.

PROOF: When (ϕ_1, ϕ_2) satisfies (3.2), a straightforward calculation shows that the evolution equation (4.1) becomes

$$\begin{cases} \phi_{1t_2} = -2A^2\phi_1 - <\phi_1,\phi_1><\phi_2,\phi_2>\phi_1 - 2<\phi_1,\phi_1>A\phi_2 \\ \qquad\quad -2<A\phi_1,\phi_1>\phi_2 - 2<\phi_1,\phi_1><\phi_1,\phi_2>\phi_2 \\ \phi_{2t_2} = 2<\phi_2,\phi_2>A\phi_1 + 2<\phi_1,\phi_2><\phi_2,\phi_2>\phi_1 + 2<A\phi_2,\phi_2>\phi_1 \\ \qquad\quad + 2A^2\phi_2 + <\phi_1,\phi_1><\phi_2,\phi_2>\phi_2 \end{cases}$$ (4.3)

It is easily obtained from (4.3) that

$$\phi_{1t_2} = -\frac{\partial H_2}{\partial \phi_2} = -\frac{\delta H_2}{\delta \phi_2}, \quad \phi_{2t_2} = \frac{\partial H_2}{\partial \phi_1} = \frac{\delta H_2}{\delta \phi_1}$$

which shows that (ϕ_1, ϕ_2) satisfies (4.2).

Noticing that $H_2 = K_2 + \frac{1}{2}K_0 K_1 + \frac{1}{8}K_0^3$ and $\{K_s\}_{s=0}^{\infty}$ is an involutive system, we can obtain

$$\{K_s, H_2\} = <\frac{\partial K_s}{\partial \phi_1}, \frac{\partial H_2}{\partial \phi_2}> - <\frac{\partial K_s}{\partial \phi_2}, \frac{\partial H_2}{\partial \phi_1}> = 0 \quad .$$

Therefore,

$$\frac{\partial}{\partial t_2}\int_{-\infty}^{\infty} K_s \, dx = \int_{-\infty}^{\infty} \frac{\partial}{\partial t_2} K_s \, dx$$
$$= \int_{-\infty}^{\infty} (<\frac{\delta K_s}{\delta \phi_1}, \frac{\partial \phi_1}{\partial t_2}> + <\frac{\delta K_s}{\delta \phi_2}, \frac{\partial \phi_2}{\partial t_2}>) \, dx$$
$$= \int_{-\infty}^{\infty} (-<\frac{\partial K_s}{\partial \phi_1}, \frac{\partial H_2}{\partial \phi_2}> + <\frac{\partial K_s}{\partial \phi_2}, \frac{\partial H_2}{\partial \phi_1}>) \, dx$$
$$= \int_{-\infty}^{\infty} \{H_2, K_s\} \, dx = 0, \quad s \geq 0.$$

In addition,

$$\{K_i, K_j\}_J = \int (\frac{\delta K_i}{\delta \phi_1}, \frac{\delta K_i}{\delta \phi_2}) J (\frac{\delta K_j}{\delta \phi_1}, \frac{\delta K_j}{\delta \phi_2})^T dx = -\int \{K_i, K_j\} dx = 0, \quad i,j \geq 0.$$

Thus $\{K_s\}_{s=0}^{\infty}$ is a series of involutive integrals of motion of (4.2). #

Let $n = 3$. We know that

$$V^{(3)}(\lambda) = V_3 + V_2\lambda + V_1\lambda^2 + V_0\lambda^3$$
$$= \begin{bmatrix} -2\lambda^3 + qr\lambda + \frac{1}{2}(qr_x - rq_x) & 2q\lambda^2 - q_x\lambda + \frac{1}{2}q_{xx} - q^2r \\ 2r\lambda^2 + r_x\lambda + \frac{1}{2}r_{xx} - qr^2 & 2\lambda^3 - qr\lambda - \frac{1}{2}(qr_x - rq_x) \end{bmatrix}.$$

Therefore the evolution equation (3.3) becomes

$$\begin{bmatrix} \phi_1 \\ \phi_2 \end{bmatrix}_{t_3} = W^{(3)}\bigg|_B \begin{bmatrix} \phi_1 \\ \phi_2 \end{bmatrix}, \quad (4.4)$$

where

$$W^{(3)} = \begin{bmatrix} -2A^3 + qrA + \frac{1}{2}(qr_x - rq_x)E_{N\times N} & 2qA^2 - q_xA + (\frac{1}{2}q_{xx} - q^2r)E_{N\times N} \\ 2rA^2 + r_xA + (\frac{1}{2}r_{xx} - qr^2)E_{N\times N} & 2A^3 - qrA - \frac{1}{2}(qr_x - rq_x)E_{N\times N} \end{bmatrix}.$$

Similarly, we can show the following theorem.

THEOREM 4. *If (ϕ_1, ϕ_2) satisfies (3.2), then the evolution equation (4.4) can be transformed into the Hamiltonian system with the Hamiltonian function $H_3 = K_3 + \frac{1}{2}K_0K_2 + \frac{1}{4}K_1^2 + \frac{3}{8}K_0^2K_1 + \frac{5}{64}K_0^4$ and the Hamiltonian operator $J = \begin{bmatrix} 0 & -E_{N\times N} \\ E_{N\times N} & 0 \end{bmatrix}$, and possesses a series of involutive integrals of motion $\{K_s\}_{s=0}^{\infty}$, which shows (4.4) is integrable in the sense of Liouville.*

It is easily proved that when (ϕ_1, ϕ_2) satisfies the equation (3.2), the right expression of the evolution equation (3.3) is only a polynomial of (ϕ_1, ϕ_2), not including any derivative items of (ϕ_1, ϕ_2). We conjecture that under (3.2), the evolution equation (3.3) for arbitrary natural number n may be transformed into a Hamiltonian system generated by a polynomial of $K_0, K_1, \cdots$ as a Hamiltonian function. Thus they may possess a series of common involutive integrals of motion $\{K_s\}_{s=0}^{\infty}$ and commute with each other. Therefore they may be Liouville integrable.

In addition, we know from Theorem 3,4 that under (3.2), the evolution equations (4.1),(4.4) are Liouville integrable. But we haven't yet known whether (4.1),(4.4) themselves are Liouville integrable or not. Notice that there exist a number of relations between the time evolution equation (3.3) and AKNS hierarchy. It is possible that the equations (3.3) for $n \geq 0$ themselves are also Liouville integrable. This needs a further investigation.

Acknowledgements: The author is greatly indebted to Professor Tu Gui-zhang for his enthusiastic guidance. The author is also grateful to Dr. Zhang Lin-bo and Mr. Xue Ju-kui for valuable suggestions.

REFERENCES

[1] C.W. Cao, Nonlinearization of the Lax system for AKNS hierarchy, to appear in Scientia Sinica A
[2] ———, Chinese Quarterly J. of Math., Vol.3, No.1, 90-96(1988)
[3] ———, Stationary Harry-Dym equation and its relation with geodesics on ellipsoid, to appear in Acta Math. Sinica
[4] Y.B. Zeng & Y.S. Li, Three kinds of constraints of potential for KdV hierarchy (preprint)
[5] ——————, The constraints of potential and the finite-dimensional integrable systems (preprint)
[6] H. Flaschka, in Proceedings of RIMS symposium on nonlinear integrable systems, World Science Publishing Co., Singapore(1983), 219-240
[7] P.D. Lax, Comm. Pure Appl. Math., Vol.21, 467-490(1968)
[8] Y.B. Zeng & Y.S. Li, On complete integrability of some finite-dimensional integrable systems (preprint)
[9] J. Moser, in Proceedings of the 1983 Beijing symposium on differential geometry and differential equation, Science Press, Beijing(1986), 157-229
[10] C.W. Cao, Henan Science, Vol.5, No.1, 1-10(1987) (in Chinese)
[11] J. Moser, in Progress in Mathematics 8, Birkhauser, Boston(1980), 233-289
[12] G.Z. Tu, J. Math. Phys., Vol.30, No.2, 330-338(1989)

Two Kinds of Finite-Dimensional Systems Related to the Generalized Schrödinger Equation

Zeng Yunbo and Li Yishen

Department of Mathematics, University of Science and Technology of China, Hefei 230026, People's Rep. of China

There are some ways to restrict a infinite-dimensional integrable Hamiltonian system to finite-dimensional invariant submanifolds of their phase space in order to obtain finite-dimensional integrable Hamiltonian systems (see, for example, ref [1-6]). We have proposed in ref [7] a straightforward way to obtain a hierarchy of finite-dimensional integrable Hamiltonian systems by restricting a hierarchy of integrable evolution equations to the invariant subspace of their recursion operator. The independent integrals of motion in involution for these Hamiltonian systems can be constructed by using the recursion operator. Then this hierarchy of integrable Hamiltonian systems are shown to commute with each other. In some cases, we may not be able to write the finite-dimensional systems we obtained into canonical Hamiltonian systems. However, using the method presented in ref [8], we can still get a natural constraint on potential by restricting potential to the invariant subspace of the recursion operator. Under this constraint condition, the two finite-dimensional systems obtained from the Lax pair possess lower order and can be shown to be consistent. The solution to these two system solves the evolution equation related to the Lax pair.

In present paper we consider the generalized Schrödinger equation [9]

$$L\phi = [\sum_{i=0}^{m-1} \lambda^i(\epsilon_i D^2 + u_i)]\phi = \lambda^m \phi, \quad \epsilon_i = 0 \text{ or } 1, \qquad (1)$$

which isospectral flows are shown to possess (m+1) compatible Hamiltonian structures [9]. The time evolution equation for ϕ is taken to be

$$\phi_{t_n} = -\frac{1}{2}B_x^{(n)}\phi + B^{(n)}\phi_x, \qquad (2)$$

where

$$B^{(n)} = \lambda^n + \frac{1}{2}\sum_{k=1}^{n} \tilde{R}_{k-1}\lambda^{n-k}, \qquad (3a)$$

$$R_k = (R_k^{(0)}, R_k^{(1)}, \ldots, R_k^{(m-2)}, \check{R}_k)^T, \quad u = (u_0, u_1, \ldots, u_{m-1})^T,$$

$$R_k = LR_{k-1} = L^k u, \qquad (3b)$$

$$L = \begin{pmatrix} 0 & \cdots & 0 & J_0 \\ 1 & \cdots & 0 & J_1 \\ \cdots & \cdots & \cdots & \cdots \\ 0 & \cdots & 1 & J_{m-1} \end{pmatrix}, \quad J_i = \frac{1}{4}\epsilon_i D^2 + u_i - \frac{1}{2}D^{-1}u_{ix},$$

$$D = \frac{\partial}{\partial x}, \quad DD^{-1} = D^{-1}D = I.$$

Here the integral constant of D^{-1} is defined to be zero. Then the hierarchy of evolution equations [9] obtained from the solvability condition of (1) and (2) can be rewritten as

$$u_{t_n} = DL^n u. \tag{4}$$

Also it is found in [9] that if ϕ is a solution of (1), then

$$D \sum_{i=0}^{m-1} \lambda^i J_i \phi^2 = \lambda^m (\phi^2)_x. \tag{5}$$

We now consider the following system instead of (1)

$$\sum_{i=0}^{m-1} \lambda_j^i (\epsilon_i D^2 + u_i) \phi_j = \lambda_j^m \phi_j, \quad j = 1, \ldots, N, \tag{6}$$

where $\lambda_j \neq \lambda_k$ when $j \neq k$. Using (5), it is easy to varify that if $\Phi = (\phi_1, \ldots, \phi_N)^T$ is a solution to (6), then

$$DL\Psi_j = \lambda_j \Psi_{jx}, \quad j = 1, \ldots, N, \tag{7a}$$

where $\Psi_j = (\psi_j^{(0)}, \psi_j^{(1)}, \ldots, \psi_j^{(m-1)})^T$,

$$\psi_j^{(0)} = \lambda_j^{m-1} \phi_j^2 - \sum_{i=1}^{m-1} \lambda_j^{i-1} J_i \phi_j^2,$$

$$\psi_j^{(m-k)} = \lambda_j^{k-1} \phi_j^2 - \sum_{i=1}^{k-1} \lambda_j^{k-1-i} J_{m-i} \phi_j^2, \tag{7b}$$

$$\psi_j^{(m-1)} = \phi_j^2.$$

Since no boundary condition on u is imposed, (7a) leads to

$$L\Psi_j = \lambda_j \Psi_j + \sum_{i=0}^{m-1} \beta_i e_i, \tag{8}$$

where β_i are some constants, $e_0 = (1, 0, \ldots, 0)^T, \ldots, e_{m-1} = (0, \ldots, 0, 1)^T$. Notice that

$$L \sum_{i=0}^{m-1} \beta_i e_i = \sum_{i=0}^{m-2} \beta_i e_{i+1} + \frac{1}{2} \beta_{m-1} u, \tag{9}$$

and take

$$u = \sum_{j=1}^{N} \alpha_j \Psi_j + C e_0, \tag{10a}$$

then we find from (8) and (9) that the linear space M spanned by $\{\Psi_1, \ldots, \Psi_N, e_0, \ldots, e_{m-1}\}$ is the invariant subspace of L. This property plays important role in our approach.

Proposition 1. The constraint on potential obtained from (10a) is of the form

$$u_{m-k} = \sum_{j=1}^{k} a_j \sum_{l_1+\ldots+l_j=k-j} <\Lambda^{l_1}G\Phi,\Phi> \ldots <\Lambda^{l_j}G\Phi,\Phi> +$$

$$\sum_{i=1}^{k-1} \epsilon_{m-i} \sum_{j=1}^{k-i} b_j \sum_{l_1+\ldots+l_j=k-i-j} [j<\Lambda^{l_1}G\Phi,\Phi>_{xx}<\Lambda^{l_2}G\Phi,\Phi> \ldots <\Lambda^{l_j}G\Phi,\Phi>$$

$$+ \frac{j(j-1)}{2} <\Lambda^{l_1}G\Phi,\Phi>_x <\Lambda^{l_2}G\Phi,\Phi>_x <\Lambda^{l_3}G\Phi,\Phi> \ldots <\Lambda^{l_j}G\Phi,\Phi>] + C\delta_{k,m},$$

$$k = 1,\ldots,m,$$
(10b)

where $l_i \geq 0$, $\Lambda = diag(\lambda_1,\ldots,\lambda_N)$, $G = diag(\alpha_1,\ldots,\alpha_N)$, $<.,.>$ is the inner product in R^N and

$$a_1 = 1, \quad a_j = -\frac{j+1}{2j}a_{j-1}, \quad b_1 = -\frac{1}{4}, \quad b_j = -\frac{j-1}{2j}b_{j-1}, \quad j = 1,2,\ldots. \quad (10c)$$

Proof. Using the following kind of identities

$$\sum_{l=1}^{k} \alpha_{k-l} \sum_{j=1}^{l} \beta_j \gamma_{l,j} = \sum_{j=1}^{k} \beta_j \sum_{l=0}^{k-j} \alpha_l \gamma_{k-l,j},$$

$$\sum_{l=0}^{k} <\Lambda^l G\Phi,\Phi>_x \sum_{l_1+\ldots+l_j=k-l} <\Lambda^{l_1}G\Phi,\Phi> \ldots <\Lambda^{l_j}G\Phi,\Phi>$$

$$= \sum_{l_1+\ldots+l_{j+1}=k} <\Lambda^{l_1}G\Phi,\Phi> \ldots <\Lambda^{l_j}G\Phi,\Phi><\Lambda^{l_{j+1}}G\Phi,\Phi>_x$$

$$= \frac{1}{j+1}D \sum_{l_1+\ldots+l_{j+1}=k} <\Lambda^{l_1}G\Phi,\Phi> \ldots <\Lambda^{l_{j+1}}G\Phi,\Phi>,$$

a straightforward calculation gives (10b) by induction.

Under the constraint condition (10), (6) becomes

$$\sum_{i=0}^{m-1} \lambda_j^i(\epsilon_i D^2 + u_i|_A)\phi_j = \lambda_j^m \phi_j, \quad j = 1,\ldots,N, \quad (11)$$

where subscript A means to substitute (10) into the expression. It is clear from (10b) that (11) is second order differential equations. Assuming Φ solves (11), then we have from (7a)

$$DLu|_A = \sum_{j=1}^{N} \lambda_j \alpha_j \Psi_{jx},$$

which yields

$$Lu|_A = \sum_{j=1}^{N} \lambda_j \alpha_j \Psi_j + 2C_1 e_{m-1} + \sum_{i=0}^{m-2} \beta_i^{(1)} e_i, \quad (12b)$$

where C_1, $\beta_i^{(1)}$ are constants. Using (7) and (9) repeatedly, we get

$$DL^k u|_A = \sum_{j=1}^{N} \lambda_j^k \alpha_j \Psi_{jx} + \sum_{i=1}^{k-1} C_i \sum_{j=1}^{N} \lambda_j^{k-i-1} \alpha_j \Psi_{jx}, \qquad (13a)$$

$$L^k u|_A = \sum_{j=1}^{N} \lambda_j^k \alpha_j \Psi_j + \sum_{i=1}^{k-1} C_i \sum_{j=1}^{N} \lambda_j^{k-i-1} \alpha_j \Psi_j + 2C_k e_{m-1} + \sum_{i=0}^{m-2} \beta_i^{(k)} e_i, \qquad (13b)$$

where C_i, $\beta_i^{(k)}$ are some constants. We find from (13b), (7b) and (3b) immediately

$$\check{R}_k|_A = <\Lambda^k G\Phi, \Phi> + \sum_{i=1}^{k-1} C_i <\Lambda^{k-1-i} G\Phi, \Phi> + 2C_k. \qquad (14)$$

It is clear that C_i are the constants of the motion for system (11). Indeed (14) can be used as a recursion formula to calculate C_k from $\check{R}_k$ and $C_1, \ldots, C_{k-1}$. Furthermore, C_k can be expressed in term of Φ, Φ_x and Λ by using (11).

Proposition 2. If Φ is a solution to (11), then u given by (10) satisfies a certain higher order stationary equation

$$DL^N u + \sum_{k=0}^{N-1} d_k DL^k u = 0, \qquad (15)$$

where d_k are some constants determined by $\lambda_1, \ldots, \lambda_N, C_1, \ldots, C_N$.

Proof. Set $(\lambda - \lambda_1)\ldots(\lambda - \lambda_N) = \lambda^N + \sum_{k=1}^{N} g_k \lambda^{N-k}$. Taking $d_N = 1$, and

$$d_{N-k} = g_k - \sum_{m=1}^{k} C_m d_{N-k+m}, \quad k = 1, \ldots, N,$$

then it is easy to varify (15) by direct calculation.

Under the constraint condition (10) and (11), we have another system from (2) and (14)

$$\phi_{j t_n} = -\frac{1}{4} \sum_{k=1}^{n} [<\Lambda^{k-1} G\Phi, \Phi>_x + \sum_{i=1}^{k-2} C_i <\Lambda^{k-2-i} G\Phi, \Phi>_x] \lambda_j^{n-k} \phi_j$$

$$+ \lambda_j^n \phi_{jx} + \frac{1}{2} \sum_{k=1}^{n} [<\Lambda^{k-1} G\Phi, \Phi> + \sum_{i=1}^{k-2} C_i <\Lambda^{k-2-i} G\Phi, \Phi> + 2C_k] \lambda_j^{n-k} \phi_{jx},$$

$$j = 1, \ldots, N,$$

(15)

which is first order partial differential equations. The evolution equation (4) is deduced from the solvability condition of (1) and (2), and (11) and (15) are obtained by substituting (10) into (1) and (2), respectively. This implies that

Proposition 3. If Φ satisfies both systems (11) and (15), then u given by (10) is a solution of equation (4).

This offers a way to solves partial differential equation (4) with high order by solving systems (11) and (15). Indeed, exactly following the procedure given in [8], the systems (11) and (15) can be shown to be consistent. More precisely, we have

Proposition 4. Let $\Phi(x,t)$ be a solution of (15) and $\Phi(x,0)$ be a solution of (11). Then $\Phi(x,t)$ satisfies (11) and u given by (10) satisfies (14).

Finally, we can find two important constraint on potential from (10).

1. Let $\alpha_1 = 1$, $C = 0$. Then (10b) and (14) become

$$u_{m-k} = \sum_{j=1}^{k} a_j \sum_{l_1+...+l_j=k-j} <\Lambda^{l_1}\Phi,\Phi> ... <\Lambda^{l_j}\Phi,\Phi>$$

$$+ \sum_{i=1}^{k-1} \epsilon_{m-i} \sum_{j=1}^{k-i} b_j \sum_{l_1+...+l_j=k-i-j} [j<\Lambda^{l_1}\Phi,\Phi>_{xx}<\Lambda^{l_2}\Phi,\Phi>...<\Lambda^{l_j}\Phi,\Phi>$$

$$+ \frac{1}{2}j(j-1)<\Lambda^{l_1}\Phi,\Phi>_x<\Lambda^{l_2}\Phi,\Phi>_x<\Lambda^{l_3}\Phi,\Phi>...<\Lambda^{l_j}\Phi,\Phi>], \quad k=1,...,m, \tag{16}$$

$$\tilde{R}_k|_A = <\Lambda^k\Phi,\Phi> + \sum_{i=1}^{k-1} C_i <\Lambda^{k-i-1}\Phi,\Phi> + 2C_k. \tag{17}$$

2. Let $G = 2\Lambda$. Then (10b) reads

$$u_{m-k} = \sum_{j=1}^{k} 2^j a_j \sum_{l_1+...+l_j=k-j} <\Lambda^{l_1+1}\Phi,\Phi> ... <\Lambda^{l_j+1}\Phi,\Phi> +$$

$$\sum_{i=1}^{k-1} \epsilon_{m-i} \sum_{j=1}^{k-i} 2^j b_j \sum_{l_1+...+l_j=k-i-j} [j<\Lambda^{l_1+1}\Phi,\Phi>_{xx}<\Lambda^{l_2+1}\Phi,\Phi>...<\Lambda^{l_j+1}\Phi,\Phi>$$

$$+ \frac{1}{2}j(j-1)<\Lambda^{l_1+1}\Phi,\Phi>_x<\Lambda^{l_2+1}\Phi,\Phi>_x<\Lambda^{l_3+1}\Phi,\Phi>...<\Lambda^{l_j+1}\Phi,\Phi>]$$

$$+ C\delta_{k,m}, \quad k=1,...,m, \tag{18}$$

where C is undetermined. Assume that

$$<\Phi,\Phi> = 1, \tag{19}$$

we have from (11) that

$$\sum_{i=0}^{m-1} \epsilon_i <\Lambda^i\Phi_{xx},\Phi_x> + \sum_{k=1}^{m-1} u_{m-k}|_A <\Lambda^{m-k}\Phi_x,\Phi> = <\Lambda^m\Phi_x,\Phi>. \tag{20}$$

Using (18) for k=1,...,m-1 and the identities used in the proof of Proposition 1, it follows from (20) that

$$\frac{d\tilde{C}}{dx} = 0, \tag{21}$$

$$\check{C} = -<\Lambda^m\Phi,\Phi> + \sum_{j=2}^{m} \frac{2^{j-1}}{j} a_{j-1} \sum_{l_1+\ldots+l_j=m-j} <\Lambda^{l_1+1}\Phi,\Phi> \ldots <\Lambda^{l_j+1}\Phi,\Phi>$$

$$+ \sum_{i=0}^{m-1} \epsilon_i <\Lambda^i\Phi_x,\Phi_x> +$$

$$\sum_{i=1}^{m-2} \epsilon_{m-i} \sum_{j=2}^{m-i} 2^{j-2}(j-1) b_{j-1} \sum_{l_1+\ldots+l_j=m-i-j} <\Lambda^{l_1+1}\Phi,\Phi>_x <\Lambda^{l_2+1}\Phi,\Phi>_x$$

$$\cdot <\Lambda^{l_3+1}\Phi,\Phi> \ldots <\Lambda^{l_j+1}\Phi,\Phi>. \tag{22}$$

Under the condition (19), we also find from (11) by using (18) for k=1,...,m-1 that

$$u_0 = <\Lambda^m\Phi,\Phi> - \sum_{i=0}^{m-1} \epsilon_i <\Lambda^i\Phi_{xx},\Phi> - \sum_{k=1}^{m-1} <\Lambda^{m-k}\Phi,\Phi> u_{m-k}$$

$$= <\Lambda^m\Phi,\Phi> - \sum_{j=2}^{m} 2^{j-1} a_{j-1} \sum_{l_1+\ldots+l_j=m-j} <\Lambda^{l_1+1}\Phi,\Phi> \ldots <\Lambda^{l_j+1}\Phi,\Phi> \tag{23}$$

$$- \sum_{i=0}^{m-1} \epsilon_i <\Lambda^i\Phi_{xx},\Phi> - \sum_{i=1}^{m-2} \epsilon_{m-i} \sum_{j=2}^{m-i} 2^{j-1} b_{j-1}(j-1) \sum_{l_1+\ldots+l_j=m-i-j}$$

$$[<\Lambda^{l_1+1}\Phi,\Phi>_{xx} <\Lambda^{l_2+1}\Phi,\Phi> \ldots <\Lambda^{l_j+1}\Phi,\Phi>$$

$$+ \frac{1}{2}(j-2) <\Lambda^{l_1+1}\Phi,\Phi>_x <\Lambda^{l_2+1}\Phi,\Phi>_x <\Lambda^{l_3+1}\Phi,\Phi> \ldots <\Lambda^{l_j+1}\Phi,\Phi>].$$

Comparing (23) with (18) for k=m yields $C = \check{C}$. Indeed (19), (23) and (18) with $C = \check{C}$ consist of the Neumann type constraint on potential of (1). Then (14) reads

$$\check{R}_k|_A = 2 <\Lambda^{k+1}\Phi,\Phi> + 2\sum_{i=1}^{k-1} C_i <\Lambda^{k-i}\Phi,\Phi> + 2C_k.$$

In sum, we can obtain two important kinds of finite-dimensional systems (11) and (15) under the constraint condition (16) and above Neumann type constraint condition, respectively. These two kinds of systems possess the property given by the Proposition 2-4.

For the special case of (1) by taking $\epsilon_1 = \ldots = \epsilon_{m-1} = 0$, (11) and (15) can be written as Hamiltonian systems which are completely integrable in the sense of Liouville[10] and commute with each other. The N independent integrals of motion in involution can be constructed by using the recursion operator L (see [11]).

ACKNOWLEDGEMENTS

This work was supported by the National Natural Science Foundation of China.

REFERENCE

1. J. Moser, Various aspects of integrable Hamiltonian systems, in Progress in Mathematics (Birkhäuser), V.3, p.233.
2. H.P. Mckean, Jr., Springer Lecture Notes in Mathematics, V. 755.
3. D.V. Choodnovsky, G.V. Choodnovsky, Nuovo Cim. B 40, 399(1977).
4. H. Airault, H.P. Mckean, Jr., J. Moser, Comm. Pure Appl. Math. 30, 95(1977).
5. H. Flaschka, Relations between infinite-dimensional and finite-dimensional isospectral equations, in Proceedings of RIMS Symposium on Nonlinear Integrable Systems Classical Theory and Quantum Theory, Tyoto, 1981, Ed M. Jimbo and T. Miwa, World Science Publishing Co., Singapore, 1983, 219-240.
6. Cao Cewen, Chinese Quarterly J. of Math., 3:1, 90(1988).
7. Yunbo Zeng and Yishen Li, An approach to the integrability of Hamiltonian systems obtained by reduction, to appear in J. Phys. A.
8. Yunbo Zeng and Yishen Li, J. Math. Phys., 30, 1679(1989).
9. M. Antonowicz and A.P. Fordy, Nonlinear Evolution Equations and Dynamical Systems (NEEDS' 87), 145-160, ed. J. Leon. World Scientific, Singapore, 1988.
10. V.I. Arnold, Mathematical Methods of Classical Mechanics, MIR (Moscow, 1975), Springer, New York, 1978.
11. Yunbo Zeng and Yishen Li, A hierarchy of integrable Hamiltonian systems related to polynomial eigenvalue problem, to be published.

Nonlinearization of the Lax Pair for the KdV Equation and Integrable Hamiltonian Systems

Zhuang Dawei and Lin Yuanqu

Department of Mathematics, Beijing University, People's Rep. of China

The Lax pair for KdV equation

$$\frac{\partial^2 y}{\partial x^2} + (\lambda - u)y = 0, \qquad \frac{\partial y}{\partial t} = -4\frac{\partial^3 y}{\partial x^3} + bu\frac{\partial y}{\partial x} + 3u_x y$$

can be nonlinearized as couple of integrable Hamiltonian systems. Since the two Hamiltonian flows are commutative, it yields a class of solutions of KdV equation which are almost periodic in x, t respectively.

1. Introduction

Since GGKM [1] found out the inverse scatttering method in 1967, a lot of work on KdV Eq. Has been done and the associated spectral problem and inverse spectral one for Schrödinger operator has been studied deeply. The inverse spectral theory for periodic and almost periodic potential has been studied by Mc Kean [2], Novikov [3], P.D.Lax [4] and J.Moser [5] etc. Especially, Moser has reduced the inverse spectral problem for finite band potentials to a Neumann problem, namely a integrable Hamiltonian system. It presents an approach to find a class of almost periodic solutions of KdV equation.

Lately, by means of nonlinearization of Lax pair for KdV Eq. Cao Cewen [6] presents a new way to solve KdV Eq. And he has considered nonlinearization of Lax pair for the nonlinear evolution equations [7].

In this paper we follow the way and perfect it. Lax pair for KdV Eq. Will be nonlinearized as a couple of integrable Hamiltonian systems, and their Hamiltonian flows are commutative. From these we can obtain a class of solutions of KdV Eq. which are alomost periodic in x, t respectively.

2. Notation and some results in [5]

The neumann problem is the differential equations

$$q'' = -Aq + uq \qquad (2.1)$$

where $A = diag(\alpha_1, \cdots, \alpha_n)$, $\alpha_1 < \alpha_2 < \cdots < \alpha_n$, $q \in \mathbb{R}^n$ and

$$\langle q, q \rangle = 1 \tag{2.2}$$

$$u = \langle Aq, q \rangle - \langle q', q' \rangle \tag{2.3}$$

$\langle \cdot, \cdot \rangle$ denotes the standard inner-product in $\mathbb{R}^n$, " ' " denotes "$\frac{\partial}{\partial x}$".
This problem can be extended to a system in $\mathbb{R}^{2n}$

$$\begin{cases} p' = \langle p, q \rangle p - Aq - \langle p, q \rangle q, \\ q' = \langle q, q \rangle p - \langle p, q \rangle q, \end{cases} \tag{2.4}$$

with the constraint (2.2). The system (2.4) just is a Hamiltonian system in the symplectic manifold $(\mathbb{R}^{2n}, d\omega = dp \wedge dq)$

$$q' = \frac{\partial H}{\partial P}, \quad P' = -\frac{\partial H}{\partial q} \tag{2.5}$$

with Hamiltonian function

$$H = \frac{1}{2}(\langle Aq, q \rangle + \langle q, q \rangle \langle p, p \rangle - \langle q, p \rangle^2). \tag{2.6}$$

Let the poisson bracket $\{F, G\}$ for two functions F, G be

$$\{F, G\} = \sum_{j=1}^{n} \left(\frac{\partial F}{\partial q_j} \frac{\partial G}{\partial p_j} - \frac{\partial F}{\partial p_j} \frac{\partial G}{\partial q_j} \right) \tag{2.7}$$

Then the confocal quadrics

$$E_k(p, q) = q_k^2 + \sum_{\substack{j=1 \\ j \neq 1}}^{n} \frac{(p_k q_j - p_j q_k)^2}{\alpha_k - \alpha_j}, \quad k = 1, \cdots, n \tag{2.8}$$

are a confocal involution system, i.e.

$$\{E_i, E_j\} = 0, \quad i, j = 1, 2, \cdots, n \tag{2.9}$$

Set

$$F_0 = \langle q, q \rangle$$

$$F_s = \sum_{k=1}^{n} \alpha_k^s E_k(p, q)$$

$$= \langle A^s q, q \rangle + \sum_{i+j=s-1} \begin{vmatrix} \langle A^i p, p \rangle & \langle A^i p, q \rangle \\ \langle A^j q, p \rangle & \langle A^j q, p \rangle \end{vmatrix}$$

$$s = 1, 2, \cdots, n. \tag{2.10}$$

It is evident that

$$H = \frac{1}{2} F_1 \tag{2.11}$$

Proposition 2.1 $F_1, \cdots, F_n$ are the integrals for the system (2.5) and in involution each other, i.e.

$$\{F_i, F_j\} = 0, \qquad i, j = 1, \cdots, n; \qquad (2.12)$$

and

$$\{F_0, F_j\} = 0, \qquad j = 1, \cdots, n. \qquad (2.13)$$

Proposition 2.2 Hamiltonian system (2.5) is integrable in Liouville sense.
Considering a differential submanifold

$$M_c = \{(p.q) \mid F_0 = 1, \ F_1 = c_1, \cdots, F_n = c_n, \ c_1, c_2, \cdots, c_n > 0\}, \qquad (2.14)$$

Then M_c is compact and connected if M_c is a connected leaf of its foliations. From (2.12), (2.13) and Arnold theorem [9], it follows

Proposition 2.3 M_c is an invariant manifold for the system (2.5), and then the phase flow with the Hamiltonian function H determines a conditionally periodic motion on M_c.

Consequently, if (p, q) is a solution of the Hamiltonian system (2.5) on M_c, then by (2.3),

$$u = \langle Aq, q\rangle - \langle q, q\rangle\langle p, q\rangle + \langle p, q\rangle^2 \qquad (2.15)$$

is necessarily almost periodic in x.

3. Nonlinearization of Lax pair for KdV Eq. and integrable Hamiltonian system

It is well-known that for KdV Eq.

$$u_t - 6uu_x + u_{xxx} = 0 \qquad (3.1)$$

there is Lax pair

$$L = -\partial^2 + u, \qquad B = -4\partial^3 + 6u\partial + 3u_x \qquad (3.2)$$

here "∂" denotes "$\frac{\partial}{\partial x}$", and it can be read an overdetermined system of equations

$$Ly = \lambda y \qquad \frac{\partial y}{\partial t} = By. \qquad (3.3)$$

Suppse that ψ_j ($j = 1, \cdots, n$) satisfy the system (3.3)

$$L\psi_j = \alpha_j \psi_j, \qquad \frac{\partial}{\partial t}\psi_j = B\psi_j, \quad j = 1, \cdots, n \qquad (3.4)$$

or in matrix form

$$\partial^2 \psi + A\psi - u\psi = 0, \qquad (3.5)$$

$$\frac{\partial}{\partial t}\psi = -4\partial^3\psi + 6u\partial\psi + 3u_x\psi \qquad (3.6)$$

here
$$\psi = (\psi_1, \cdots, \psi_n)^T, \qquad A = diag(\alpha_1, \cdots, \alpha_n) \qquad (3.7)$$

Under the constranits
$$u = \langle Aq, q \rangle - \langle q', q' \rangle, \qquad (3.8)$$
$$\langle q, q \rangle = 1, \qquad q = \psi, \qquad (3.9)$$

by §2 the neumann problem (3.5) can be nonlinearization as an integrable Hamiltonian system

$$\begin{cases} p' = \langle p, q \rangle p - Aq - \langle p, q \rangle q = -\dfrac{\partial H}{\partial q} \\ q' = \langle q, q \rangle p - \langle p, q \rangle q = \dfrac{\partial H}{\partial p} \end{cases} \qquad (3.10)$$

Substituting (3.5), (3.8-3.10) in (3.6), the equatiom (3.6) is nonlinearized as a Hamiltonian system in symplectic manifold ($\mathbb{R}^{2n}$, $d\omega = dp \wedge dq$)

$$\dot{p} = -\frac{\partial \tilde{H}}{\partial q}, \qquad \dot{q} = \frac{\partial \tilde{H}}{\partial p} \qquad (3.11)$$

with the Hamiltonian function
$$\tilde{H} = 2F_0 F_1 - \frac{1}{2} F_1^2, \qquad (3.12)$$

here "$\cdot$" denotes "$\frac{\partial}{\partial t}$".

Proposition 3.1 $\tilde{H}$ and H given by (3.12), (2.6) are in involution.

Proof. By a calculation [8], it is easy to see
$$\{\tilde{H}, H\} = 0, \qquad (3.13)$$

i.e., $\tilde{H}$ and H are in involution.

Corollary 1. $\tilde{H}$, $F_1, \cdots, F_n$ and F_0 are in involution each other; and $F_1, \cdots, F_n$ are integrables of the system (3.11), and M_c in (2.14) is an invariant manifold for the system (3.11) too.

By [9], it is easy from (3.13) to see

Corolary 2. The Hamiltonian phase flows g_H^x and $g_{\tilde{H}}^t$ are commutative.

By [5] or [8], from corollary 1 we have

Proposition 3.2 The Hamiltonian system (3.11) is integrable in Liouville sense, and its phase flows also determine a conditionally periodic motion on M_c.

Thus if (p, q) is a solution of system (3.11) on M_c, then the function u given by (2.15) is necessarily almost periodic in t.

Setting
$$y = \begin{pmatrix} p_0 \\ q_0 \end{pmatrix} \in M_c \qquad (3.14)$$

from corollary 2 it yields
$$g_{\tilde{H}}^t(g_H^x y) = g_H^x(g_{\tilde{H}}^t y) = \begin{pmatrix} p(x, t) \\ q(x, t) \end{pmatrix} \in M_c \qquad (3.15)$$

which satisfies either (3.10) and (3.11). And $\psi = q(x, t)$ also satisfies the systems

(3.5) and (3.6) with
$$u = \langle A\psi, \psi \rangle - \langle \psi', \psi' \rangle, \qquad \langle \psi, \psi \rangle = 1,$$

Since KdV Eq. is equivalent to the compatible condition

$$\frac{\partial^2 \psi}{\partial t \partial x} = \frac{\partial^2 \psi}{\partial x \partial t} \tag{3.16}$$

for the overdetermined system (3.3), i.e., the systems (3.5), (3.6), it follows by a calculation that
$$u = \langle Aq, q \rangle - \langle q, q \rangle \langle p, p \rangle - \langle p, q \rangle^2$$
is a solution of KdV Eq., which is almost periodic in x, t respectively where p, q are given by (3.15).

In general, Lax pair of a nonlinear evolution equation can be nonlinearized as a couple of Hamiltonian system which commutative and integrable in Liouville sense. Solving these systems, one can get a class of solutions of the nonlinear evolution equation, especially the periodic and almost periodic solutions. It presents an approach to solve nonlinear evolution equations. This is of universal significance, for example, a class of conditionally periodic solutions of Nonlinear Schrodinger Eq. is obtained by Pan Tao [10].

Reference

[1] C.S. Gardner, J.M. Green, M.D. Kruskal, R.M. Mliura, Method for solving the Korteweg-de Vries equation, Phys. Rev. Lett., 19 (1967) 1059-1097

[2] H.P. Mc Kean, P. Van Moerbeke, The spectrum of Hill's qeuation, Invent. Math., 30 (1975) 217-274

[3] B.A. Dubrovin, V.B. Matveev, S.P. Novikov, Nonlinear equations of Kortewg-de Vries type, finite zoned linear operators and Abelian varieties, Russian Math. Surveys, 31 (1967) 59-146

[4] P.D. Lax, Periodic solutions of the KdV equation, Comm. Pure Appl. Math., 28 (1975) 141-188

[5] J. Moser, Integrable Hamiltonian systems and spectral theory, In proc. of 1983 Beijing (1986) 157-229

[6] Cao Cewen, A cubic system which generates Bargnann potential and N-gap potential, preprinting (1988)

[7] Cao Cewen, Nonlinearization of the Lax system for AKNS hierachk, Preprinting (1988)

[8] Cao Cewen, Classical integrable systems, Preprinting (1988)

[9] V.I. Arnold, Mathematical methods of classical mechanics, Springer-Verlag (1978)

[10] Pan Tao, Qian Min, The inverse spectral problem with finite bands for Dirac operator and conditional periodic solutions of Nonlinear Schrodinger equation, Preprinting (1989)

Part III

Quantum Aspects and Statistical Mechanics

Quantum and Classical Statistical Mechanics of the Integrable Models in 1 + 1 Dimensions

R.K. Bullough[1], *D.J. Pilling*[1], *J. Timonen*[2], *Yi Cheng*[3], *and Yu-Zhong Chen*[4]

[1]Department of Mathematics, UMIST, P.O. Box 88, Manchester M60 1QD, UK
[2]Department of Physics, University of Jyväskylä, SF-40100 Jyväskylä, Finland
[3]UMIST and University of Science and Technology of China,
Hefei, Anhui, People's Rep. of China
[4]UMIST and Institute for Theoretical Physics, Beijing, People's Rep. of China

Abstract In a short but remarkable paper Yang and Yang [1] showed that the free energy of a model system consisting of N bosons on a line with repulsive δ-function interactions was given by a set of coupled integral equations. The Yangs' chosen model is in fact the repulsive version of the *quantum* Nonlinear Schrödinger (NLS) model. We have shown that with appropriate extensions and different dispersion relations and phase shifts similar formulae apply to 'all' of the integrable models *quantum or classical*. These models include the sine-Gordon (s-G) and sinh-Gordon (sinh-G) models, the two NLS models (attractive and repulsive), the Landau-Lifshitz (L-L') model which includes all four previous models, and so on. A significant feature is the bose-fermi equivalence of these models: the classical limit of the bose theories yields the classical statistical mechanics of the models in a simple way but in e.g. s-G fermion features remain even in classical limit. The models also include the integrable lattices and the Toda lattice is one of these. We comment on the statistical mechanics of the Toda lattice at the end of the paper. For simplicity the sinh-G model is used throughout this paper to exemplify the method of calculating the free energy, the method of 'generalised Bethe ansatz'.

1 Introduction

One of the simplest classical integrable models with real physical interest is the 'repulsive' NLS model in one space and one time (1+1) dimensions. The NLS has equation of motion

$$-i\phi_t = \phi_{xx} - 2c\phi^*\phi^2 \quad , \tag{1}$$

where $\phi(x,t) \in \mathcal{C}$, $x \in \mathcal{R}$, and ϕ^* is complex conjugate of ϕ. This classical system is a Hamiltonian system with Hamiltonian

$$H[\phi] = \int dx [\phi_x^*\phi_x + c(\phi^*)^2\phi^2] \tag{2}$$

and Poisson bracket $\{\phi,\phi^*\} = i\delta(x-x')$. Hamilton's equations $\phi_t = \{H,\phi\}$ are equation (1). The parameter c is a real-valued coupling constant, and, for the 'repulsive' NLS model, $c>0$ ($c<0$ is 'attractive'). Evidently this is consistent with the repulsive character of the non-linear self-potential $c(\phi^*)^2\phi^2$, which is positive when $c>0$. The NLS model is 'completely integrable' [2,3], i.e. it has a complete set of independent constants, a continuous infinity of them, commuting under the bracket.

The history of the classical integrable models is that they were first solved with ϕ, defined for $x \in \mathcal{R}$, vanishing exponentially fast as $x \to \pm\infty$ [2,3]. However, we shall find that it is periodic boundary conditions (b.c.s), rather than these decaying b.c.s, which play a fundamental role in the statistical mechanics.

At first sight the 'attractive' classical NLS ($c<0$) is more interesting than the repulsive case: for the former has soliton solutions. However, when $c<0$ the quantum NLS has no lower bound to its ground state. Thus starting with the simpler cases we should first of all be concerned with the repulsive case of the quantum or classical NLS.

Still, in a sense which becomes plain once we transform to action-angle variables (these are available for completely integrable systems and they are essential to our approach to the statistical mechanics) the two NLS models are non-relativistic forms of two *covariant* models: the attractive NLS corresponds to the s-G model

$$\phi_{xx} - \phi_{tt} = m^2 \sin\phi, \quad H[\phi] = \gamma_0^{-1} \int dx \left[\frac{1}{2}\gamma_0^2 \Pi^2 + \frac{1}{2}\phi_x^2 + m^2(1-\cos\phi)\right] \quad , \quad (3)$$

where real positive γ_0 is a coupling constant, and the Poisson bracket is $\{\Pi, \phi\} = \delta(x-x')$. The repulsive NLS model corresponds to the sinh-G model

$$\phi_{xx} - \phi_{tt} = m^2 \sinh\phi, \quad H[\phi] = \gamma_0^{-1} \int dx \left[\frac{1}{2}\gamma_0^2 \Pi^2 + \frac{1}{2}\phi_x^2 + m^2(\cosh\phi - 1)\right] \quad (4)$$

with the same bracket as s-G. By canonical transformation $\Pi \to \Pi\gamma_0^{-1/2}$, $\phi \to \phi\gamma_0^{1/2}$, and analytic continuation $\gamma_0 \to -\gamma_0$ sinh-G is found from s-G. Note that this analytic continuation is by no means trivial. The s-G model has three kinds of soliton solutions, kinks, antikinks and breathers; the sinh-G like the repulsive NLS has no soliton solutions. We shall find in consequence that we can develop the statistical mechanics of sinh-G wholly in parallel with that of the repulsive NLS. Because of the infrared cut-off provided by the "mass" m in the sinh-G model (4), the statistical mechanics of the sinh-G is in fact more tractable than that of the repulsive NLS, and therefore we shall use sinh-G here to describe the method of 'generalised Bethe ansatz' we introduced [4-10] recently.

As noted above, the main reason for considering the statistical mechanics of *integrable* models is the possibility of using their action-angle variables as canonical variables. We are now concerned with integrable fields so their Hamiltonians can be expressed in terms of a continuous infinity of commuting action variables, and motion is on tori and generalised tori. For the s-G model for example the Hamiltonian in terms of action variables for decaying b.c.s on the real line is [2,3]

$$H[p] = \sum_{i=1}^{N_K+N_R} (M^2 + p_i^2)^{1/2} + \sum_{j=1}^{N_B} (4M^2 \sin^2\theta_j + p_j^2)^{1/2} + \int_{-\infty}^{\infty} \omega(k) P(k) dk \quad , \quad (5)$$

where $M = 8m\gamma_0^{-1}$, $\omega(k) = (m^2 + k^2)^{1/2}$, and we do not make a distinction between kinks and antikinks. Action variables $4\gamma_0^{-1}\theta_j$ form the action variables for the internal oscillation modes of the breathers. It is important to notice that the term

$$H = \int_{-\infty}^{\infty} \omega(k) P(k) dk \quad (6)$$

in (5), for which the action variable $0 \le P(k) < \infty$ appears as a density of modes with dispersion $\omega(k)$, is also the Hamiltonian in terms of action variables for the sinh-G model. It is also the Hamiltonian for the linear Klein-Gordon (KG) model with the same dispersion, while it becomes the Hamiltonian of both the repulsive NLS model and its linearised form when $\omega(k) = k^2$. Thus it might appear that the sinh-G model for example has the same free energy as the bunch of harmonic oscillators described by the linear KG model. This is not the case, however. As we shall see below, nonlinear integrable models cannot be described in terms of noninteracting harmonic modes except in their excitation spectrum. The answer to this apparent paradox lies in the fact that the action variables, and the related Hamiltonians, are found for decaying b.c.s for fields defined on the whole real axis. The number *densities* $N_j L^{-1}$ of the particles described by the action variables in Hamiltonians like (5) or (6) are then strictly zero ($L \to \infty$), and the thermodynamic limit cannot be sensibly defined. We shall show in the subsequent sections how this problem can be solved by imposing periodic b.c.s. An effective use of the thermodynamic limit $L \to \infty$ then results in infinite line action variables which are *coupled* through a condition induced by the periodic b.c.s.

For the *quantum* forms the NLS models for example involve operators $\hat{\phi}$, $\hat{\phi}^\dagger$, and $\hat{H}$. They are normally ordered so that, for quantum repulsive NLS anyway, there is a well defined ground state. Heisenberg's equations of motion and commutators are

$$-i\hat{\phi}_t = \hat{\phi}_{xx} - 2c\hat{\phi}^\dagger \hat{\phi}^2 \quad ; \quad [\hat{\phi}, \hat{\phi}^\dagger] = \delta(x - x') \quad . \tag{7}$$

This is the second quantised form, with $\hat{N} = \int dx \hat{\phi}^\dagger \hat{\phi}$, of the Schrödinger wave mechanical problem

$$\sum_{i=1}^{N} \left[-\frac{\partial^2}{\partial x_i^2} + c \sum_{j \neq i} \delta(x_i - x_j) \right] \Psi(\{x_i\}) = E \Psi(\{x_i\}) \quad . \tag{8}$$

The wave mechanical problem is solved by the *Bethe ansatz* (BA) wave functions [11]

$$\Psi(\{x_j\}) = \sum_P a(P) \, e^{i \sum_{j=1}^{N} k_{P_j} x_j} \quad ; \quad x_1 < x_2 < \ldots < x_N, \tag{9}$$

in which $a(P) = \Pi e^{i\Delta_{jj'}}$ and $e^{i\Delta_{jj'}}$ is the 2-body S-matrix phase shift for each of the transpositions $j \leftrightarrow j'$ making up the permutation P. Bose symmetry is imposed so that $\Psi(\{x_j\})$ is also defined for $x_2 < x_1 < \ldots < x_N$, etc. .

The phase shifts

$$e^{i\Delta_{jj'}} = -\frac{c - i(k_j - k_{j'})}{c + i(k_j - k_{j'})} \quad ; \quad \Delta_{ij} = -2 \tan^{-1} \frac{c}{k_i - k_j}. \tag{10}$$

The smooth branch $-2\pi < \Delta_{ij} < 0$ is taken — see below. Periodicity on $x_j \to x_j + L$ then requires

$$L k_i = 2\pi n_i + \sum_{j \neq i} \Delta_{ij} \quad ; \quad i = 1, \ldots, N \quad , \tag{11}$$

and the n_i are integers.

The wave numbers k_i solving (11) are the only wave numbers (N of them) allowed. For $L \to \infty$ and $N \to \infty$ with $NL^{-1} > 0$, a *finite density* limit, both the k_i and the wave numbers $2\pi n_i L^{-1}$ are dense and distinct. When $c < 0$, additional 'string' solutions arise generalising (11) [2,3]. The BA wave functions (9) solve (8) exactly [11]: $E = \sum_{i=1}^{N} k_i^2$ and the k_i label this eigenstate. Yang and Yang [1] used the same BA method to solve the quantum statistical mechanics (SM). Evidently the conditions (11) are crucial to both of these calculations.

Notice that the choice of the smooth branch of the $\tan^{-1}$ for Δ_{ij} in (10) means that in both BA and quantum inverse method the boson problem is *described* in terms of fermions. However, we shall show below that there is an equivalent description in terms of bosons.

With reference to classical integrability it is unnecessary to say very much here. The reader is referred to [2,3,12]. But since it is relevant to our method of 'generalised Bethe ansatz' for both classical *and* quantum SM we give a quick summary next.

The spectral transform (ST) method (inverse scattering method) [2,3,12,13] for solving a classical integrable model in 1+1 dimensions on the real line finds spectral data S(0) from initial data, follows its evolution to S(t), t>0, and inverts. The r-matrix allows description of the brackets of the data S(t) in the form

$$\{T(\lambda) \overset{\otimes}{,} T(\mu)\} = [T(\lambda) \otimes T(\mu), r(\lambda, \mu)] \tag{12}$$

The left side has, for AKNS-ZS-Dirac type models [2], 16 Poisson brackets constructed from the 4×4 elements of spectral data defined through the 'monodromy' matrix

$$T(\lambda) = \begin{bmatrix} A(\lambda) & B(\lambda) \\ C(\lambda) & D(\lambda) \end{bmatrix} \quad ; \tag{13}$$

for the NLS and sinh-G models the r-matrix is

$$r(\lambda, \mu) = \begin{bmatrix} f(\lambda,\mu) & 0 & 0 & 0 \\ 0 & 0 & g(\lambda,\mu) & 0 \\ 0 & g(\lambda,\mu) & 0 & 0 \\ 0 & 0 & 0 & f(\lambda,\mu) \end{bmatrix}, \tag{14}$$

and $f = g = c(\lambda - \mu)^{-1}$ for NLS, $f = -\frac{1}{8}\gamma_0 \coth(\lambda - \mu)$, $g = -\frac{1}{8}\operatorname{cosech}(\lambda - \mu)$ for sinh-G, while $\gamma_0 \to -\gamma_0$ to get the r-matrix for s-G. It is important to recognize that periodic b.c.s are assumed. Our methods of functional integration and generalised BA [4-10] connect in thermodynamic limit, i.e., at finite density as $L \to \infty$, action-angle variables under periodic b.c.s with period L to action-angle variables under vanishing b.c.s at $\pm\infty$.

Stay here with repulsive NLS for simplicity: $T(\lambda) = \widehat{\prod_{n}} L_n(\lambda)$ in terms of elementary monodromy matrices.

$$L_n(\lambda) = \begin{bmatrix} 1 - \frac{1}{2}i\lambda a_0 & -i\sqrt{c}\phi_n^* \\ i\sqrt{c}\phi_n & 1 + \frac{1}{2}i\lambda a_0 \end{bmatrix} \quad ; \tag{15}$$

T is ordered from $n = -\frac{1}{2}N$ on the right to $\frac{1}{2}N - 1$ on the left. Periodic b.c.s are assumed and (15) shows the continuous NLS is discretized to a lattice, spacing a_0, period $L = Na_0$. In thermodynamic limit $L \to \infty$ at finite density and, for continuum models like NLS and sinh-G, $a_0 \to 0$. The second step involves an ultraviolet divergence problem in the classical SM of both sinh-G and s-G. but this is easily handled [4,10].

Staying with NLS, if $x = na_0$, and $\phi_n \to \phi(x)$ as $a_0 \to 0$ in (15) the lattice monodromy problem $v_{n+1} = L_n(\lambda)v_n$ (v_n is a 2-spinor) becomes the usual AKNS-ZS-Dirac spectral problem [2,3,12]

$$Lv = \lambda v \quad , \quad \frac{\partial v}{\partial t} = Av \quad , \quad L = \begin{bmatrix} i\frac{\partial}{\partial x} & -\sqrt{c}\phi^*(x) \\ \sqrt{c}\phi(x) & -i\frac{\partial}{\partial x} \end{bmatrix} \tag{16}$$

with A as given in e.g. [2,3,14]. Compatibility of the Lax pair A, L means $L_t = [A, L]$ is the equation of motion (1) for the NLS models. Since L is self-adjoint for $c>0$ there are no solitons for the repulsive NLS model and this simplifies the SM.

The sequence here is through a series of *non*-integrable lattices as $a_0 \to 0$, converging on the integrable NLS model in the limit. But there is an integrable NLS *lattice* with the same r-matrix, Eqn. (14) [16]:

$$L_n(\lambda) = \begin{bmatrix} 1 - \frac{1}{2}i\lambda a_0 + \frac{1}{2}c\phi_n^*\phi_n a_0^2 & -i\sqrt{c}\phi_n^*\rho_n a_0 \\ i\sqrt{c}\rho_n\phi_n a_0 & 1 + \frac{1}{2}i\lambda a_0 + \frac{1}{2}c\phi_n^*\phi_n a_0^2 \end{bmatrix}, \tag{17a}$$

$$v_{n+1} = L_n(\lambda)v_n \quad ; \quad \rho_n = \left[1 + \frac{1}{4}c\phi_n^*\phi_n a_0\right]^{\frac{1}{2}} \quad , \tag{17b}$$

and $\{\phi_n^*, \phi_m\} = -ia_0^{-1}\delta_{nm}$. This also converges to the usual AKNS-ZS-Dirac spectral problem [2,3] as $a_0 \to 0$. Since the matrix trace of (12) is the Poisson bracket $\{\Delta(\lambda), \Delta(\mu)\} = 0$, $\Delta(\lambda) = TrT(\lambda) = A(\lambda) + D(\lambda) = A(\lambda) + A^*(\lambda)$, (12) is an integrability condition and $\ell n\Delta(\lambda)$ is a generator of classical Hamiltonians.

Quantisation is now straightforward: periodic b.c.s and thermodynamic limit at finite density are retained and the bracket $\{\phi, \phi^*\} = i\delta(x - x') \to [\hat{\phi}, \hat{\phi}^+] = \delta(x - x')$. This induces *operator* spectral data and monodromy matrices $\hat{T}(\lambda)$ through the elementary operators $\hat{L}_n(\lambda)$. Then (12) is replaced by

$$R\hat{T}(\lambda) \otimes \hat{T}(\mu) = \hat{T}(\mu) \otimes \hat{T}(\lambda)R \tag{18}$$

and R is the R-matrix. For repulsive NLS and sinh-G (in one convention [7])

$$R(\lambda, \mu) = \begin{bmatrix} f(\lambda, \mu) & 0 & 0 & 0 \\ 0 & 1 & g(\lambda, \mu) & 0 \\ 0 & g(\lambda, \mu) & 1 & 0 \\ 0 & 0 & 0 & f(\lambda, \mu) \end{bmatrix} \quad ; \tag{19}$$

$f(\lambda, \mu) = 1 + ic(\lambda - \mu)^{-1}$, $g(\lambda, \mu) = ic(\lambda - \mu)^{-1}$ for repulsive NLS; $f(\lambda, \mu) = -\sinh(\lambda' - \mu' - i\mu_0)/\sinh(\lambda' - \mu')$, $g(\lambda, \mu) = i\sin\mu_0/\sinh(\lambda' - \mu')$ for sinh-G: λ', μ' are $\ell n\lambda$, $\ell n\mu$ and $\mu_0 = \pi(1 + \gamma_0/8\pi)$ [7]. Since (18) means $[\hat{\Delta}(\lambda), \hat{\Delta}(\mu)] = 0$, $\hat{\Delta}(\lambda) \equiv Tr\hat{T}(\lambda)$, (18) is a quantum integrability condition and $\ell n\hat{\Delta}(\lambda)$ is a generator of operator Hamiltonians.

In our 'generalised Bethe ansatz' method we begin with the classical action, do not introduce operators, and appear to use only the classical r-matrix. However, the R-matrix and the Yang-Baxter relations [7,15] underlie the theory: the commutation relations (18) put conditions on the allowed modes (labelled by λ, μ). The conditions from the quantum inverse method [3,17] are (11) for models like the NLS and sinh-G and these coincide with those found by quantum Bethe ansatz [7,18]: they must be extended for e.g. s-G [4,14]. The 'generalised Bethe ansatz' method derives quantum or *classical* conditions for the allowed modes generalising the quantum inverse method and Bethe ansatz results.

2 The Generalised Bethe Ansatz Method: Use of Action-Angle Variables

As noted in the Introduction, to define properly the thermodynamic limit of the sinh-G model for example we have to impose periodic b.c.s with period L (say). To this end we use the classical r-matrix formulation of the inverse method which we introduced briefly above. Working in the large L limit it is relatively easy to show [8,9,19] for the integrable sinh-G lattice that, with $\omega(k) = (m^2 + k^2)^{1/2}$,

$$L^{-1}H[p] = L^{-1} \sum_{n=-\frac{1}{2}N}^{\frac{1}{2}N} \omega(\tilde{k}_n)P_n + O(L^{-1}) \quad , \tag{20}$$

and if the l.h.s. is $O(1)$ as $L \to \infty$ as required by the thermodynamic limit this means that $L^{-1}H[p] = O(1) + O(L^{-1})$ as $L \to \infty$ and the action variables $P_n = O(1)$. Note that with period $L = Na_0$, $n = \frac{1}{2}N$ is strictly speaking excluded from (20): the actual count is merely convenient as is the choice N =even.

When we impose periodic b.c.s on the (integrable) lattice we can use Floquet theory to prove [13,19,20] that

$$tr\ T(\lambda) \equiv tr\ T_N(\lambda) = 2 \tag{21}$$

must be satisfied by periodic solutions ($tr\ T_N(\lambda) = -2$ for antiperiodic solutions). By expressing $T_N(\lambda)$ in terms of the scattering data and transforming λ to wave vectors $\tilde{k}$, Eqn. (21) means that the *allowed* wave vectors $\tilde{k}_n$ must satisfy [13,14,19]

$$\tilde{k}_n = k_n - L^{-1} \sum_{m \neq n} \Delta_c(\tilde{k}_n, \tilde{k}_m) P_m \ , \tag{22}$$

where $k_n = 2\pi n L^{-1}$, and the classical phase shift

$$\Delta_c(k,k') = -\frac{1}{4}\gamma_0 m^2 \big(k\omega(k') - k'\omega(k)\big)^{-1} \ . \tag{23}$$

Note how (22) generalises (11) to the classical problem, and in particular, how $\omega(\tilde{k}_n)$ in $H[p]$, (20), depends on the P_n through (22). Thus from this viewpoint, the oscillators making up (20) are *nonlinear* oscillators ('large amplitude phonons' [6,8]). With (22) $H[p]$ and hence the partition function Z can be evaluated by iterating $\omega(\tilde{k}_n) = \omega(k_n) +$ terms in P_n. Apparently this iteration is formal only for classical repulsive NLS; there are no problems for the classical sinh-G model [5].

The classical action variables P_n in (20) and (22) are numbers of modes labelled by $\tilde{k}_n$ and when quantised become the occupation numbers. If the quantised particles obey Bose-Einstein statistics then $P_j = 0, 1, 2, \ldots$, and if they obey Fermi-Dirac statistics, $P_j = 0, 1$. Furthermore the classical phase shifts must be replaced by their quantum S-matrix counterparts which in the boson case are [5,8]

$$\Delta_b(k,k') = -2\tan^{-1}[m^2 \sin(\gamma_0''/8)/(k\omega(k') - k'\omega(k)] \ , \tag{24}$$

where $\gamma_0'' = \gamma_0(1 + \gamma_0/8\pi)^{-1}$ and $\Delta_b(+\infty) = 0$ so that it has a jump of 2π at $k - k'$. The fermion form of the phase shift is the continuous branch of Δ_b,

$$\Delta_f(k,k') = \Delta_b(k,k') - 2\pi\theta(k' - k) \ , \tag{25}$$

where $\theta(x)$ is the unit step function, and $\Delta_f = 0$ for $k \to \infty$, $\Delta_f = -2\pi$ for $k \to -\infty$. Notice that it is the bosonic phase shift Δ_b which tends to the classical expression Δ_c in the limit $\gamma_0 \to 0$. The continuous branch for the fermion form of the phase shift is needed [18] in order to satisfy the Pauli principle (no two particles can have the same momentum [18]). The fermion form of (22) is exactly the same as the Bethe ansatz result (11).

To make better contact with the pioneering work of Yang and Yang [1] we now take the exact thermodynamic limit $L \to \infty$. Since P_n is a particle number in $\tilde{k}_n$, $P_n L^{-1}$ is a number density, and $P_n L^{-1} \to \rho(\tilde{k})d\tilde{k}$ as $L \to \infty$. Note that $\rho(\tilde{k})$ are $O(1)$ because P_n are $O(1)$: we thus have a finite density limit. Now relabel the wave vectors $\tilde{k} \to k$, and define $h(k) = \lim_{L \to \infty}(2\pi n L^{-1})$ ($= \lim_{L \to \infty} k_n$ in the previous notation in (22)), then as $L \to \infty$ (22) is [21]

$$h(k) = k + \int_{-\infty}^{\infty} \Delta_c(k,k')\rho(k')dk' \ . \tag{26}$$

The allowed modes k define a density of allowed states $f(k)$ such that

$$f(k) = (2\pi)^{-1}\frac{dh(k)}{dk} \tag{27}$$

so (26) means

$$2\pi f(k) = 1 + \int_{-\infty}^{\infty} \frac{d}{dk}\Delta_c(k,k')\rho(k')dk' \quad . \tag{28}$$

From (20) one also has the energy E (Hamiltonian)

$$EL^{-1} = \int_{-\infty}^{\infty} \omega(k)\rho(k)dk \quad . \tag{29}$$

Notice that Eqns. (28) and (29) are *classical* results; there are corresponding results for bosons and fermions with their repective densities replacing the classical density $\rho(k)$, and phase shifts Δ_b and Δ_f replacing Δ_c.

The procedure now is to calculate an entropy S and so the free energy F in terms of the density of states and the density of particles: this procedure generalises the work of Yang and Yang [1] on the repulsive NLS to a boson description and to the classical case. We find [8] that in the classical case

$$S_{cl}L^{-1} = \int_{-\infty}^{\infty} \rho(k)[\ell n(f(k)/\rho(k)) + 1] \quad , \tag{30}$$

and for bosons and fermions

$$SL^{-1} = \int_{-\infty}^{\infty} [(f+\rho)\ell n(f+\rho) - f\ell n f - \rho \ell n \rho]dk \quad . \tag{31}$$

The equilibrium free energy is now found by minimising $FL^{-1} = EL^{-1} - \beta^{-1}SL^{-1}$, i.e. we set $\delta(FL^{-1})/\delta\rho(k) = 0$. From (29) we readily find that $\delta(EL^{-1})/\delta\rho(k) = \omega(k)$. For the variation of entropy we need the variation of the density of states $f(k)$: from (28) one finds

$$\frac{\delta f(k)}{\delta \rho(k')} = \frac{1}{2\pi}\frac{d}{dk}\Delta_c(k,k') \quad , \tag{32}$$

and similar expressions for bosons and fermions. The minimising condition is then an integral equation involving the density of states and the density of particles. These are not independent because of (28). One finds they can be related through certain excitation energies at finite temperatures. These excitation energies $\epsilon(k)$ are to be defined such that

$$f\rho^{-1} = \beta\epsilon(k), \quad \text{for classical particles,}$$
$$f\rho^{-1} = e^{\beta\epsilon(k)} - 1, \quad \text{for bosons,} \tag{33}$$
$$f\rho^{-1} = e^{\beta\epsilon(k)}, \quad \text{for fermions.}$$

In this way the integral equations are expressed in terms of finite temperature excitation energies $\epsilon(k)$ alone. Together with the corresponding minimal free energies, the results are

$$\epsilon(k) = \omega(k) + (2\pi\beta)^{-1}\int_{-\infty}^{\infty} \frac{d}{dk}\Delta_c(k,k')\ell n(\beta\epsilon(k'))dk'$$

$$\lim_{L\to\infty} FL^{-1} = (2\pi\beta)^{-1}\int_{-\infty}^{\infty} \ell n(\beta\epsilon(k))dk \tag{34}$$

for classical particles,

$$\epsilon(k) = \omega(k) + (2\pi\beta)^{-1} \int_{-\infty}^{\infty} \frac{d}{dk}\Delta_b(k,k') ln(1 - e^{-\beta\epsilon(k')})dk'$$

$$\lim_{L\to\infty} FL^{-1} = (2\pi\beta)^{-1} \int_{-\infty}^{\infty} ln(1 - e^{-\beta\epsilon(k)})dk \tag{35}$$

for bosons, and

$$\epsilon(k) = \omega(k) - \mu + (2\pi\beta)^{-1} \int_{-\infty}^{\infty} \frac{d}{dk}\Delta_f(k,k') ln(1 + e^{-\beta\epsilon(k')})dk'$$

$$\lim_{L\to\infty} FL^{-1} = \mu\bar{n} - (2\pi\beta)^{-1} \int_{-\infty}^{\infty} ln(1 + e^{-\beta\epsilon(k)})dk \tag{36}$$

for fermions. For the latter it is convenient to introduce chemical potential μ and particle density $\bar{n}$ so that we have in fact minimised the negative pressure = $EL^{-1} - \beta^{-1}SL^{-1} - \mu NL^{-1}$.

There are two interesting free particle limits of (36) for sinh-G: both arise where $\sin(\gamma_0''/8) \to 0$ but this is where $\gamma_0/(1 + \gamma_0/8\pi) = 0$ or 8π, so $\gamma_0 = 0$ or $\gamma_0 \to \infty$. In both cases $\epsilon(k) = \omega(k) - \mu$, just as for a gas of free bosons: since $\gamma_0 \to 0$ is the free KG this is expected, but the result when $\gamma_0 \to \infty$ is not expected. In fact the repulsive NLS (1) ($c>0$) has the same integral equations like (36) in fermion form except that $\omega(k) = k^2$ and $\Delta_f = -2\tan^{-1}(c/(k-k'))$, smooth branch [1]. When the coupling constant $c \to \infty$, $\epsilon(k) = k^2 - \mu$, a gas of free fermions. This is the impenetrable bose gas [11].

Notice that the fermion result (36) and the boson result (35) are indeed equivalent: one can be transformed to the other through $ln(1 + e^{-\beta\epsilon}) = -ln(1 - e^{-\beta\epsilon})$, and $\frac{d}{dk}\Delta_f(k,k') = \frac{d}{dk}\Delta_b(k,k') + 2\pi\delta(k'-k)$. Also the classical result (34) is the natural classical limit of the *boson* result (35).

Yang and Yang proved [1] that the fermion form (36) for the repulsive NLS iterates: similar arguments apparently apply to sinh-G. We have not found analytic expressions for the resulting integrals, however. The classical result (34) for sinh-G also iterates (for repulsive NLS it does not iterate). We find that

$$\lim_{L\to\infty} FL^{-1} = -m\beta^{-1}\Big[\frac{1}{4}t - \frac{1}{8}t^2 + \frac{3}{16}t^3 - \frac{53}{128}t^4 + \frac{594}{2^9}t^5 \\ - \frac{7922}{2^{11}}t^6 + \cdots - \frac{478935069186}{2^{23}}t^{12} + \cdots\Big] + F_{KG} , \tag{37}$$

where $t = (M\beta)^{-1}$, $M = 8m\gamma_0^{-1}$ and F_{KG} is the free energy of the KG lattice. Note that because of the classical model (17) there is a classical ultraviolet divergence: $\lim a_0 \to 0$ cannot be taken in the free energy. In fact, instead of $\omega(k) = (m^2 + k^2)^{1/2}$ we have to use a discretised version, e.g. $\omega(k) = (m^2 + \sin^2(\frac{1}{2}ka_0)/(\frac{1}{2}a_0)^2)^{1/2}$. However, the first iterate of (34) gives the free particle result F_{KG} where $a_0 \to 0$ cannot be taken; in the remaining terms which give the power series contribution in (37), $a_0 \to 0$ can be and has been taken. The *nonlinear* contributions to the free energy are independent of the discretization procedure. This is true for all integrable fields. The result (37) checks identically with that found [5] by the the transfer integral method [22].

Even though the classical system (34) does not iterate for the repulsive NLS, we have found [23] from these equations and also by applying the transfer integral method to the corresponding functional integral that

$$\lim_{L \to \infty} FL^{-1} = (\beta a_0)^{-1} \ell n(\beta a_0^{-2}) + Kc^{1/3}\beta^{-4/3} , \qquad (38)$$

where a_0^{-2} is an energy in our units. The constant K is determined by a numerical solution for the lowest eigenvalue of the radial Schrödinger equation derived through the transfer integral method as $K = 1.49 \times 2^{-2/3}$ [23]. Remarkably enough [14] we also find (38) for the attractive NLS which does have soliton solutions: in this case $K \to -K$. It is open at present whether $|K|(c/\beta)^{1/3} \propto T^{1/3}$ forms an estimate of the soliton number density. The attractive NLS in any case needs special discussion [14].

3 Results for Other Integrable Models and Comments

The generalised BA method in the form in which it was applied to the repulsive NLS model and the sinh-G model in the previous section, must be significantly extended when applied to integrable models which admit soliton solutions. The models which admit soliton solutions and which we have solved so far include the s-G models [5], the massive Thirring model [14,24], the attractive NLS model [14,24], the Landau-Lifshitz model [14,24], the isotropic ferromagnet [25], and with some provisos, the Toda lattice [10]. One of the most significant new features related to the *continuum* models listed above is the difference between their classical and quantum thermodynamic limits [6]. The breather solutions or the pulse soliton solutions of these models form a discrete set of bound states when quantised. Each of these bound states is a distinct particle and has a well-defined finite density thermodynamic limit. In contrast with this, the classical thermodynamic limit at finite density forces these solutions to become dense in the phase space only in the vanishing amplitude limit. As this thermodynamic limit is also such that the ordinary small oscillations ('phonons') give no contribution to the energy density EL^{-1} of the model [6,8,14], it means [6,8,14] that the breather-like solutions replace the phonon contribution to give a 'large amplitude phonon' contribution with the *same* dispersion $\omega(k)$. These 'large amplitude phonons' interact through mutual phase shifts unlike the ordinary phonons.

For the s-G model, for example, the classical energy density EL^{-1} is [4,6,8]

$$\lim_{L \to \infty} EL^{-1} = \int_{-\infty}^{\infty} \omega(k)\rho(k)dk + 2\int_{-\infty}^{\infty} \rho_s(p)E_s(p)dp , \qquad (39)$$

where $E_s(p) = (M^2 + p^2)^{1/2}$, and the factor 2 multiplying the kink contribution, the second term on the right side of (39), accounts for the antikinks. Eqn. (39) should be compared with the *zero density* result (5). Note that the breathers now make no contribution as such and only 'large amplitude phonons' do so.

Extending the analysis of the previous section to this case, we then find [4] the classical result is

$$\lim_{L \to \infty} FL^{-1} = (2\pi\beta)^{-1}\int_{-\infty}^{\infty} dk \ell n(\beta\epsilon_{ph}(k)) - 2(2\pi\beta)^{-1}\int_{-\infty}^{\infty} dp\, e^{-\beta\epsilon_s(p)}$$

$$\epsilon_{ph}(k) = \omega(k) + (2\pi\beta)^{-1}\int_{-\infty}^{\infty} dk' \frac{d}{dk}\Delta(k,k')\ell n(\beta\epsilon_{ph}(k'))$$

$$+ 2(2\pi\beta)^{-1} \int_{-\infty}^{\infty} dp \frac{d}{dk} \Delta_s(k,p) e^{-\beta\epsilon_s(p)}$$

$$\epsilon_s(p) = E_s(p) + 2(2\pi\beta)^{-1} \int_{-\infty}^{\infty} dk \frac{d}{dp} \Delta_s(p,k) \ell n(\beta\epsilon_{ph}(k)) \qquad (40)$$

$$+ 2(2\pi\beta)^{-1} \int_{-\infty}^{\infty} dp' \frac{d}{dp} \Delta_{ss}(p,p') e^{-\beta\epsilon_s(p')} \quad ,$$

while the corresponding *semi*classical results are obviously $\ell n(\beta\epsilon_{ph}) \to \ell n(1-\exp(-\beta\epsilon_{ph}))$ and $\exp(-\beta\epsilon_s) \to \ell n(1+\exp(-\beta\epsilon_s))$. In some contrast the exact fully quantised fermion result coincides with that [26,27] of the usual BA method, but we have not yet worked out the details for the boson version. The phase shifts are conveniently expressed in terms of rapidities [4] so that for Eqn. (40) and semiclassical limit ($p \to M \sinh y$, $k \to m \sinh x$)

$$\Delta(x,x') = \frac{1}{4}\gamma_0(\sinh(x-x'))^{-1} \quad , \quad \Delta_s(x,y) = -2\tan^{-1}\left[(\sinh(x-y))^{-1}\right]$$

$$\frac{\partial}{\partial y}\Delta_{ss}(y,y') = 8\gamma_0^{-1} \ell n \left[\frac{\cosh(y-y')-1}{\cosh(y-y')+1}\right] \quad . \qquad (41)$$

The system (40) with (41) now iterates. Again all the integrals can be done and the result is [4]

$$\lim_{L\to\infty} = F^{(1)} + F^{(2)} + F^{(3)} + \cdots + F_{KG} \quad ;$$

$$F^{(1)} = -m\beta^{-1}(8/\pi t)^{1/2} e^{-1/t} \left[1 - \frac{7}{8}t - \frac{59}{128}t^2 - \frac{897}{1024}t^3 - \cdots\right]$$

$$- m\beta^{-1}\left[\frac{1}{4}t + \frac{1}{8}t^2 + \frac{3}{16}t^3 + \frac{53}{128}t^4 + \frac{594}{2^9}t^5 + \cdots\right] \quad ,$$

$$F^{(2)} = \frac{8m}{\pi} M e^{-2/t} \{\ell n\left(\frac{4C}{t}\right) - \frac{5}{4}t\left[\ell n\left(\frac{4C}{t}\right)+1\right] - \cdots\} \quad , \qquad (42)$$

$F^{(q)} = O(e^{-q/t})$, $q = 3, 4, \cdots$; F_{KG} is as before (with the same ultraviolet divergence problem associated with it); $\ell n C = 0.5771 \cdots$ (Euler's constant). Note that all of the expansions in (42) are strictly asymptotic.

It is evident that (42) is the analytic continuation in γ_0 of (37). Because there are now soliton solutions the $F^{(q)}$ have the q-kink-plus-antikink contributions depending on $\exp(-q/t)$. It is relevant to the analysis of the Toda lattice for example, which has *zero* mass corresponding to m, that the development in multisoliton contributions is not possible and a different development arises — see below. Note that it is the finiteness of the mass M which controls the convergence of (42) as $T \to 0$. The robust features of both (37) and (42) seem to us quite extraordinary given the apparent delicacy of the means of finding these asymptotic expansions. Note further that all of the classical phase shifts like (41) emerge through the analytical properties of the classical spectral data and all of them are found in this way. Thus the classical r-matrix (14) in (12) has determined everything.

Another interesting model is the (classical) isotropic Heisenberg ferromagnet with equations of motion

$$\vec{S}_t = \vec{S} \times \vec{S}_{xx} + J\vec{S} \times \vec{S}, \quad \vec{S} = (S_1, S_2, S_3), \quad |\vec{S}| = 1 \quad ; \qquad (43)$$

which is related to attractive NLS by gauge transformation [28]. We find [25,29] for the classical SM of this model when a *transverse* magnetic field h is imposed

$$\lim_{L\to\infty} FL^{-1} = (2\pi\beta)^{-1}\int_{-\infty}^{\infty} dk \ln(\beta^*\epsilon(k)) - J(1+h) \;,$$

$$\epsilon(k) = k^2 + h + (2\pi\beta)^{-1}\int_{-\infty}^{\infty} dk' \frac{d}{dk}\Delta_c(k,k')\ln(\beta^*\epsilon(k')) \;,$$

$$\Delta_c(k,k') = J^{-1}kk'(k-k')^{-1} \;, \tag{44}$$

where $\beta^* = J\beta$. Note that the classical magnon-magnon phase shift Δ_c has coupling constant J^{-1} and does not depend on $(k-k')$ alone. The system (44) iterates. With the scaled temperature $s = (\beta J h^{1/2})^{-1}$ (compare with t in (42) for s-G and (37) for sinh-G) (44) yields the low temperature asymptotic expansion

$$FL^{-1} = \beta^{-1}a_0^{-1}[\ell n(\pi^2 J\beta a_0^{-2})-2] - J(1+h) + Jh(s - \frac{1}{4}s^2 - \frac{1}{64}s^3 + \frac{5}{512}s^4 + \cdots) \;, \tag{45}$$

which agrees with, and extends, the result [30] found by the transfer integral method.

The corresponding results for attractive NLS follow formally from (44) for $h \to 0$ (with a shift of energy): however Δ_c for the attractive NLS model is not the same as (44) for the ferromagnet: $\Delta_c(k,k') = 2c(k-k')^{-1}$, $c>0$. The asymptotic expansion of (44), namely (45), is not defined for $h \to 0$ since s is not defined: we have not found any low temperature asymptotic expansion for the attractive NLS. Note that the classical SM of the attractive NLS is given formally by Eqns. (34) with the different Δ_c. This is exactly what we find for the classical SM of the repulsive NLS except that $c \to -c$ in the phase shift Δ_c. It is remarkable that both attractive NLS with breather-like solitons and phonons and repulsive NLS with no solitons and only phonons have formally equivalent classical SM. A similar situation obtains for s-G as we have seen above.

Our results for the classical Toda lattice are still not complete so our results as reported here take the form of a comment. By choosing suitable units of energy and time, the classical Hamiltonian of the Toda lattice can be expressed in the dimensionless form [10]

$$H[Q_n] = ab^{-1}\sum_{n=-N}^{N}\left[\frac{1}{2}b^2 a^{-2}P_n^2 + e^{-(Q_n-Q_{n-1})} - 1\right] \;, \tag{46}$$

where a and b are the parameters of Toda's potential [31] and ba^{-1} acts as the coupling constant; $\{P_n, Q_{n'}\} = \delta_{nn'}$. We can find action-angle variables for this periodic lattice such that [14]

$$H[p] = 2\int_0^\pi 2\sin(\frac{1}{2}k)P(k)dk + ab^{-1}\sum_{j=1}^{2N_s}[\sinh(2ba^{-1}p_j) - 2ba^{-1}p_j] \tag{47}$$

with $p_j > 0$. The classical phase shifts can be found in the same way as for the other integrable lattices and we find [14] for example

$$\Delta_s(k,p) = -2\tan^{-1}\left[\cot(\frac{1}{2}k)\tanh(\frac{1}{2}ba^{-1}p)\right] \;,$$

$$\Delta_{ss}(p,p') = -4\tanh^{-1}\left[\tanh(\frac{1}{2}ba^{-1}p)\coth(\frac{1}{2}ba^{-1}p')\right] \;. \tag{48}$$

The phonon-phonon shift Δ is given in [14]. It is important that there are actually two phase shifts Δ_s, and two Δ_{ss}'s depending on whether pairs of excitations move in the same or opposite directions [10,14,31]. This means that we have to extend the integral equations like (40) which we otherwise find for the classical SM of this model. We have not yet solved these extended integral equations.

There is a complication surrounding the Hamiltonian (47) in action variables which we have not found in any other model: the action variables $P(k)$ and p_j are not independent but satisfy the condition [3]

$$\int_0^{2\pi} \frac{1}{\omega(k)} P(k) dk - \sum_{j=1}^{2N_s} 2p_j = 0 \quad . \tag{49}$$

There is a corresponding change in the usual brackets which gain extra factors depending on k and p_j in a rather complicated fashion [3,10]. This complication makes the calculation for example of the soliton number density difficult, and all the ramifications related to this problem are still being explored. Note that to find a quantum theory of BA type all such problems must reoccur [7].

We note briefly that there is a Hamiltonian formulation of Feynman's functional integration method by which the free energies of the integrable models can be calculated [4,5,7,10]. We have shown that this method and the method of 'generalised BA' described in this paper are wholly equivalent; both methods make use of the action-angle variables, and periodic b.c.s are imposed so that the action variables are coupled through periodicity conditions like (22) or its extensions. It is in fact these periodicity conditions where all the 'action' resides: otherwise the partition functions would be factorisable, and the systems they describe would behave like ideal gases. The role of the functional integral is only to minimise the free energy: in the 'generalised BA' method described in this paper minimisation is made explicit.

We notice finally that in principle the methods described here are not restricted to 1+1 dimensional models, but it remains to be seen how difficult the finite density limit and the periodicity conditions turn out to be in the 2+1 dimensional case. So far the 2+1 dimensional models which have been considered under vanishing b.c.s are trivial in the sense that although the action-angle variables can be found [32], there are no phase shifts [32]. Phase shifts have been crucial to all of the calculations in 1+1 dimensions.

References

[1.]   Yang CN, Yang CP. J.Math.Phys. 10:115 (1969)

[2.]   Bullough RK, Caudrey PJ. Solitons. (Springer, Heidelberg 1980)

[3.]   Faddeev LD, Takhtajan LA. Hamiltonian methods in the theory of solitons. (Springer, Berlin 1987)

[4.]   Timonen J, Stirland M, Pilling DJ, Cheng Yi, Bullough RK. Phys.Rev.Lett. 56:2233 (1986)

[5.]   Bullough RK, Pilling DJ, Timonen J. J.Phys. A19:L955 (1986)

[6.]   Timonen J, Chen YZ, Bullough RK. Nucl.Phys. B, 5A:58 (1988)

[7.]   Bullough RK, Pilling DJ, Timonen J. In: Lakshmanan M (ed.) Solitons. (Springer, Heidelberg 1988)

[8.]   Timonen J, Pilling DJ, Chen YZ, Bullough RK. In: Bar'yakhtar VG et al. (eds.) Plasma theory and nonlinear and turbulent processes in physics. (World Scientific, Singapore 1988)

[9.] Bullough RK, Pilling DJ, Timonen J. In: Christiansen PL, Parmentier RD (eds.) Structure, coherence and chaos. (Manchester UP, Manchester 1988)

[10.] Bullough RK, Timonen J, Chen YZ, Cheng Yi, Stirland M. In: Fordy AP, Degasperis A (eds.) Nonlinear evolution equations: Integrability and spectral methods. (Manchester UP, Manchester 1989)

[11.] Lieb EH, Liniger W. Phys.Rev. 130:1605 (1963)

[12.] Caudrey PJ. Physica 6D:51 (1982); In: Fordy AP (ed.) Soliton theory a survey of results. (Manchester UP, Manchester 1989)

[13.] Bullough RK, Pilling DJ, Timonen J. In: Claro (ed.) Nonlinear phenomena in physics (Springer, Berlin 1985)

[14.] Chen Yu-zhong. Ph.D. Thesis. University of Manchester (submitted March 1989)

[15.] Bullough RK, Olafsson S. In: Solomon A (ed.) Proc. of the 17th Intl. Conf. on Differential Geometric Methods in Theoretical Physics, Chester 1988 (World Scientific, Singapore 1989)

[16.] Izergin AG, Korepin VE. Lett.Math.Phys. 5:199 (1981)

[17.] Sklyanin EK, Takhtajan LA, Faddeev LD. Teor.Mat.Fiz. 40:194 (1979) [Theor.Math.Phys. 40:688 (1980)]

[18.] Thacker HB. Rev.Mod.Phys. 53:253 (1981)

[19.] Cheng Yi. Ph.D. Thesis. University of Manchester (1987)

[20.] Flaschka H, McLaughlin DW. Prog.Theor.Phys. 55:438 (1976)

[21.] There is a sign error in Eqns. (7) and (10) of Ref. [5] (k_n and $\bar{k}_n$ were accidentally interchanged), but the subsequent equations there are correct

[22.] Stone M, Reeve J. Phys. Rev. D18:4746 (1976); Dingle RB, Müller HJW. J. Reine und Angew. Math. 211:11 (1962)

[23.] Bullough RK, Pilling DJ, Timonen J. In: Takeno S (ed.) Dynamical problems in soliton systems. (Springer, Berlin 1985)

[24.] Chen YZ, Bullough RK, Timonen J. To be published

[25.] Chen YZ, Bullough RK, Timonen J, Tognetti V, Vaia R. To be published

[26.] Fowler M, Zotos X. Phys.Rev. B24:2634 (1982); B25:5806 (1982); Imada M, Hida K, Ishikawa M. Phys.Lett. 90A:79 (1982); J.Phys. C16:36 (1983)

[27.] Chen NN, Johnson MD, Fowler M. Phys.Rev.Lett. 56:907 (1986); 56:1427(E) (1986); Phys.Rev. B34:1851 (1986); Timonen J, Bullough RK, Pilling DJ. Phys.Rev. B34:6525 (1986)

[28.] Lakshmanan M. Phys.Lett. 61A:53 (1977); Takhtajan LA. Phys.Lett. 64A:235 (1977)

[29.] Bullough RK, Chen YZ, Olafsson S, Timonen J. In: Lochak P, Balabane M, Sulem C (eds.) Integrable systems and applications (Springer, Heidelberg, to appear 1989)

[30.] Nakamura K, Sasada T. J.Phys. C11:L171 (1987)

[31.] Toda M. Theory of nonlinear lattices. (Springer, Berlin 1981)

[32.] Jiang Z, Bullough RK, Manakov SV. Physica 18D:305 (1980); Jiang Zhuhan. Ph.D. Thesis. University of Manchester (1987); Bakurov VG, Manakov SV, Jiang Z, Bullough RK. To be published

Link Polynomials and Exactly Solvable Models

M. Wadati[1], Y. Akutsu[2], and T. Deguchi[1]

[1] Institute of Physics, College of Arts and Sciences, University of Tokyo, Komaba 3-8-1, Meguro-ku, Tokyo 153, Japan
[2] Institute of Physics, Kanagawa University, Rokkakubashi, Kanagawa-ku, Yokohama 221, Japan

Link polynomial, topological invariant for knots and links, is constructed from an exactly solvable model in statistical mechanics. A general theory consists of two steps. First, representation of the braid group is made from the Boltzmann weights of the exactly solvable model. Second, Markov trace is defined on the braid group representation. Sufficient conditions for the existence of the Markov trace are explicitly given. The knot theory based on exactly solvable models also includes braid-monoid algebra, graphical approach and two-variable extension of link polynomials. In addition, application of the theory to graph-state models is presented.

1 Knots, Links and Link Polynomials

We begin with a brief summary of knot theory [1,2]. A string is a one-dimensional object. Strings assume a variety of configurations in three-dimensional space. A knot is a closed string which does not cross with itself. An assembly of closed strings with mutual entanglements is called a link. Link is often a general terminology for knot and link. A problem we shall consider is a classification of knots and links. This problem is known as one of the most fundamental problems in topology.

Suppose that two knots (or links) are given, we like to judge whether they are different or not. Two knots (or links) are defined to be topologically equivalent or ambient isotopic when they are transformed each other by continuous deformations without tearing the string(s). To classify knots and links in a systematic manner, it is significant to have topological invariant, that is, a quantity which does not change under continuous deformations of strings. When the invariant is in a form of (Laurant) polynomial with continuous parameter(s), it is called link polynomial.

We introduce braid and braid group to describe knots and links. Prepare two horizontal bars and choose n points on each of them. Braid is formed when n points on the upper bar are connected by n strings to n points on another bar directly below the first n points (Fig.1). Trivial n-braid is a configuration where no intersection between the strings is present. Let us name a string from the i-th point on the upper bar i-th string. An operation of making an intersection, where i-th string passes over (under) $i+1$-th string is denoted by b_i (b_i^{-1}), $i = 1, 2, \ldots, n-1$ (Fig.2). The operation b_i^{-1} is the inverse of b_i.

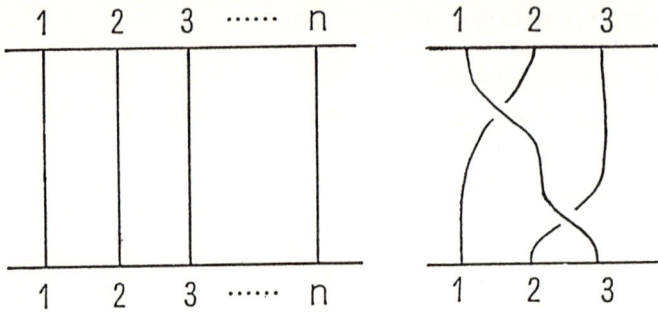

Fig.1: Trivial n-braid (left) and a non-trivial 3-braid represented by $b_1 b_2$ (right).

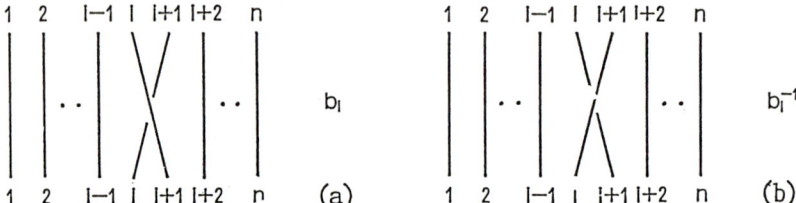

Fig.2: Operations b_i and b_i^{-1}. Other strings except i-th and $i+1$-th strings remain fixed.

A product of two braid operations, say b_1 and b_2, is written as $b_1 b_2$ (Fig.1). An n-braid in general is composed from the trivial n-braid by successive applications of $\{b_i\}$ and $\{b_i^{-1}\}$. The operators $\{b_i, i = 1, 2, \ldots, n-1\}$ define a group, the braid group, which is denoted by B_n. By regarding the trivial n-braid as the identity operation in B_n, we can identify any element in B_n as an n-braid. However, expression of a braid in terms of braid group elements is not unique. Topological equivalence between different expressions of a braid is guaranteed by the following relations (Fig.3):

$$b_i b_j = b_j b_i, \qquad |i - j| \geq 2 \qquad (1.1a)$$
$$b_i b_{i+1} b_i = b_{i+1} b_i b_{i+1}. \qquad (1.1b)$$

This is the defining relation of the braid group by Artin. Under the relation (1.1), each topologically equivalent class (isotopy class) of braids is identified with an element in B_n.

Given a braid, one may tie opposite ends to form a link (closed braid, Fig.4). A closed braid represents an oriented link meaning that each string in the link has a definite orientation. Conversely, according to Alexander's theorem, any oriented link is represented by a closed braid. However, the representation of a link as a closed braid is highly non-unique: infinitely many braids give the same link when they are closed. The equivalent braids expressing the same link are mutually transformed by a finite sequence of two types of operations, the Markov moves I and II (Fig.5):

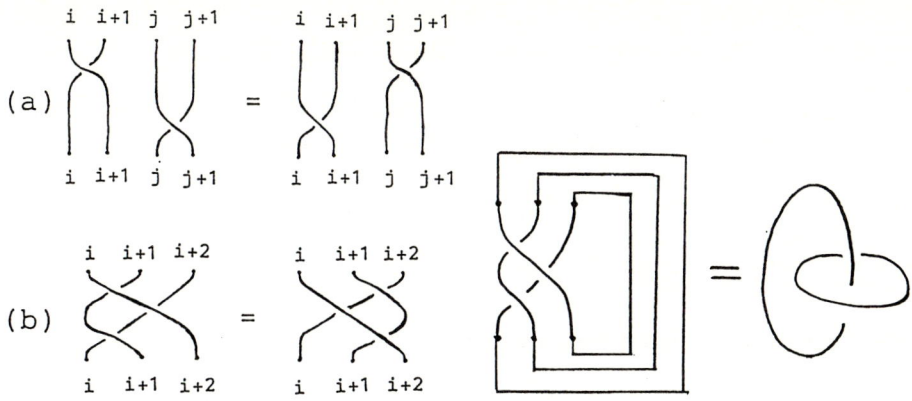

Fig.3: Defining relation of the braid group.   Fig.4: A link as a closed braid.

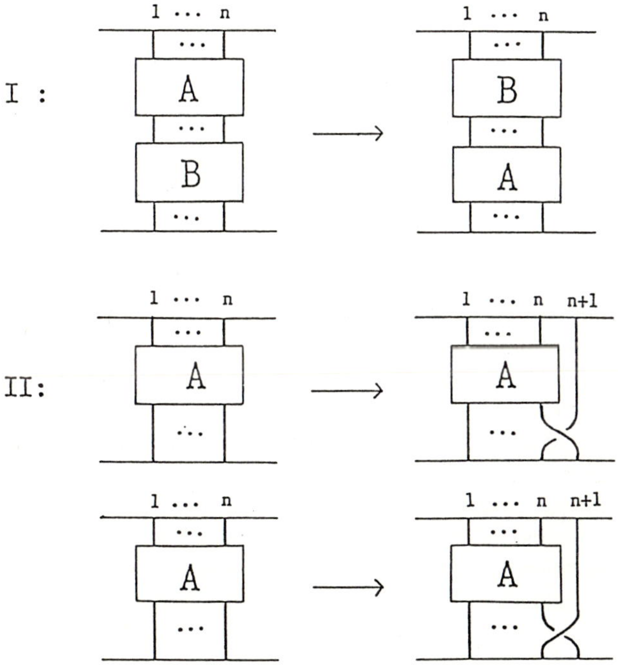

Fig.5: Markov moves, I and II.

I. $AB \to BA$ $\quad (A, B \in B_n)$
II. $A \to Ab_n,$
    $A \to Ab_n^{-1}$ $\quad (A \in B_n,\ b_n \in B_{n+1}).$ \hfill (1.2)

Hence, we can obtain a link polynomial in the following scheme. We first make a representation of B_n. And then we find a quantity defined on the representation which is invariant under the Markov moves.

Let us denote the representation of b_i by G_i and a link polynomial by $\alpha(\cdot)$. The link polynomial $\alpha(\cdot)$ should satisfy the conditions:

I. $\alpha(AB) = \alpha(BA), \quad (A, B \in B_n)$ (1.3a)
II. $\alpha(AG_n) = \alpha(AG_n^{-1}) = \alpha(A), \quad (A \in B_n, \ G_n \in B_{n+1})$. (1.3b)

This quantity is obtained if we can find a functional $\phi(\cdot)$ called Markov trace, which have the following properties (the Markov properties):

I. $\phi(AB) = \phi(BA), \quad (A, B \in B_n)$ (1.4a)
II. $\phi(AG_n) = \tau\phi(A),$
    $\phi(AG_n^{-1}) = \bar{\tau}\phi(A) \quad (A, B \in B_n),$ (1.4b)

where the constants τ and $\bar{\tau}$ are defined by

$$\tau = \phi(G_i), \quad \bar{\tau} = \phi(G_i^{-1}) \quad \text{for all } i. \tag{1.5}$$

With the Markov trace $\phi(\cdot)$, the link polynomial $\alpha(\cdot)$ is given by a formula

$$\alpha(A) = (\tau\bar{\tau})^{-(n-1)/2} \left(\frac{\bar{\tau}}{\tau}\right)^{e(A)/2} \phi(A), \quad A \in B_n. \tag{1.6}$$

Here $e(A)$ is the exponent sum of b_i's in the braid A. For instance, if $A = b_2^3 b_3^{-1} b_1 b_2^{-2}$, then $e(A) = 3 - 1 + 1 - 2 = 1$.

2 Yang-Baxter Relation

A general method has been developed to construct link polynomials from exactly solvable models in statistical mechanics. Complementary review papers are found in [3,4,5,6,7]. As explained in §1, to construct a link polynomial, we have to make a representation of the braid group and then find the Markov trace defined on the representation. We shall show that both steps are accomplished in a natural and systematic way by the theory of exactly solvable models.

The Yang-Baxter relation [8,9], a sufficient condition for the solvability, takes different forms depending on the models under consideration. For 1+1 dimensional field theory, the Yang-Baxter relation is the factorization condition for the many-body S-matrices and is often called factorization equation [10]. Let us denote the scattering amplitude (Fig.6) for the process $(i, j) \to (k, \ell)$ by $S_{jl}^{ik}(u)$, where the indices i, j, k and ℓ specify inner degrees of freedom such as charge, spin and flavor. The parameter u, referred to as spectral parameter, is the rapidity difference of two colliding particles. The factorization equation reads as

$$\sum_{\alpha\beta\gamma} S_{\gamma\tau}^{\beta q}(v) S_{k\gamma}^{\alpha p}(u+v) S_{j\beta}^{i\alpha}(u) = \sum_{\alpha\beta\gamma} S_{\beta q}^{\alpha p}(u) S_{\gamma\tau}^{i\alpha}(u+v) S_{k\gamma}^{j\beta}(v), \tag{2.1}$$

which is schematically explained in Fig.7.

For two-dimensional statistical mechanics, we have two types of models [11], the vertex models and the IRF (Interaction Round a Face) models (Fig.6). As was pointed out by Zamolodchikov, any factorized S-matrix can be interpreted as the

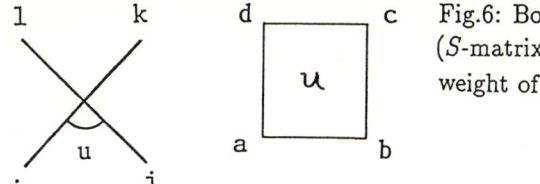

Fig.6: Boltzmann weight of vertex model (S-matrix) $S^{ik}_{jl}(u)$ (left) and Boltzmann weight of IRF model $w(a,b,c,d;u)$ (right).

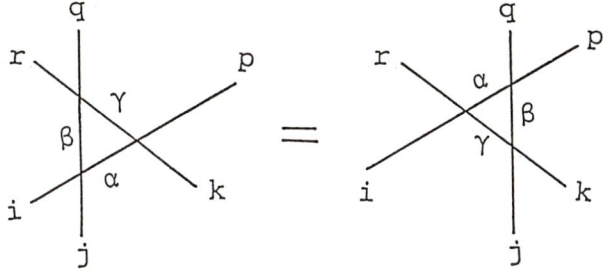

Fig.7: Factorization equation (Yang-Baxter relation for S-matrice and vertex model).

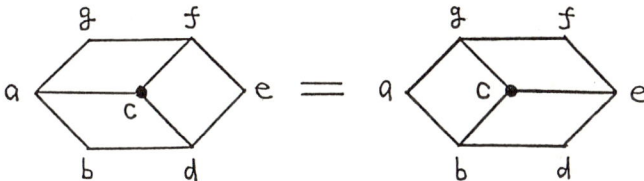

Fig.8: Star-triangle relation (Yang-Baxter relation for IRF model).

Boltzmann weight of a solvable vertex model. This equivalence is helpful, in particular, for a pictorial understanding of the knot theory based on exactly solvable models. For the vertex model, the factorization equation (2.1) is the commutability condition of the transfer matrices. We introduce IRF models. State variables are located on the sites of a square lattice. The Boltzmann weight is assigned to each state variable configuration round a unit square (face). By $w(a,b,c,d;u)$ we denote the Boltzmann weight for the configuration (a,b,c,d) round the face. For the IRF models, the commutability condition of the transfer matrices is called star-triangle relation and reads as (Fig.8)

$$\sum_c w(b,d,c,a;u)w(a,c,f,g;u+v)w(c,d,e,f;v)$$
$$= \sum_c w(a,b,c,g;v)w(b,d,e,c;u+v)w(c,e,f,g;u). \qquad (2.2)$$

The Boltzmann weights of most of the exactly solvable models satisfy the following relations in addition to the Yang-Baxter relation.

1) *standard initial condition*

$$S^{ik}_{jl}(0) = \delta(i,\ell)\delta(j,k), \tag{2.3a}$$
$$w(a,b,c,d;0) = \delta(a,c), \tag{2.3b}$$

where $\delta(\cdot,\cdot)$ is the Kronecker's delta.

2) *inversion relation (unitarity condition)*

$$\sum_{m,p} S^{mk}_{p\ell}(u)S^{ip}_{jm}(-u) = \rho(u)\rho(-u)\delta(i,\ell)\delta(j,k), \tag{2.4a}$$
$$\sum_{e} w(e,c,d,a;u)w(b,c,e,a;-u) = \rho(u)\rho(-u)\delta(b,d), \tag{2.4b}$$

where $\rho(u)$ is a model-dependent function.

3) *second inversion relation (second unitarity condition)*

$$\sum_{m,p} S^{im}_{p\ell}(\lambda - u)S^{kp}_{mj}(\lambda + u) \cdot \frac{r(p)r(m)}{r(k)r(j)} = \rho(u)\rho(-u)\delta(i,j)\delta(k,\ell) \tag{2.5a}$$

$$\sum_{e} w(c,e,a,b;\lambda-u)w(a,e,c,d;\lambda+u) \cdot \frac{\psi(e)\psi(b)}{\psi(a)\psi(c)}$$
$$= \rho(u)\rho(-u)\delta(b,d). \tag{2.5b}$$

We call λ crossing parameter (or crossing point) and $r(i)$ and $\psi(a)$ crossing multipliers.

4) *crossing symmetry*

$$S^{ik}_{k\ell}(u) = S^{j\ell}_{\bar{k}\bar{i}}(\lambda - u) \cdot \left[\frac{r(i)r(\ell)}{r(j)r(k)}\right]^{1/2}, \tag{2.6a}$$

$$w(a,b,c,d;u) = w(b,c,d,a;\lambda-u) \cdot \left[\frac{\psi(a)\psi(c)}{\psi(b)\psi(d)}\right]^{1/2}. \tag{2.6b}$$

For vertex model, we have introduced the notation $\bar{j} = -j$ etc. for "charge conjugation". We assume that

$$r(\bar{j}) = 1/r(j). \tag{2.7}$$

5) *charge (or spin) conservation condition*

$$S^{ik}_{j\ell}(u) = 0 \quad \text{unless } i+j = k+\ell. \tag{2.8}$$

For IRF models, this condition corresponds to the single-valuedness of the state variables around a face.

We are now going to relate exactly solvable models and knot theory. The Yang-Baxter relation (Fig.7) indicates an invariance under displacement of one of three lines over an intersection of the other two lines. Interestingly, this scattering diagram is similar to the graphical illustration of the braid group (Fig.3).

To make the above observation concrete, we define Yang-Baxter operator (Fig.9) for vertex models by [12,13,14]

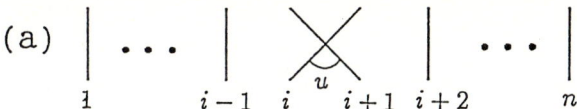

Fig.9: Yang-Baxter operator $X_i(u)$; (a) vertex model, (b) IRF model.

$$X_i(u) = \sum_{k,\ell,m,p} S_{\ell p}^{km}(u) I^{(1)} \otimes \cdots \otimes e_{pk}^{(i)} \otimes e_{m\ell}^{(i+1)} \otimes \cdots \otimes I^{(n)}, \qquad (2.9)$$

where $\otimes$ means a tensor product, $I^{(k)}$ an identity matrix on k-th position, e_{pk} a matrix such that $(e_{pk})_{ab} = \delta(p,a)\delta(k,b)$. For IRF models, we define Yang-Baxter operator (Fig.9) by [15,16]

$$\begin{aligned}{}[X_i(u)]_{\ell_0\cdots\ell_n}^{p_0\cdots p_n} &= \prod_{j=0}^{i-1} \delta(p_j,\ell_j) \cdot w(\ell_i,\ell_{i+1},p_i,\ell_{i-1};u) \prod_{j=i+1}^{n} \delta(p_j,\ell_j) \\ &\qquad \text{where } p_{j+1} \sim p_j, \ell_{j+1} \sim \ell_j \text{ for all } j, \\ &= 0, \text{otherwise}. \end{aligned} \qquad (2.10)$$

The definition (2.10) may require explanations. For IRF models, there are constraints on nearest-neighbouring state variable pairs. Suppose a state variable d is assigned on a lattice site. On the nearest-neighbouring sites, possible values of state variable b are restricted by the defining condition of the model. When b takes one of the allowed values, the state variable b is said to be allowable or admissible to d and this relation is expressed as $b \sim d$. Hereafter, when we consider IRF model, we always assume that suffices $\{p_j\}$ and $\{\ell_j\}$ satisfy the condition: $p_{j+1} \sim p_j, \ell_{j+1} \sim \ell_j$ for all j.

It is readily shown that the Yang-Baxter operators $\{X_i(u)\}$ satisfy the following relations (Yang-Baxter algebra)[11,12,13]:

$$X_i(u)X_j(v) = X_j(v)X_i(u), |i-j| \geq 2 \qquad (2.11a)$$
$$X_i(u)X_{i+1}(u+v)X_i(v) = X_{i+1}(v)X_i(u+v)X_{i+1}(u). \qquad (2.11b)$$

We can see close similarity between (1.1) and (2.11): the only difference is that (2.11) has spectral parameters as the arguments. If we set $u = u+v = v$, (2.11) reduces to (1.1). This suggests that we can get braid group representation $\{G_i\}$ from the Yang-Baxter operator. A possibility such that $u = u+v = v$ is $u = v = \infty$. In this way, we obtain representation of the braid group from the Yang-Baxter operator $X_i(u)$:

$$G_i = \lim_{u \to \infty} X_i(u)/\rho(u), \qquad (2.12a)$$

$$G_i^{-1} = \lim_{u \to \infty} X_i(-u)/\rho(-u). \tag{2.12b}$$

The limit $u \to \infty$ (with suitable normalization of the Boltzmann weights if necessary) implies a certain direction in the complex u-plane. The existence of the limit $u \to \infty$ requires that the model should be critical: the Boltzmann weights are parametrized by hyperbolic or trigonometric functions.

We introduce "weight matrix", $\sigma^{(\pm)}_{\ell k,ij}$ for vertex models (factorized S-matrices) and $\sigma(a,b,c,d;\pm)$ for IRF models:

$$\sigma^{(\pm)}_{\ell k,ij} = \lim_{u \to \infty} S^{ik}_{j\ell}(\pm u)/\rho(\pm u), \tag{2.13}$$
$$\sigma(a,b,c,d;\pm) = \lim_{u \to \infty} w(a,b,c,d;\pm u)/\rho(\pm u). \tag{2.14}$$

Then, the braid operator G_i is given for vertex models as

$$Gi = \sum_{k,\ell,m,p} \sigma^{(+)}_{pm,k\ell} I^{(1)} \otimes \cdots \otimes e^{(i)}_{pk} \otimes e^{(i+1)}_{m\ell} \otimes \cdots \otimes I^{(n)}, \tag{2.15}$$

and for IRF models as

$$[G_i]^{p_0 \cdots p_n}_{\ell_0 \cdots \ell_n} = \prod_{j=0}^{i-1} \delta(p_i,\ell_j) \sigma(\ell_i,\ell_{i+1},p_i,\ell_{i-1};+) \prod_{j=i+1}^{n} \delta(p_j,\ell_j). \tag{2.16}$$

Another possibility for the condition $u = u + v = v$ is $u = v = 0$. Due to the standard initial condition, if we set $u = 0$ in $X_i(u)$ we have

$$X_i(0) = I_n, \tag{2.17}$$

where I_n is the identity operator in the representation of B_n. From the inversion relation (2.4), we have

$$G_i \cdot G_i^{-1} = G_i^{-1} \cdot G_i = I_n. \tag{2.18}$$

The formula (2.12) is applicable to any solvable model at criticality. Corresponding to an exactly solvable model, we obtain a representation of the braid group.

3 Markov Trace and General Theory

We construct Markov trace $\phi(\cdot)$ on the representation of the braid group. With the Markov trace $\phi(\cdot)$, (1.6) yields link polynomial.

We first consider the vertex model (the factorized S-matrix). We assume the following form of a trace $\phi(\cdot)$[12,14]:

$$\phi(A) = \text{Tr}(H^{(n)}A)/\text{Tr}(H^{(n)}), \quad A \in B_n, \tag{3.1}$$

where

$$H^{(n)} = h^{(1)} \otimes h^{(2)} \otimes \cdots \otimes h^{(n)}, \tag{3.2}$$

and $h^{(i)} = h$ is a diagonal matrix whose elements are

$$h_{pq} = r^2(p)\delta(p,q). \tag{3.3}$$

Here $r(p)$ is the crossing multiplier of the model. We have normalized $\phi(\cdot)$ as $\phi(I) = 1$. The trace $\phi(\cdot)$ becomes the Markov trace when the following conditions hold:

$$\sigma^{(\pm)}_{pm,k\ell} \neq 0 \quad \text{only when } p + m = k + \ell, \tag{3.4}$$
$$r^2(p)r^2(m) = r^2(\ell)r^2(k) \quad \text{when } p + m = k + \ell, \tag{3.5}$$
$$\sum_\ell \sigma^{(+)}_{k\ell,k\ell} r^2(\ell) = \chi(\lambda) \quad \text{(independent of } k\text{)}, \tag{3.6a}$$
$$\sum_\ell \sigma^{(-)}_{k\ell,k\ell} r^2(\ell) = \bar{\chi}(\lambda) \quad \text{(independent of } k\text{)}. \tag{3.6b}$$

The condition (3.4) is the charge (or spin) conservation condition (2.8). The conditions (3.4) and (3.5) are sufficient for the Markov property I. For the Markov property II, the conditions (3.4) and (3.6) are sufficient. The constants τ and $\bar\tau$ in (1.4b) are expressed as

$$\tau = \chi(\lambda)/\xi(\lambda),$$
$$\bar\tau = \bar\chi(\lambda)/\xi(\lambda), \tag{3.7}$$

where

$$\xi(\lambda) = \sum_p r^2(p). \tag{3.8}$$

We next consider the IRF model. We assume that a trace $\phi(\cdot)$ has a form [15,16,17]:

$$\phi(A) = \hat{\mathrm{Tr}}(H^{(n)}A)/\hat{\mathrm{Tr}}(H^{(n)}), \quad A \in B_n, \tag{3.9a}$$
$$[H^{(n)}]^{p_0 \cdots p_n}_{\ell_0 \cdots \ell_n} = \delta(p_0,\ell_0)\delta(p_1,\ell_1) \cdots \delta(p_n,\ell_n) \cdot \frac{\psi(\ell_n)}{\psi(\ell_0)}. \tag{3.9b}$$

Here $\psi(\ell)$ is the crossing multiplier and $\hat{\mathrm{Tr}}(\cdot)$ is the constrained trace defined by

$$\hat{\mathrm{Tr}}(A) = \widetilde{\sum_{\substack{\ell_1 \cdots \ell_n \\ \ell_0 : \text{fixed}}}} A^{\ell_0 \cdots \ell_n}_{\ell_0 \cdots \ell_n}, \quad A \in B_n. \tag{3.9c}$$

The symbol $\widetilde{\sum}$ means the summation under the constraint imposed on the model. From (3.9), we have

$$\hat{\mathrm{Tr}}(H^{(n)}A) = \widetilde{\sum_{\substack{\ell_1 \cdots \ell_n \\ \ell_0 : \text{fixed}}}} A^{\ell_0 \cdots \ell_n}_{\ell_0 \cdots \ell_n} \cdot \psi(\ell_n)/\psi(\ell_0). \tag{3.10}$$

The trace $\phi(\cdot)$ becomes the Markov trace when the following conditions hold:

$$\sum_{b \sim a} \sigma(a,b,a,c;+)\frac{\psi(b)}{\psi(a)} = \chi(\lambda) \quad \text{(independent of } c, a\text{)}, \tag{3.11a}$$

$$\sum_{b \sim a} \sigma(a,b,a,c;-)\frac{\psi(b)}{\psi(a)} = \bar{\chi}(\lambda) \qquad \text{(independent of } c, a\text{)}, \tag{3.11b}$$

$$\sum_{b \sim a} \psi(b)/\psi(a) = \xi(\lambda) \qquad \text{(independent of } a\text{)}. \tag{3.12}$$

Here the summations are over all values of b admissible to a. The trace $\phi(\cdot)$ automatically satisfies the Markov property I. The conditions (3.11) and (3.12) are sufficient for the Markov property II. The constants τ and $\bar{\tau}$ are expressed as

$$\tau = \chi(\lambda)/\xi(\lambda),$$
$$\bar{\tau} = \bar{\chi}(\lambda)/\xi(\lambda). \tag{3.13}$$

Thus, sufficient conditions for the existence of the Markov trace (and then link polynomial) have been explicitly given.

The conditions (3.6) for the vertex models ((3.11) and (3.12) for the IRF models) can be written compactly as

$$\sum_{\ell} S^{k\ell}_{\ell k}(u) r^2(\ell) = H(u;\lambda)\rho(u), \qquad \text{(independent of } k\text{)}, \tag{3.14a}$$

$$\sum_{b \sim a} w(a,b,a,c;u)\frac{\psi(b)}{\psi(b)} = H(u;\lambda)\rho(u), \qquad \text{(independent of } a,c\text{)} \tag{3.14b}$$

We call (3.14) extended Markov property and $H(u;\lambda)(= H(u))$ characteristic function. The τ factors in the Markov property II are given by

$$\tau = \lim_{u \to \infty} \frac{H(u;\lambda)}{H(0;\lambda)},$$
$$\bar{\tau} = \lim_{u \to \infty} \frac{H(-u;\lambda)}{H(0;\lambda)}. \tag{3.15}$$

What is significant in this theory is that the extended Markov property (and the charge conservation condition for vertex models) is sufficient for the existence of the Markov trace.

Furthermore, when the model has the crossing symmetry, the projection relation [16]

$$X_i(\lambda)X_i(u) = \beta(u)X_i(\lambda),$$
$$\beta(u) = H(\lambda - u)\rho(\lambda - u), \tag{3.16}$$

is equivalent to the extended Markov property. Thus, the existence of the Markov trace for the models with the crossing symmetry can be examined simply by proving the extended Markov property (3.14) or equivalently the projection relation (3.16).

In the previous and presnet sections, a general theory has been presented to construct link polynomials from the exactly solvable models. Application to various solvable models was studied in detail. We shall only write down the results.

(1) N-state vertex model [12,13,14]

From the N-state vertex model asymmetrized by the symmetry breaking transformation [12] we get the braid operator which satisfies the N-th order reduction relation:

$$(G_i - C_1)(G_i - C_2)\cdots(G_i - C_N) = 0 \tag{3.17a}$$

where for $j = 1, 2, \cdots, N$

$$C_j = (-1)^{j+N} t^{\frac{1}{2}N(N-1) - \frac{1}{2}j(j-1)}, \quad t = e^{2\lambda}. \tag{3.17b}$$

In the $N = 2$ case (6-vertex model) the braid operator is a representation of the generator of the Hecke algebra (see (7.1) and (7,2)). The crossing multiplier for the asymmetrized N-state vertex model is

$$r(k) = e^{-\lambda k} = t^{-k/2}, \quad k = -s, -s+1, \cdots, s, \tag{3.18}$$

where

$$s = (N-1)/2. \tag{3.19}$$

The extended Markov property (3.14a) is satisfied with the characteristic function given as

$$H(u; \lambda) = \frac{\sinh(N\lambda - u)}{\sinh(\lambda - u)}. \tag{3.20}$$

From (3.15), the constants τ and $\bar{\tau}$ are

$$\begin{aligned}\tau &= 1/(1 + t + \cdots + t^{N-1}), \\ \bar{\tau} &= t^{N-1}/(1 + t + \cdots + t^{N-1}).\end{aligned} \tag{3.21}$$

It is remarkable that there exists an infinite sequence of link polynomials corresponding to the N-state vertex models ($N = 2, 3, 4, 5, \cdots$) [12,13]. The N=2 case corresponds to the Jones polynomial [18]. In the $N \geq 3$ cases we have new link polynomials. From the reduction relation, we obtain the skein relations (the Alexander-Conway relations) for the link polynomials:

$$\begin{aligned}\alpha(L_+) &= (1-t)t^{\frac{1}{2}}\alpha(L_0) + t^2\alpha(L_-), \quad (N=2) \tag{3.22a}\\ \alpha(L_{2+}) &= t(1 - t^2 + t^3)\alpha(L_+) + (t^4 - t^5 + t^7)\alpha(L_0) \\ &\quad - t^8\alpha(L_-), \quad (N=3) \tag{3.22b}\\ \alpha(L_{3+}) &= t^{3/2}(1 - t^3 + t^5 - t^6)\alpha(L_{2+}) + t^6(1 - t^2 + t^3 + t^5 - t^6 + t^8)\alpha(L_+) \\ &\quad + t^{25/2}(-1 + t - t^3 + t^6)\alpha(L_0) - t^{20}\alpha(L_-), \quad (N=4). \tag{3.22c}\end{aligned}$$

In (3.22a), by L_+, L_0 and L_- we have denoted links which have the configuration of b_i, b_i^0 and b_i^{-1}, at an intersection. Similarly, L_{2+}, L_+, L_0 and L_- in (3.22b) and L_{3+}, L_{2+}, L_+, L_0 and L_- in (3.22c) should be understood.

(2) ABCD IRF models [17]

The IRF model corresponding to affine Lie algebra $A^{(1)}_{m-1}$ ($B^{(1)}_m$, $C^{(1)}_m$, $D^{(1)}_m$) is called $A^{(1)}_{m-1}$ ($B^{(1)}_m$, $C^{(1)}_m$, $D^{(1)}_m$) model. The crossing parameter λ and the sign factor σ are defined as

$$\begin{aligned}
\lambda &= m\omega/2, & \sigma &= 1 & &\text{for } A^{(1)}_{m-1}, \\
\lambda &= (2m-1)\omega/2, & \sigma &= 1 & &\text{for } B^{(1)}_m, \\
\lambda &= (m+1)\omega, & \sigma &= -1 & &\text{for } C^{(1)}_m, \\
\lambda &= (m-1)\omega, & \sigma &= 1 & &\text{for } D^{(1)}_m,
\end{aligned} \quad (3.23)$$

where ω is a parameter. The reduction relations are

$$(G_i - 1)(G_i + \gamma^2) = 0 \quad \text{for } A^{(1)}_{m-1}, \quad (3.24)$$

$$(G_i - 1)(G_i - \beta)(G_i + \gamma^2) = 0 \quad \text{for } B^{(1)}_m, C^{(1)}_m \text{ and } D^{(1)}_m, \quad (3.25)$$

with

$$\gamma = e^{-i\omega} \quad \text{for } A^{(1)}_{m-1}, B^{(1)}_m, C^{(1)}_m \text{ and } D^{(1)}_m, \quad (3.26)$$

$$\beta = \sigma e^{-i[2\lambda + \omega(1+\sigma)]} \quad \text{for } B^{(1)}_m, C^{(1)}_m \text{ and } D^{(1)}_m. \quad (3.27)$$

The extended Markov property (3.14b) is proved and the characteristic functions are calculated as

$$H(u) = \frac{\sin(m\omega - u)}{\sin(\omega - u)} \quad \text{for } A^{(1)}_{m-1}, \quad (3.28)$$

$$H(u) = \frac{\sigma \sin(2\lambda - u)\sin(\sigma\omega + \lambda - u)}{\sin(\lambda - u)\sin(\omega - u)} \quad \text{for } B^{(1)}_m, C^{(1)}_m \text{ and } D^{(1)}_m, \quad (3.29)$$

(The explicit forms of the crossing multipliers are given in [17]). Using the reduction relations and the Markov traces, we obtain the (generalized) skein relations:

$$\begin{aligned}
\alpha(L_+) &= (1-t)t^{(m-1)/2}\alpha(L_0) + t^m \alpha(L_-) & &\text{for } A^{(1)}_{m-1}, & (3.30)\\
\alpha(L_{2+}) &= (1-t+\beta)e^{-i(2\lambda+\omega(\sigma-1))} \cdot \alpha(L_+) & & & \\
&\quad + (t+\beta t - \beta)e^{-2i(2\lambda+\omega(\sigma-1))} \cdot \alpha(L_0) & & & \\
&\quad - t\beta e^{-3i(2\lambda+\omega(\sigma-1))} \cdot \alpha(L_-), & &\text{for } B^{(1)}_m, C^{(1)}_m \text{ and } D^{(1)}_m, & (3.31)
\end{aligned}$$

where

$$t = e^{-2i\omega}. \quad (3.32)$$

For $A^{(1)}_{m-1}$ model, the Alexander polynomail is obtained by the limit $m \to 0$, while $m = 2$ corresponds to the Jones polynomial.

Link polynomials thus obtained are one-variable invariants for each fixed m. It is noted that m is independent of t. We now have two variables t and m. The link polynomial constructed from $A^{(1)}_{m-1}$ model corresponds to the two-variable extension [19,20] of the Jones polynomial. The link polynomails from $B^{(1)}_m$, $C^{(1)}_m$, $D^{(1)}_m$ models correspond to the Kauffman polynomial [2,21]. We thus have explicit realizations of the Kauffman polynomial and the two-variable extension of the Jones polynomial (HOMFLY polynomial). The link polynomials constructed by

Turaev [22] correspond to the vertex-model analog of the present link polynomials constructed from $A^{(1)}_{m-1}$, $B^{(1)}_m$, $C^{(1)}_m$, $D^{(1)}_m$ IRF models.

4 Braid-Monoid Algebra

We shall discuss an algebra and its role in the knot theory associated with solvable models [16]. In the following we use the factorized S-matrices (vertex models). The discussion goes parallel for IRF models [16,17]. From the crossing symmetry and the standard initial condition, we have

$$S^{ik}_{j\ell}(\lambda) = [\frac{r(i)r(\ell)}{r(j)r(k)}]^{1/2} S^{j\ell}_{ki}(0) = r(i)r(\ell)\delta(\ell,\bar{k})\delta(i,\bar{j}). \tag{4.1}$$

Recall that λ is the crossing parameter (also referred to as crossing point) and $r(\bar{j}) = 1/r(j)$. We introduce an operator E_i by

$$E_i = X_i(\lambda). \tag{4.2}$$

Putting (4.1) into (2.9), we get the following relations:

$$E_i E_j = E_j E_i, \quad |i-j| \geq 2, \tag{4.3}$$
$$E_i E_{i\pm 1} E_i = E_i, \tag{4.4}$$
$$E_i^2 = q^{1/2} E_i, \tag{4.5}$$

where

$$q^{1/2} = \sum_k r^2(k). \tag{4.6}$$

Relations (4.3)-(4.6) are the defining relations of the Temperley-Lieb algebra.

We can associate the Temperley-Lieb operator E_i with the monoid diagram (Fig.10). We then sometimes call the Temperley-Lieb operator monoid operator. This view-point gives a basis of a graphical approach to knot theory. The detail will be discussed in §5.

We have derived braids and monoids from the Yang-Baxter operator. For the factorized S-matrices, the S-matrices at $u = \infty$ (high energy limit) correspond to braid operators, while the S-matrices at $u = \lambda$ (crossing point) correspond to monoid operators. From the models which satisfy the crossing symmetry and the Markov property II, we obtain an algebra which we call braid-monoid algebra:

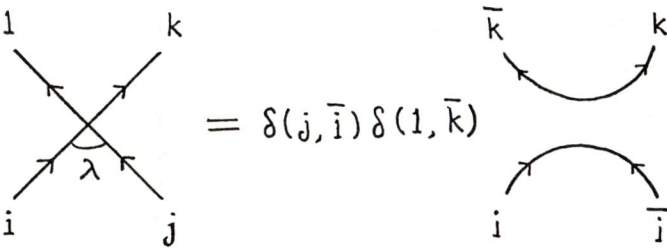

Fig.10: Monoid diagram.

$$f(G_i) = 0, \tag{4.7a}$$
$$E_i = g(G_i), \tag{4.7b}$$
$$E_i^2 = q^{1/2} E_i, \tag{4.7c}$$
$$G_i E_i = E_i G_i = c E_i, \tag{4.7d}$$
$$G_i G_{i+1} G_i = G_{i+1} G_i G_{i+1}, \tag{4.7e}$$
$$E_i E_{i\pm1} E_i = E_i, \tag{4.7f}$$
$$E_i G_{i\pm1} G_i = G_{i\pm1} G_i E_{i\pm1} = E_i E_{i\pm1}, \tag{4.7g}$$
$$G_{i\pm1} E_i G_{i\pm1} = G_i^{-1} E_{i\pm1} G_i^{-1}, \tag{4.7h}$$
$$G_{i\pm1} E_i E_{i\pm1} = G_i^{-1} E_{i\pm1}, \tag{4.7i}$$
$$E_{i\pm1} E_i G_{i\pm1} = E_{i\pm1} G_i^{-1}, \tag{4.7j}$$
$$E_i G_{i\pm1} E_i = c^{-1} E_i, \tag{4.7k}$$
$$G_i G_j = G_j G_i, \quad |i-j| \geq 2 \tag{4.7l}$$
$$E_i E_j = E_j E_i, \quad |i-j| \geq 2. \tag{4.7m}$$

Here $f(G_i)$, $g(G_i)$, $q^{1/2}$ and c are model-dependent. In the derivation of the relations (4.7), we have defined the braid operator G_i by

$$G_i^{\pm 1} = \lim_{u \to \infty} \frac{X_i(\pm u)}{[\rho(\lambda \mp u)\rho(\pm u)]^{1/2}}. \tag{4.8}$$

Remark that the Yang-Baxter operator may have an overall arbitrary factor. The quantities $q^{1/2}$ and c are related to the characteristic function $H(u) = H(u; \lambda)$ of the extended Markov property as

$$q^{1/2} = (\tau \bar{\tau})^{-1/2} = H(0), \tag{4.9}$$
$$c = \lim_{u \to \infty} \frac{H(\lambda - u)\rho(\lambda - u)}{[\rho(u)\rho(\lambda - u)]^{1/2}}. \tag{4.10}$$

The model-dependent part of the braid-monoid algebra is given as follows.
(1) **N-state vertex model** [16]

$$(G_i - c_1 I)(G_i - c_2 I) \cdots (G_i - c_N I) = 0, \tag{4.11a}$$
$$E_i = \frac{q^{1/2}}{(c_1 - c_2)(c_1 - c_3)\cdots(c_1 - c_N)}(G_i - c_2 I)(G_i - c_3 I)\cdots(G_i - c_N I), \tag{4.11b}$$

where

$$c_r = (-1)^{N-1}(-1)^{r+1} t^{\frac{1}{4}(N^2-1)-\frac{1}{2}r(r-1)}, \tag{4.11c}$$
$$c = (-1)^{N-1}(-1)^{r+1} t^{\frac{1}{4}(N^2-1)}, \tag{4.11d}$$
$$q^{1/2} = \frac{\sinh N\lambda}{\sinh \lambda} = t^{\frac{1}{2}(N-1)} + \cdots + t^{-\frac{1}{2}(N-1)}. \tag{4.11e}$$

(2) **B, C, D IRF models** [17]

$$(G_i - \gamma^{-1} I)(G_i + \gamma I)(G_i - \gamma^{-1}\beta I) = 0, \tag{4.12a}$$
$$E_i = \frac{\gamma}{\beta(\gamma - \gamma^{-1})}(G_i - \gamma^{-1} I)(G_i + \gamma I), \tag{4.12b}$$

where

$$c = \beta\gamma^{-1} = \sigma e^{-i(2\lambda+\sigma\omega)}, \tag{4.12c}$$

$$q^{1/2} = \frac{\sin 2\lambda \cdot \sin(\sigma\omega + \lambda)}{\sin\lambda \cdot \sin\omega} \tag{4.12d}$$

For the $N = 3$ case and B, C and D IRF models, the braid-monoid algebra reduces to the Birman-Wenzl-Murakami (BWM for short) algebra [23,24] with

$$\ell = \sqrt{-1}t^{-2}, \quad m = \sqrt{-1}(t - t^{-1}) \quad \text{for the } N = 3 \text{ case,} \tag{4.13a}$$

$$\ell = \sqrt{-1}\gamma\beta^{-1}, \quad m = \sqrt{-1}(\gamma - \gamma^{-1}) \quad \text{for } B_m^{(1)}, C_m^{(1)} \text{ and } D_m^{(1)}. \tag{4.13b}$$

The BWM algebra was derived from the Kauffman polynomial and ℓ and m in (4.13) are constants in their notation.

5 Graphical Approach

The theory shown in §2 and §3 may be called an "algebraic approach" since it mainly uses algebraic objects such as braid and braid group. By emphasizing the diagrammatical interpretation of braid and monoid, we shall here develop a "graphical approach"[25].

We call a projection of link L onto a plane link diagram $\hat{L}$. It does not have multiple points but double points. Link diagrams expressing ambient isotopic links are transformed into each other by a finite sequence of the Riedemeister moves I, II and III (Fig.11). Thus, link invariant may also be defined as a Reidemeister move invariant. The following is noteworthy. The Reidemeister move I corresponds to the Markov property II while the Riedemeister moves II and III are nothing but the inversion relation (unitarity) and the Yang-Baxter relation, respectively.

Link diagram contains crossings (intersections). We associate a sign $\epsilon(C)$ to crossing C as depicted in Fig.12. The writhe $w(\hat{L})$ of the link diagram $\hat{L}$ is defined as a sum of signs for all crossings of the diagram:

$$w(\hat{L}) = \sum_C \epsilon(C). \tag{5.1}$$

The writhe is invariant under the Reidemeister moves II and III, but is changed by the move I. Definition: link diagrams are regularly isotopic if and only if they are transformed each other by a finite sequence of the Reidemeister moves II and III. Thus, the writhe is a regular isotopy invariant. Let $\hat{L}$ be a link diagram which is equivalent to the closed braid of a braid A. The exponent sum of the closed braid is, except for a sign convention, the writhe of the link diagram:

$$e(A) = -w(\hat{L}). \tag{5.2}$$

To formulate link plynomial directly on the link diagram, we consider the factorized S-matrices (the vertex model) with the crossing symmetry. We regard a string as a trajectory (an orbit) of particle and a braid as orbits of particles. We choose the time direction upward. A particle with state variabel $\bar{j}$ going upwards is assumed to be the antiparticle with state variable j going downwards.

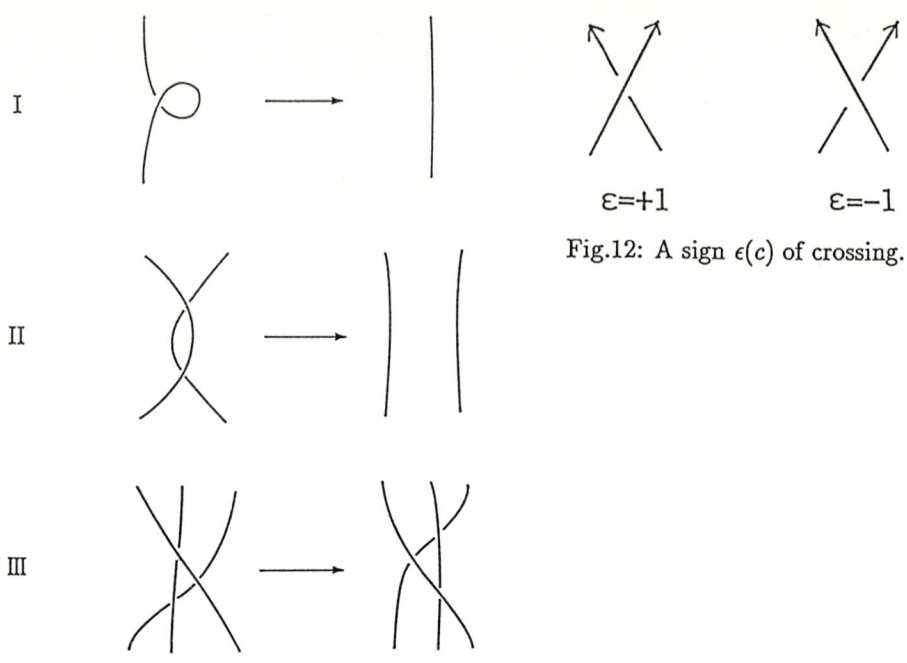

Fig.12: A sign $\epsilon(c)$ of crossing.

Fig.11: Reidemeister moves, I, II and III.

To each part of a link diagram $\hat{L}$, we assign state variables. The value of the state variable may change only at crossing points. The link diagram consists of the following elements: annihilation, creation, braid, inverse braid and line diagrams. To them, we assign weights $r(i)\delta(i,\bar{j})$, $r(\ell)\delta(\ell,\bar{k})$, $\sigma^{(+)}_{\ell k,ij}$, $\sigma^{(-)}_{\ell k,ij}$ and $\delta(j,k)$ in this order (Fig.13). We have three important observations:

(1) S-matrix at $u = \lambda$, $S^{ik}_{jl}(\lambda)$, is separated into two parts, $r(i)\delta(i,\bar{j})$ and $r(\ell)\delta(\ell,\bar{k})$, which respectively correspond to the creation and annihilation processes of particle-antiparaticle pair.

(2) Weights $\sigma^{(+)}_{\ell k,ij}$ and $\sigma^{(-)}_{\ell k,ij}$ correspond to braiding and inverse braiding operations. Invariance under the Reidemeister moves II and III is guaranteed by the inversion relation and the Yang-Baxter relation at $u = \pm\infty$ satisfied by $\sigma^{(+)}_{\ell k,ij}$ and $\sigma^{(-)}_{\ell k,ij}$.

(3) Evaluation of the link polynomial by use of the Markov trace corresponds to that of weighted sum defined on a link diagram.

These key observations set a basis of the graphical approach. The monoid diagram and the weights for the creation and annihilation diagrams were originally introduced by L.H. Kauffman for the bracket polynomial which gives a graphical calculation of the Jones polynomial [25]. We shall derive the graphical calculation from the view point of the solvable models in a unified way. The link polynomials constructed from the solvable (IRF and vertex) models with the crossing symmetry are in this way graphically calculable [16].

The normalization of the weight matrices $\sigma^{(\pm)}_{\ell k,ij}$ is defined by (4.8). Taking the charge conservation condition into account, we perform a summation over all

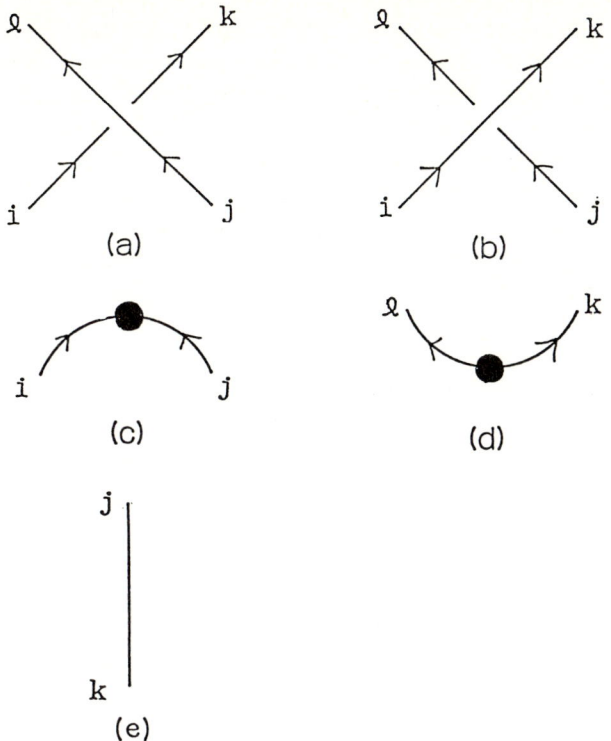

Fig.13: Elementary diagrams for vertex models. (a) braid diagram $\sigma^{(+)}_{\ell k,ij}$, (b) inverse braid diagram $\sigma^{(-)}_{\ell k,ij}$, (c) annihilation diagram $r(i)\delta(i,\bar{j})$, (d) creation diagram $r(\ell)\delta(\ell,\bar{k})$, (e) line diagram $\delta(j,k)$. Solid circle means that the crossing multiplier is multiplied there.

possible state variables for the link diagram. We denote the sum by $\text{Tr}(\hat{L})$. This sum is a regular isotopy invariant. We define a constant c by

$$c = \sum_{\ell} \sigma^{(+)}_{k\ell,k\ell} r^2(\ell). \tag{5.3}$$

We multiply the factor $c^{-w(\hat{L})}$, which is also a regular isotopy invariant, to $\text{Tr}(\hat{L})$. It is obvious that $c^{-w(\hat{L})}\text{Tr}(\hat{L})$ is a regular isotopic invariant. Further, this combination is invariant under the Reidemeister move I. Then we arrive at a conclusion that a link polynomial is expressed as

$$\alpha(L) = c^{-w(\hat{L})}\text{Tr}(\hat{L})/\text{Tr}(\hat{K}_0). \tag{5.4}$$

Here $\text{Tr}(\hat{K}_0)$ is the sum for the trivial knot (a loop). The link polynomial is normalized so that $\alpha(K_0) = 1$ for the trivial knot K_0. Since the trivial knot consists of annihilation and creation diagrams, we have

$$\text{Tr}(\hat{K}_0) = \sum_j r^2(j) = q^{1/2}. \tag{5.5}$$

From a view point of the braid-monoid algebra, the formula (5.4) can also be verified as follows [16]. The relation (4.7e) $\sim$ (4.7j) implies that elements of the braid-monoid algebra are invariant under the Reidemeister moves II and III. The effect of the Reidemeister move I is, as seen from (4.7d) and (4.7k), a constant factor c given by (4.10). This change in the trace cancels with a change of $c^{-w(\hat{L})}$ and then (5.4) beceoms a topological invariant of the link L. Because of the extended Markov property, two expressions of c, (4.10) and (5.3), are the same.

For the IRF models, a graphical approach can be introduced in a similar manner. The monoid diagram corresponds to $E_i = X_i(\lambda)$. Since

$$w(a,b,c,d;\lambda) = \delta(b,d) \cdot [\psi(a)\psi(c)/\psi(b)\psi(d)]^{1/2}, \tag{5.6}$$

we may decompose the monoid diagram into two parts, $[\psi(a)/\psi(b)]^{1/2}$ and $[\psi(c)/\psi(b)]^{1/2}$. We assign weights $[\psi(a)/\psi(b)]^{1/2}$ and $[\psi(c)/\psi(b)]^{1/2}$ to the annihilation and creation diagrams, respectively (Fig.14). To the braid and inverse braid diagrams, we give the weight matrices $\sigma(a,b,c,d;+)$ and $\sigma(a,b,c,d;-)$ whose nor-

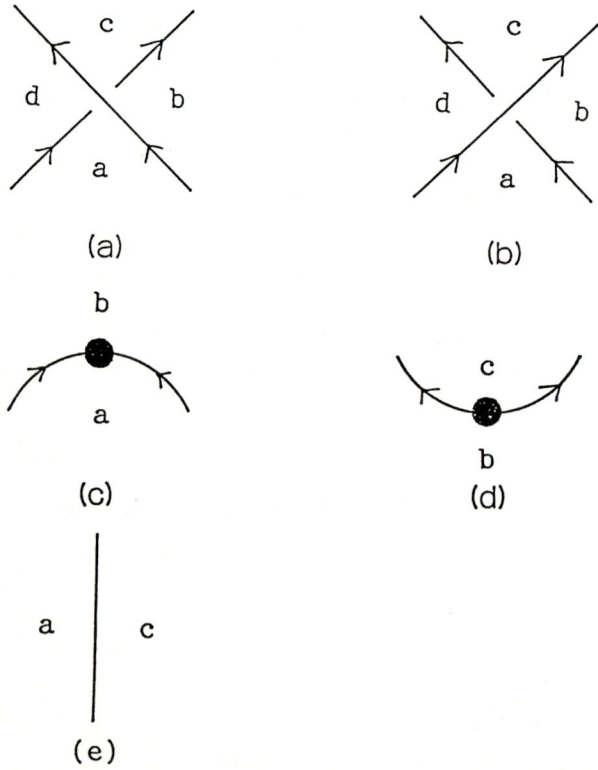

Fig.14: Elementary diagrams for IRF models. (a) braid diagram $\sigma(a,b,c,d;+)$, (b) inverse braid diagram $\sigma(a,b,c,d;-)$, (c) annihilation diagram $[\psi(b)/\psi(a)]^{1/2}\delta(b,d)$, (d) creation diagram $[\psi(c)/\psi(b)]^{1/2}\delta(b,d)$, (e) line diagram $\delta(a \sim c)$ meaning that 1 when a is admissible to c and 0 otherwise. Solid circle denotes that the crossing multiplier is multiplied there.

malization is given by (4.8). The similar argument leads to the formula (5.4), where

$$\text{Tr}(\hat{K}_0) = \sum_{b \sim a} \psi(b)/\psi(a) = q^{1/2}. \tag{5.7}$$

To summarize, we conclude that an exactly solvable model with the crossing symmetry gives a graphical construction of link polynomials.

6 Application to Graph-State Models

In this section we shall add some other examples which give interesting application of the knot theory. Special emphasis is paid on the crossing symmetry and crossing multiplier.

The constraint of an IRF model can be expressed by a graph. Points in a graph express the values of the state variables of the model. If a state a is admissible to a state b, then the point for a is connected by a line (bond) to the point for b. In general, graphs may be oriented. There may be such a case that a is admissible to b but b is not admissible to a. In the case, we put an arrow from a to b. Thus, the graph contains all the information on states and admissibility relations of the model. In the following, we shall restrict ourselves to graphs which are not oriented.

Let us introduce graph-state IRF models. We start from a graph. The crossing multipliers $\{\psi(j)\}$ of the model are determined by

$$\sum_{b \sim a} \psi(b) = q^{1/2} \psi(a), \tag{6.1}$$

where the summation runs over all states admissible to a. It is interesting to notice that (6.1) is considerd as eigenvalue equation for $\{\psi(j)\}$ defined on the graph. We may rewrite (6.1) as

$$C\Psi = q^{1/2}\Psi. \tag{6.2}$$

We term a matrix C which expresses admissibility condition admissibility matrix. Introducing the crossing parameter λ

$$\cos \lambda = q^{1/2}, \tag{6.3}$$

we obtain Yang-Baxter operator

$$X_i(u) = \frac{\sin(\lambda - u)}{\sin \lambda}[I + \frac{\sin u}{\sin(\lambda - u)} U_i], \tag{6.4}$$

where U_i is the Temperley-Lieb operator whose matrix elements are given by

$$[U_i]_{p_0 p_1 \cdots p_n}^{\ell_0 \ell_1 \cdots \ell_n} = \prod_{j=0}^{i-1} \delta(\ell_j, p_j) \cdot \delta(\ell_{i-1}, \ell_{i+1}) \cdot \frac{[\psi(\ell_i)\psi(p_i)]^{1/2}}{\psi(\ell_{i-1})} \cdot \prod_{j=i+1}^{n} \delta(\ell_j, p_j). \tag{6.5}$$

It is easy to prove that the Boltzmann weights of the graph-state model satisfy the crossing symmetry

$$w(a,b,c,d;u) = [\frac{\psi(a)\psi(c)}{\psi(b)\psi(d)}]^{1/2} w(b,c,d,a;\lambda-u). \tag{6.6}$$

In this way, a solvable IRF model with the crossing symmetry (6.6) can be constructed from a given graph. For a graph with a finite size it is a restricted model. For a graph with an infinite size, it is an unrestricted model.

We shall construct graph-state models for graphs (a) and (b) given in Figure 15.

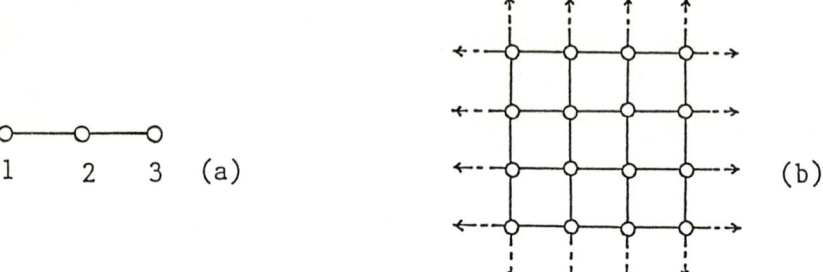

Fig.15: Graph-state models.

(a) Graph (a) represents the constraint condition of the restricted 8VSOS (8 vertex solid-on-solid) model with three states 1, 2, 3. Let us consider a sequence of "spins" $\{\sigma_i\}$ where $\sigma_{i\pm 1}$ are the nearest-neighbor spins for a spin σ_i. The graph indicates that for $\sigma_i = 2$, then $\sigma_{i\pm 1}$ take two spin values 1 and 3; for $\sigma_i = 1$ or 3, then $\sigma_{i\pm 1}$ take only 2. For the restricted 8VSOS model, the admissibility matrix C is given by

$$C = \begin{pmatrix} 0 & 1 & 0 \\ 1 & 0 & 1 \\ 0 & 1 & 0 \end{pmatrix}. \tag{6.7}$$

The eigenvalues of the matrix C is $q^{1/2} = 0, \pm\sqrt{2}$. Because of positivity of the Boltzmann weights, we choose an eigenvalue $\sqrt{2}$, and the crossing multipliers are found to be

$$\psi(a) = \sin(\frac{\pi}{4}a), \quad \text{for } a = 1,2,3. \tag{6.8}$$

(b) The graph (b) is a two-dimensional square lattice. The crossing multiplier is given by

$$\psi(\vec{a}) = \sin(\vec{a}\cdot\vec{\nu} + \omega_0), \tag{6.9}$$

where the points on the two-dimensional square lattice are represented by

$$\vec{a} = (z_1, z_2), \quad z_1, z_2 \in \mathbf{Z} \tag{6.10a}$$

and $\vec{\nu}$ is an arbitrary real vector

$$\vec{\nu} = (\nu_1, \nu_2). \tag{6.10b}$$

The eigenvalue $q^{1/2}$ is

$$q^{1/2} = \cos \nu_1 + \cos \nu_2. \tag{6.11}$$

From the above examples, it is clear how to construct a solvable model for an arbitrary graph. From the graph-state models we can construct link polynomials according to the theory presented in §2 and §3. The characteristic function $H(u; \lambda)$ for the graph state models are found to be

$$H(u; \lambda) = q^{1/2} + \frac{\sin u}{\sin(\lambda - u)}, \tag{6.12}$$

where the crossing point λ is determined from the relation (6.3). The braid-monoid algebra and graphical formulation are also straightforwardly introduced.

7 Two-Variable Extension

The general theory for construction of link polynomials has been extended to include two-variable link polynomials [26,27]. We start from the generators $\{g_i, i = 1, 2, \cdots, n-1\}$ of the Hecke algebra $H(t,n)$:

$$g_i g_j = g_j g_i, \quad |i-j| \geq 2, \tag{7.1a}$$
$$g_i g_{i+1} g_i = g_{i+1} g_i g_{i+1}, \tag{7.1b}$$
$$g_i^2 = (1-t)g_i + t. \tag{7.2}$$

Note that the braid operator made from the 6-vertex model (and the 8VSOS model, the A_{m-1} vertex and IRF models) satisfies the defining relations (7.1) and (7.2) of the Hecke algebra.

We form a composite string by combining $(N-1)$ strings and attaching a projector $P^{(N)}$ at each end (Fig.16). This means the following. We make a multiplet of $(N-1)$ particles and extract a given symmetric component by the projectors. To make a discussion simple, we shall here consider a fully symmetric projection (general case has been discussed in [27,5,6]). For the case of spin 1/2 particles, the fully symmetric component has the spin $(N-1)/2$. In this sense, the N-state vertex model describes the composition of spin 1/2 particles into a spin $s = (N-1)/2$ particle. Explicit form of the projector $P_i^{(N)}$ is derived through a recursion formula:

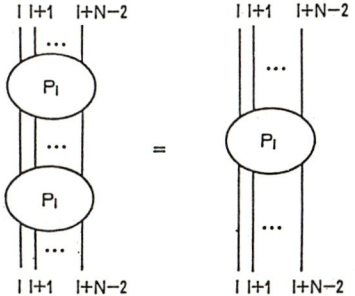

Fig.16: A composite string. Two diagrams are equivalent since $P_i^2 = P_i$.

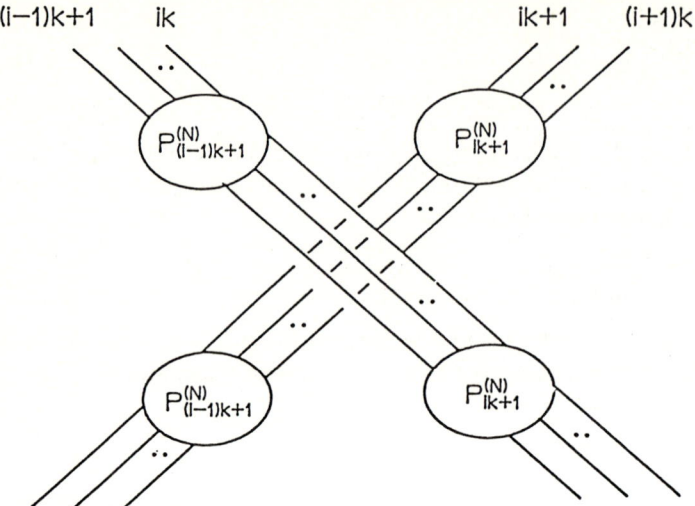

Fig.17: Generator G_i of $B_n^{[s]}$. Note that $k \equiv N - 1 = 2s$.

$$P_i^{(N)} = P_i^{(N-1)} h_{i+N-3}^{(N)} P_i^{(N-1)}, \qquad N \geq 3,$$
$$P_i^{(2)} \equiv 1 \tag{7.3}$$

where

$$h_j^{(N)} = \frac{\tau_{N-2}}{\tau_{N-1}}\left(\frac{t^{N-2}}{\tau_{N-2}} + g_i\right), \tag{7.4}$$

$$\tau_m = 1 + t + t^2 + \cdots + t^m, \tag{7.5}$$

Using the composite string, we introduce generators $\{G_i; i = 1, 2, \cdots, n-1\}$ in "spin s" representation $B_n^{[s]}$ of the braid group B_n. We prepare n sets of $k = N - 1$ strings and combine each set of k strings into a composite string with projectors at both ends. The generator G_i in $B_n^{[s]}$ is defined by (Fig.17):

$$G_i = P_{(i-1)k+1}^{(N)} P_{ik+1}^{(N)} \bar{G}_i^{(N)} P_{(i-1)k+1}^{(N)} P_{ik+1}^{(N)}, \qquad G_i \in B_n^{[s]} \tag{7.6}$$

where

$$\bar{G}_i^{(N)} = g_i^{(1)} g_i^{(2)} \cdots g_i^{(N-1)}$$
$$g_i^{(\ell)} = g_{ik+1-\ell} g_{ik+2-\ell} \cdots g_{(i+1)k-\ell}, \qquad \ell = 1, 2, \cdots, N-1. \tag{7.7}$$

It is easy to see that the generators $\{G_i\}$ satisfy the defining relation of the braid group:

$$G_i G_j = G_j G_i, \qquad |i - j| \geq 2, \tag{7.8a}$$
$$G_i G_{i+1} G_i = G_{i+1} G_i G_{i+1}. \tag{7.8b}$$

In addition, they satisfy the N-th order reduction relation

$$(G_i - c_i)(G_i - c_2) \cdots (G_i - c_N) = 0, \tag{7.9a}$$

where for $r = 1, 2, \cdots, N$
$$c_r = (-1)^{N+r} t^{\frac{1}{2}N(N-1) - \frac{1}{2}r(r-1)}. \tag{7.9b}$$

We now present two-variable extension of link polynomials. By $\psi(\cdot)$ we denote the Markov trace (Ocneanu's trace) associated with a representation of $B_n^{[1/2]} (= H(t,n))$. It satisfies the normalization condition
$$\psi(I) = 1, \qquad I\text{: identity in } B_n^{[1/2]}, \tag{7.10}$$
and the Markov properties
$$\text{I.} \quad \psi(AB) = \psi(BA), \qquad (A, B \in B_n^{[1/2]}), \tag{7.11a}$$
$$\text{II.} \quad \psi(Ag_n) = z\psi(A),$$
$$\psi(Ag_n^{-1}) = \bar{z}\psi(A), \qquad (A \in B_n^{[1/2]}), \tag{7.11b}$$
where
$$z = \psi(g_i),$$
$$\bar{z} = \psi(g_i^{-1}) \qquad \text{for all } i. \tag{7.12}$$

Recall that the Hecke algebra (defined by (7.1) and (7.2)) has one parameter t. This z (or $\bar{z}$) is another independent parameter. In this way, a pair (t, z) enters into the two-variable Jones polynomial. It is sometimes convenient to introduce a variable ω by
$$\omega = \bar{z}/z \tag{7.13}$$
and change a set of variables (t, z) into (t, ω).

Generalizing the Ocneanu's formulation [19] described above, we define a trace (generalized Ocneanu's trace) $\psi^{[s]}$:
$$\psi^{[s]}(A) = \psi(A) / [\psi(P_j^{(N)})]^n, \qquad A \in B_n^{[s]}. \tag{7.14}$$
Note that $A \in B_n^{[s]}$ consists of the generators in $B_{(N-1)n}^{[1/2]}$. It can be shown that $\psi^{[s]}(\cdot)$ satisfies the Markov properties:
$$\text{I.} \quad \psi^{[s]}(AB) = \psi^{[s]}(BA), \qquad (A, B \in B_n^{[s]}), \tag{7.15}$$
$$\text{II.} \quad \psi^{[s]}(AG_n) = Z\psi^{[s]}(A), \tag{7.16a}$$
$$\psi^{[s]}(AG_n^{-1}) = \bar{Z}\psi^{[s]}(A), \tag{7.16b}$$
$$(A \in B_n^{[s]}, \quad G_n \in B_{n+1}^{[s]}),$$
where
$$Z = \psi^{[s]}(G_j) = \frac{(1-t)(1-t^2)\cdots(1-t^{N-1})}{(1-\omega t)(1-\omega t^2)\cdots(1-\omega t^{N-1})},$$
$$\bar{Z} = \psi^{[s]}(G_j^{-1}) = \frac{\omega^{N-1}(1-t)(1-t^2)\cdots(1-t^{N-1})}{(1-\omega t)(1-\omega t^2)\cdots(1-\omega t^{N-1})}. \tag{7.17}$$

With the generalized Ocneanu's trace, two-variable link polynomial $\alpha_\omega^{[s]}(\cdot)$ is given by a formula:
$$\alpha_\omega^{[s]}(A) = (\bar{Z}Z)^{-(n-1)/2} \left(\frac{\bar{Z}}{Z}\right)^{e(A)/2} \psi^{[s]}(A), \qquad (A \in B_n^{[s]}), \tag{7.18}$$
where $e(A)$ is the exponent sum of the generators in $B_n^{[s]}$.

When we set $\omega = t$, the two-variable link polynomial $\alpha_\omega^{[s]}(\cdot)$ reduces to the new link polynomial for the N-state vertex model. While $\alpha_\omega^{[1/2]}$ is the two-variable Jones polynomial [19,20], $\alpha_\omega^{[s]}(\cdot)$ for $s = 1, 3/2, 2, \cdots$ are new [26,27]. It is interesting to notice that the $N = 3$ two-variable polynomial is different from the Kauffman polynomial. Since the $N = 2$ case contains the Alexander polynomial as the limit $\omega \to 1/t$, the two-variable link polynomials $\alpha_\omega^{[s]}(\cdot)$ are also considered as the generalizations of the Alexander polynomial.

The composite A_{m-1} model gives the equivalent result, in the following sense. The braid operator constructed from the composite Yang-Baxter operator for the A_{m-1} model corresponds to the composite string representation [17]. For fixed m the link polynomial from the composite A_{m-1} model is one-variable restriction of $\alpha_\omega^{[s]}(\cdot)$ with $\omega = t^{m-1}$. If we consider m as a continuous parameter independent of t, then the link polynomial from the composite A_{m-1} model is equivalent to $\alpha_\omega^{[s]}(\cdot)$.

Acknowledgements

One of the authors (M.W) would like to express his sincere thanks to Professor Gu Chaohao, for inviting him to International Conference on Nonlinear Physics, Shanghai, April 24 ~ 29, 1989.

References

[1] J.S. Birman: *Braids, Links and Mapping Class Groups* (Princeton University Press, 1974).

[2] L.H. Kauffman: *On Knots* (Princeton University Press, 1987).

[3] M. Wadati and Y. Akutsu: Prog. Theor. Phys. Suppl. **94** (1988) 1.

[4] Y. Akutsu, T. Deguchi and M. Wadati: in *Braid Group, Knot Theory and Statistical Mechanics*, ed. C.N. Yang and M.L. Ge (World Scientific Pub., 1989).

[5] M. Wadati, T. Deguchi and Y. Akutsu: Phys. Reports (in press).

[6] T. Deguchi, M. Wadati and Y. Akutsu: Adv. Stud. in pure Math. **19** (1989), Kinokuniya-Academic Press.

[7] M. Wadati, T. Deguchi and Y. Akutsu: in *Nonlinear Evolution Equations, Integrability and Spectral Methods*, ed. A. Fordy (Manchester University Press, 1989).

[8] C.N. Yang: Phys. Rev. Lett. **19** (1967) 1312.

[9] R.J. Baxter: Ann. of Phys. **70** (1972) 323.

[10] M. Karowski, H.J. Thun, T.T. Truong and P.H. Weisz: Phys. Lett. **67B** (1977) 321.
K. Sogo, M. Uchinami, A. Nakamura and M. Wadati: Prog. Theor. Phys. **66** (1981) 1284.
K. Sogo, M. Uchinami, Y. Akutsu and M. Wadati: Prog. Theor. Phys. **68** (1982) 508.

[11] R.J. Baxter: *Exactly Solved Models in Statistical Mechanics (Academic Press, 1982)*.

[12] Y. Akutsu and M. Wadati: J. Phys. Soc. Jpn. **56** (1987) 839.

[13] Y. Akutsu and M. Wadati: J. Phys. Soc. Jpn. **56** (1987) 3039.

[14] Y. Akutsu, T. Deguchi and M. Wadati: J. Phys. Soc. Jpn. **56** (1987) 3464.

[15] Y. Akutsu, T. Deguchi and M. Wadati: J. Phys. Soc. Jpn. **57** (1988) 1173.

[16] T. Deguchi, M. Wadati and Y. Akutsu: J. Phys. Soc. Jpn. **57** (1988) 1905.

[17] T. Deguchi, M. Wadati and Y. Akutsu: J. Phys. Soc. Jpn. **57** (1988) 2921.

[18] V.F.R. Jones: Bull. Amer. Math. Soc. **12** (1985) 103.

[19] P. Freyd, D. Yetter, J. Hoste, W.B.R. Lickorish, K. Millett and A. Ocneanu: Bull. Amer. Math. Soc. **12** (1985) 239.

[20] J.H. Przytycki and K.P. Traczyk: Kobe J. Math. **4** (1987) 115.

[21] L. Kauffman: Topology **26** (1987) 395.

[22] V.G. Turaev: Invent. Math. **92** (1988) 527.

[23] J.S. Birman and H. Wenzl: Trans. Amer. Math. Soc. (to appear).

[24] J. Murakami: Osaka J. Math. **24** (1987) 745.

[25] L.H. Kauffman: Statistical Mechanics and the Jones Polynomial, preprint (to appear in Proceedings of 1986 Santa Cruz Conference on the Artin Braid Group)

[26] Y. Akutsu and M. Wadati: Commun. Math. Phys. **117** (1988) 243.

[27] T. Deguchi, Y. Akutsu and M. Wadati: J. Phys. Soc. Jpn. **57** (1988) 757.

R-Matrices and Higher Poisson Brackets for Integrable Systems

W. Oevel

Fachbereich Mathematik, Universität Paderborn, Fed. Rep. of Germany

The tri-hamiltonian nature of Lax-equations is revealed: starting with an R-matrix on an associative algebra g equipped with a trace form there are 3 compatible Poisson brackets with linear, quadratic and cubic dependence on the coordinates. The invariant functions (Casimir functions) on g^* are in involution relative to these brackets, they yield a hierarchy of integrable tri-hamiltonian Lax-equations. The results can be applied to solvable PDE's such as the Korteweg-de Vries equation as well as to finite integrable systems such as the Toda lattice. In these cases the Poisson structures considered here turn out to be abstract versions of the first 3 hamiltonian operators of these equations obtained by their well-known recursion operators.

1. Introduction

This paper essentially consists of a summary of the results found in [13] by the author and O. Ragnisco.

Since the discovery of Inverse Scattering Techniques for integrable equations it is known that one of the most essential tools for solving these equations is given by their Lax representation. One may think of the Lax equation as the abstract dynamical system from which the "physical" dynamical systems are obtained by introducing suitable charts. Hence the phase space for most of these equations can be regarded as given by the set of Lax operators (taking values in some Lie algebra). In this sense for many examples of solvable equations the phase space itself is a Lie algebra and systematic constructions of integrable systems on Lie algebras have been given (e.g. a review of some of the relevant involution theorems can be found in [14]). Also infinite dimensional equations (PDE's) such as the celebrated Korteweg-de Vries can be regarded as systems on an infinite dimensional Lie algebra of pseudo-differential operators [1]. Thus Lie-algebras seem to provide a unifying picture for the algebraic treatment of integrable systems of various kinds.

It has turned out that the factorization method and the group of dressing transformations for these integrable systems can be understood in terms of Poisson-Drinfeld groups acting on the phase space [15] (see also [6] and [7] for a review and further references). In this setup dressing transformations are Poisson maps relative to the Poisson structure on the product of the group with this phase space. The Poisson structure on the group corresponds to a Lie bracket on the dual of the associated Lie algebra or a second Lie bracket on the Lie algebra, respectively. It is engendered by a linear map called the "R-matrix", which leads to a direct construction of integrable systems using the Casimir functions on the dual algebra. These dynamical systems are hamiltonian relative to the Lie-Poisson structure arising from this modified Lie bracket. In [15.1] it was shown that for "unitary" R-matrices on associative algebras with traceform there is a further Poisson bracket leading to a bi-hamiltonian formulation of the integrable equations associated to the Casimir functions.

We will generalize this result by defining a natural second (quadratic) Poisson bracket connected to an R-matrix without assuming unitarity. Furthermore, a natural third Poisson bracket with cubic dependence on the coordinates is found, leading to a tri-hamiltonian formulation of the Lax-equations. While tri-hamiltonian formulations are known for special examples of solvable equations (e.g. [9],[12]) we thus have managed to give a more general "explanation" for this phenomenon by finding the abstract tri-hamiltonian formulation of the class of Lax-equations associated to R-matrices.

2. Basic Definitions

Different geometric concepts are found in the literature ([3],[15]) leading to slightly different definitions for R-matrices. In any case the basic idea is to to start with a Lie-algebra g and introduce an additional structure from which a modified Lie-bracket can be derived. This might be done by claiming the dual g^* to carry a Lie-algebra structure [3] satisfying a suitable compatibility condition with the Lie-algebra structure of g. Another way is to look for a modification of the Lie-bracket on g itself [15]. For the sake of simplicity we will assume g to be equipped with a symmetric, invariant nondegenerate metric $<,>$, i.e.

$$<a,b> = <b,a>, \quad <a,[b,c]> = <b,[c,a]>, \quad a,b,c \in g, \qquad (2.1)$$

such that g can be identified with its dual g^*. For semi-simple Lie-algebras the Cartan-Killing form on g will provide such a metric, another important class of examples is given by nondegenerate trace forms (such as the usual trace on the algebra of matrices). In this case the different concepts of R-matrices coincide and are given by the following definition.

Definition 1: An R-matrix is a linear map from a Lie algebra g to itself such that the bracket $[a, b]_R := [Ra, b] + [a, Rb]$ is a second Lie product on g.

We will in general not assume the "unitarity" condition [15] for R, i.e. we do not automatically assume R to be skew-symmetric relative to the above pairing.

Looking for convenient characterizations of such a map R one easily checks that the Jacobi-identity for $[,]_R$ can be rewritten as the Jacobi-identity of the expression $[[Ra, Rb] - R[a, b]_R, c]$. Hence any linear map R will have the desired properties, if the first entry in this expression is just a scalar multiple of the original bracket $[a, b]$. Hence a sufficient condition for R to be an R-matrix is given by the equation

$$[Ra, Rb] - R[a, b]_R = -\alpha [a, b], \qquad (2.2)$$

α being a scalar parameter. Following [15] we will refer to (2.2) as the (classical) **Yang-Baxter equation**, or YB(α) for short. The correspondence of this terminology to the classical tensor notation is explained in [15.1]. As to the dependence of the parameter α obviously only the two cases $\alpha = 0$ and $\alpha = 1$ have to be considered, as for any $\alpha \neq 0$ the dilatation $R \to (1/\sqrt{\alpha}) R$ maps solution of YB(α) to solutions of YB(1). The case $\alpha = 1$ is also called the modified Yang-Baxter equation.

For a given Lie-algebra g the YB-equation may be understood as an equation for the linear map R. For the special class of simple Lie-algebras a systematic scheme of solving the YB-equations (for both cases $\alpha = 0, 1$) is given in [2]. The most important class of solutions of (2.2) arises in a very simple manner: assume that the Lie-algebra g can be decomposed into two sub-algebras g_+ and g_-, i.e. let $g = g_+ \oplus g_-$. Denoting the projections to these sub-algebras by $P_\pm$, it is easily verified that

$$R := P_+ - P_- \qquad (2.3)$$

solves YB(1) and hence provides an R-matrix on g. In this case the modified Lie-bracket $[\ ,\ ]_R$ is easily calculated to be $[a,b]_R = 2[P_+a, P_+b] - 2[P_-a, P_-b]$.

3. Main results

Let g be an associative algebra equipped with the natural Lie-product $[a,b] := ab - ba$. We furthermore assume the existence of a symmetric, non-degenerate "traceform" $tr: g \to \mathbf{R}$, yielding an invariant metric

$$<L_1, L_2> := tr(L_1 L_2) \qquad (3.1)$$

on g (identified with its dual g^* by means of this duality). We now will regard $g^* = g$ as a manifold and construct several Poisson brackets from suitable R-matrices on g.

Let $R: g \to g$ be some linear map. We will be interested in the the following 3 brackets on the space $C^\infty(g^*)$ of smooth functions on $g^* = g$:

$$\begin{aligned}\{f_1, f_2\}_1(L) &:= <[L, df_1], R(df_2)> - <[L, df_2], R(df_1)> , \\ \{f_1, f_2\}_2(L) &:= <[L, df_1], R(Ldf_2 + df_2 L)> - <[L, df_2], R(Ldf_1 + df_1 L)> , \quad (3.2)\\ \{f_1, f_2\}_3(L) &:= <[L, df_1], R(Ldf_2 L)> - <[L, df_2], R(Ldf_1 L)> , \end{aligned}$$

evaluated at a point $L \in g = g^*$. Referring to the dependence on the point L we will call them the "linear", the "quadratic" and the "cubic" bracket. From the invariance (2.1) of the metric (3.1) it is obvious that the linear bracket coincides with the Lie-Poisson bracket arising from the modified Lie-product $[\ ,\ ]_R$, i.e. the linear bracket will become a Poisson bracket iff R is an R-matrix on g. There are similar conditions for the other brackets to become Poisson brackets:

Lemma 1:

a) For any R-matrix R on g the linear bracket is a Poisson bracket.

b) If R and its skew-symmetric part $\frac{1}{2}(R-R^*)$ both satisfy a Yang-Baxter equation (2.2) (with the same α), then the quadratic bracket is a Poisson bracket.

c) If R solves a Yang-Baxter equation (2.2), then the cubic bracket is a Poisson bracket.

The linear bracket has already been identified as the Lie-Poisson structure of $[\,,\,]_R$. For the other 2 brackets the proof consists of some tedious but straightforward calculations [13]. We remark that for "unitary" R-matrices, when R coincides with its skew-symmetric part, the quadratic bracket considered here coincides with the bracket derived in [15.1]. To any bracket $\{\,,\,\}$ we can associate a Poisson tensor P via

$$\{f_1, f_2\}(L) = <df_2, P(L)df_1> \qquad (3.3)$$

Using the invariance of our metric one immediately obtains the following Poisson tensors and hence the associated hamiltonian vector fields $X_f := P\,df$ corresponding to the above 3 brackets:

$$P_1(L)\,df = [L, R(df)] + R^*([L, df]) \,,$$
$$P_2(L)\,df = [L, R(Ldf + dfL)] + LR^*([L, df]) + R^*([L, df])L \,, \qquad (3.4)$$
$$P_3(L)\,df = [L, R(LdfL)] + LR^*([L, df])L \,.$$

We remark that thinking of g being given by an algebra of matrices there is a simple relation between the linear and the cubic bracket. Assuming a matrix $L \in g$ to be invertible we can apply a "coordinate transformation" $L \to L^{-1}$, which sends the linear bracket to the cubic one and vice versa via the usual transformation laws. In [13] smooth deformations were found relating all 3 brackets. Hence all brackets are compatible in the sense of [4],[5],[10], if the hypotheses of lemma 1 are satisfied. I.e., as a special case, any skew-symmetric solution of YB(α) leads to the construction of 3 compatible Poisson brackets (3.2).

These compatibility results are a strong hint that the brackets (3.2) should turn out to be interesting in the context of integrable multi-hamiltonian equations. Indeed, a set of commuting equations can be found admitting the brackets (3.2) as hamiltonian structures. Looking for a natural set of functions in involution w.r.t. (3.2), we consider the invariant functions (Casimir functions) $C \in C^\infty(g^*)$ characterized by

$$[L, dC(L)] = 0 \,. \qquad (3.5)$$

Obviously these functions are in involution relative to all 3 brackets (3.2).

Theorem: The Casimir functions on $g^* = g$ are in involution relative to the 3 brackets (3.2). The hamiltonian equations associated to a Casimir function C are Lax-equations given by

$$\frac{d}{dt} L = P_1(L) dC = [L, R(dC)],$$

$$\frac{d}{dt} L = P_2(L) dC = [L, R(2 L dC)], \qquad (3.6)$$

$$\frac{d}{dt} L = P_3(L) dC = [L, R(L^2 dC)].$$

As we have assumed a traceform on g a natural set of Casimir functions is given by the traces of powers of the "Lax-operator" $L \in g$:

$$C_{(k)}(L) := \frac{1}{k} tr(L^k), \quad dC_{(k)}(L) = L^{k-1}, \quad k = 1,2,3,\dots. \qquad (3.7)$$

Taking these $C_{(k)}$'s as Hamiltonian functions one finds a hierarchy of equations

$$\frac{d}{dt} L := [L, R(L^k)] = P_1 dC_{(k+1)} = \frac{1}{2} P_2 dC_{(k)} = P_3 dC_{(k-1)}, \qquad (3.8)$$

which are obviously tri-hamiltonian relative to the 3 Poisson structures (3.2). As the $C_{(k)}$'s are in involution w.r.t. all brackets the above hierarchy of equations commutes and each equation admits all Casimir functions on g^* as conservation laws. In this sense the above equations shall be called "integrable": they admit a set of globally defined conservation laws in involution.

It is well known that those integrable equations arising in a bi-hamiltonian scheme [4],[5],[10] with 2 compatible Poisson operators usually admit infinitely many hamiltonian formulations: assuming invertibility of one of the Poisson structures one obtains a hereditary [4] recursion operator that generates a hierarchy of compatible Poisson tensors. For the examples worked out in [13] and indicated later-on we found that the 3 brackets considered here coincide with the first 3 Poisson structures obtained from the well-known recursion operators for those equations. In this sense our choice of Poisson brackets (3.2) seems to be justified. The surprising observation here is that on the abstract algebraic level 3 hamiltonian formulations for the the Lax-hierarchy (3.8) of integrable equations can be identified without assuming non-degeneracy for any of the brackets under consideration, i.e. without using a recursion operator. It remains open whether further Poisson brackets (maybe of fourth, fifth and higher order in L) can be identified on the abstract algebraic level "explaining" the known hierarchy of hamiltonian formulations for certain integrable equations of mathematical physics.

4. Applications

4.1. Integrable PDE's and the algebra of pseudo-differential operators

Following Adler [1] we will consider the algebra of pseudo-differential operators $g = g_+ \oplus g_-$ given by

$$g_+ := \bigcup_{m=0}^{\infty} \{ \sum_{k=0}^{m} a_k(x) D^k \}, \quad g_- := \{ \sum_{k=0}^{\infty} I^{k+1} b_k(x) \}, \qquad (4.1)$$

where $D = \partial/\partial x$ is the differential operator and $I = (\partial/\partial x)^{-1}$ is a formal integration operator.

The variable coefficients $a_k(x)$ and $b_k(x)$ are to be regarded as multiplication operators, the algebraic multiplication rules for D and the formal integration I are given by

$$DI = ID = identity \ , \ Da = a_x + aD \ , \quad (4.2)$$

$$Ia = aI - a_x I^2 + a_{xx} I^3 - \dots \ , \quad aI = Ia + I^2 a_x + I^3 a_{xx} + \dots \ ,$$

with $a_x = \frac{\partial a}{\partial x}$. Here the last rule of how to commute integration and multiplication with some arbitrary function a is to model the rule of integration by parts. Obviously the subspaces $g_\pm$ form subalgebras of g and hence an R-matrix (2.3) is given with the above decomposition. The essential observation of [1] was the construction of a traceform on g: it turns out that the "trace" tr defined by

$$tr(L) = tr(\sum_{k=0}^{\infty} I^{k+1} b_k + \sum_{k=0}^{m} a_k D^k) := \int b_0(x) \, dx \ . \quad (4.3)$$

yields a symmetric metric $<L_1, L_2> := tr(L_1 L_2)$ on g (see [1]). Hence we have found the situation of the last section and all the results of [1] can be rephrased in terms of the R-matrix induced by the above decomposition. We observe that relative to the duality given by the above trace one has $g_+ = g_-^*$, hence R turns out to be skew-symmetric. So, by lemma 1 we have found 3 compatible Poisson brackets (3.2) on the space g of pseudo-differential operators. It is easy to see that all 3 Poisson brackets can be restricted Poisson brackets on the subalgebra $g_+ = g_-^*$ of purely differential operators. Hence a set of integrable tri-hamiltonian equations on g_+ is given with (3.8). A technical difficulty arises from the fact that one should not consider Casimir functions $C_{(k)} := tr(L^k)/k$ with integer powers $k \in \mathbb{N}$, as these restrict to trivial functions on the subspace g_+ (note that integer powers of a purely differential operator will again be purely differential operators (with vanishing trace)). But certain fractional powers of differential operators can be calculated explicitly on a purely algebraic level (see e.g. [11] for details).

E.g., consider $L_S := a_0 + D^2 \in g_+$, i.e. the Schroedinger operator with potential a_0. By explicitly calculating the square root $L_S^{1/2}$, one finds

$$\frac{d}{dt} L_S = [L_S, R(L_S^{3/2})] \quad \Leftrightarrow \quad \frac{d}{dt} a_0 = -\frac{1}{2}(a_{0xxx} + 6 a_0 a_{0x}) \ , \quad (4.4)$$

i.e. the Korteweg-de Vries equation. Of course, one should have expected to find the KdV in this approach, as the Schroedinger operator L_S is the well-known Lax-operator of the KdV. Indeed, the equations (3.8) yield the well-known hierarchy of higher KdV-equations by choosing $k = 1/2 + n$, $n \in \mathbb{N}$. The simplest way to see this are the Lenard-relations given by the tri-hamiltonian formulations (3.8). We try to calculate the 3 Poisson brackets (3.2) by evaluating the Poisson tensors (3.4) at the point $L_S \in g_+$. For technical details we refer to [11],[13], here we just briefly state the results: As a simple exercise one finds that the subspace $S^{(2)} := \{a_0 + D^2\}$ is an integral manifold of the linear bracket, hence the linear Poisson structure is restricted to $S^{(2)}$ easily. As expected, one finds the differential operator D as the first hamiltonian operator of the KdV and its higher equations. Note that for the "linear bracket" the linear dependence on the fields is destroyed by this reduction. It turns out that the quadratic bracket admits the subspace $S' := \{ a_0 + a_1 D + D^2 \}$ as integral manifold. We restrict P_2 given by (3.4) to this subspace and a suitable reduction procedure yields a canonical projection of this structure to the subspace $S^{(2)}$ of Schroedinger operators. The resulting hamiltonian operator is found to be given by $D^3 + 2 D a_0 + 2 a_0 D$, i.e. the well known second hamiltonian operator of the KdV and its higher equations.

Thus the Lenard-relations (3.8) between the hamiltonian formulations given by P_1 and P_2 lead to the well-known bi-hamiltonian structure and hence to the usual recursion operator (Lenard-operator) of the KdV when restricting the considerations to the Schroedinger operator L_S. Similarly, one can try to evaluate the cubic Poisson tensor (3.4) at the point $L = L_S$. For this tensor again a nontrivial -but still natural- reduction procedure is needed to restrict it to the subspace $S^{(2)}$ of Schroedinger operators. It turns out that the hamiltonian operator resulting from this reduction coincides with the hamiltonian operator obtained by applying the above recursion-operator to the second hamiltonian operator ([13]). We remark that although the cubic bracket is restricted to purely differential operators, there are non-localities (integrations) arising from the reduction procedure.

Of course, the abstract hierarchy (3.8) and the abstract hamiltonian operators (3.2) can be evaluated at any arbitrary point $L \in g_+$. It is simple to see that all the subspaces

$$S^{(j)} := \{ a_0 + a_1 D + ... + a_{j-2} D^{j-2} + D^j \} \tag{4.5}$$

are integral manifolds of the linear bracket and hence are invariant relative to the flows given by the hamiltonian equations (3.8). Hence the linear bracket as well as the hierarchy of equations (3.8) can be restricted to $S^{(j)}$. For a given j a reasonable choice for the Casimir functions will be $C_{(k)} = tr(L^k)/k$ with $k = n/j$, $n \in \mathbb{N}$. For all different choices of j these equations yield integrable partial differential equations in terms of the "physical variables" $a_0, a_1, ..., a_{j-2}$. E.g., for $L = a_0 + a_1 D + D^3$ (known to be a Lax-operator related to the Boussinesq equation) one finds the integrable equation

$$\frac{d}{dt} L = [L, R(L^{2/3})] \quad \Leftrightarrow \quad \frac{d}{dt} \begin{bmatrix} a_0 \\ a_1 \end{bmatrix} = \frac{2}{3} \begin{bmatrix} 2 a_{1xxx} - 3 a_{0xx} + 2 a_1 a_{1x} \\ 3 a_{1xx} - 6 a_{0x} \end{bmatrix}, \tag{4.6}$$

which can be rewritten as $3 a_{1tt} + 4 (a_{1xxx} + 4 a_1 a_{1x})_x = 0$, i.e. the Boussinesq equation in its usual form.

The subspaces (4.5) are not just integral manifolds of the linear bracket but symplectic leaves. Hence the restriction of the linear bracket to $S^{(j)}$ yields an invertible hamiltonian operator for the integrable equations (3.8) evaluated on $S^{(j)}$. The quadratic as well as the cubic bracket do not admit $S^{(j)}$ as integral manifolds, hence these brackets have to be restricted to these spaces by means of the non-trivial reduction procedure already mentioned for the KdV-example above. In any case, the reduced quadratic bracket and the invertible restriction of the linear bracket yield a recursion operator for the partial differential equations given by the restrictions of the hierarchy (3.8) to $S^{(j)}$. E.g., for $j = 3$ these reduced hamiltonian operators provide the well-known bi-hamiltonian formulation of the Boussinesq-equation in the form (4.6). Again, it turns out that the reduction of the cubic tensor to $S^{(3)}$ coincides with the hamiltonian operator obtained by applying the recursion operator resulting from P_1 and P_2 to the reduction of the quadratic tensor P_2 ([13]).

4.2. Lattice systems: the classical Toda lattice

Following [1],[8],[16] we start with the Lie-algebra $g = gl(n)$ of all $n \times n$-matrices, which shall be split ($g = g_+ \oplus g_-$) into the subalgebras g_+ given by the lower triangular matrices and $g_- = so(n)$, i.e. the skew-symmetric matrices. We identify g with its dual g^* via the nondegenerate pairing $<L_1,L_2> := tr(L_1L_2)$, where tr is the usual trace of matrices. The duals of the subalgebras $g_\pm$ are easily identified as g_+^* consisting of the symmetric matrices (annihilating the skew-symmetric matrices in g_- via the trace-pairing) and g_-^* consisting of the strictly lower matrices (annihilating the lower matrices in g_+ via the trace-pairing). The projections to the subalgebras (and their duals, respectively) corresponding to the decomposition $g = g_+ \oplus g_-$ (and the dual decomposition $g = g^* = g_+^* \oplus g_-^*$, respectively) are found to be

$$P_+L = \underline{l}(L) + \overline{u}(L)^T + d(L) \in g_+ \quad ; \quad P_-L = \overline{u}(L) - \overline{u}(L)^T \in g_- ,$$
$$P_+^*L = d(L) + \overline{u}(L) + \overline{u}(L)^T \in g_+^* \quad ; \quad P_-^*L = \underline{l}(L) - \overline{u}(L)^T \in g_-^* . \quad (4.7)$$

Here $\underline{l}(L)$ is to denote the strictly lower, $\overline{u}(L)$ the strictly upper, and $d(L)$ the diagonal part of the matrix L, the suffix ^{T} is to denote the usual transposition of matrices. From the above decomposition $g = g_+ \oplus g_-$ we find an R-matrix (2.3) solving YB(1) and its adjoint map R^* :

$$R(L) := P_+L - P_-L = \underline{l}(L) + 2\,\overline{u}(L)^T + d(L) - \overline{u}(L) ,$$
$$R^*(L) := P_+^*L - P_-^*L = -\underline{l}(L) + 2\,\overline{u}(L)^T + d(L) + \overline{u}(L) . \quad (4.8)$$

By splitting all matrices into their strictly lower, their strictly upper and their diagonal parts, it is a simple exercise to show that also the skew-symmetric part

$$\frac{1}{2}(R - R^*) : L \to \underline{l}(L) - \overline{u}(L) \quad (4.9)$$

of the above R-matrix is a solution of YB(1). Hence, according to lemma 1 we have found 3 compatible Poisson brackets (3.2) on g.

We will now look for a low-dimensional integral manifold for these brackets. We can immediately restrict ourselves to the subspace g_+^* of symmetric matrices (see [13]). By elementary arguments ([13]) it is shown that the subspace of symmetric matrices of the special form

$$S := \{ b + a\,T_+ + T_-\,a \} \quad (4.10)$$

is an integral manifold for all 3 Poisson brackets (3.2). Here $b = diag(b_1,b_2,...,b_n)$ and $a = diag(a_1,a_2,...,a_{n-1},0)$ are diagonal matrices and $T_\pm$ are the "shiftmatrices"

$$T_+ := \begin{bmatrix} 0 & 1 & 0 & .. & . & 0 \\ 0 & 0 & 1 & .. & . & . \\ 0 & 0 & 0 & .. & . & . \\ . & . & . & .. & . & . \\ . & . & . & .. & 0 & 1 \\ 0 & . & . & .. & 0 & 0 \end{bmatrix} \quad , \quad T_- := \begin{bmatrix} 0 & 0 & 0 & .. & 0 & 0 \\ 1 & 0 & 0 & .. & . & . \\ 0 & 1 & 0 & .. & . & . \\ . & . & . & .. & . & . \\ . & . & . & .. & 0 & 0 \\ 0 & . & . & .. & 1 & 0 \end{bmatrix} \quad (4.11)$$

consisting of just one line on the first super-diagonal (sub-diagonal). We remark that S consists of the the well known Lax-operator for the classical Toda lattice [9.2]. Indeed, choosing the Casimir functions $C_k := tr(L^k)/k$, $k = 2,3,4,...$, one finds the Toda lattice as the first equation of the tri-hamiltonian hierarchy (3.8):

$$\frac{d}{dt} L = P_1 \, dC_2 = [L, RL] \iff \frac{d}{dt} \begin{bmatrix} b_i \\ a_i \end{bmatrix} = \begin{bmatrix} 4 \, a_i^2 - 4 \, a_{i-1}^2 \\ 2 \, a_i \, (b_{i+1} - b_i) \end{bmatrix} \qquad (4.12)$$

with $i = 1,..,n$ and $a_0 = a_n := 0$. All 3 Poisson tensors (3.4) can be restricted to their integral manifold S easily. An explicit calculation reveals that their restrictions coincide with the 3 local hamiltonian operators of the Toda lattice found by Kupershmidt ([9.2]). Hence our choice of abstract Poisson brackets (3.2) on the Lax-operators seems to be an "explanation" for the tri-hamiltonian nature of the Toda-hierarchy.

We remark that instead of using the R-matrix (4.8) one can also start with the "unitary" R-matrix (4.9). Again (4.10) turns out to be an integral manifold for the hierarchy (3.8) and again the Toda hierarchy and its tri-hamiltonian formulation is derived. We also remark that the periodic Toda lattice can be treated in the same framework by substituting the shiftmatrices (4.11) by their cyclic counterparts and modify the notions of "upper" and "lower" diagonal parts of a matrix in a suitable way.

References

[1] M. Adler, *On a Trace Functional for Formal Pseudo-Differential Operators and the Symplectic Structure of the Korteweg-de Vries Equations*, Invent. math. 50, 219-248 (1979)

[2] A.A. Belavin, V.G. Drinfeld, *Triangle Equations and simple Lie algebras*, Mathematical Physics Reviews Vol. 4, Moscow, 94-165 (1987)

[3] V.G. Drinfel'd, *Hamiltonian structures on Lie groups, Lie bialgebras and the geometric meaning of the classical Yang-Baxter equations*, Soviet Math. Dokl. 27, 68-70 (1983)

[4] B. Fuchssteiner, A.S. Fokas, *Symplectic structures, their Baecklund transformations, and hereditary symmetries*, Physica 4D, 47-66 (1981)

[5] I.M. Gel'fand, I.Y. Dorfman, *Hamiltonian operators and algebraic structures related to them*, Funct. analy. i. Eqs. Priloz. 13, 248-262 (1979); *The Schouten bracket and hamiltonian operators*, Funct. analy. i. Eqs. Priloz. 14, 223-226 (1980)

[6] Y. Kosmann-Schwarzbach, *Poisson-Drinfeld Groups*, in "Topics in Soliton Theory and Exactly Solvable Nonlinear Equations", M. Ablowitz, B. Fuchssteiner, M. Kruskal (eds.), World Scientific Publ., Singapore, 191-215 (1987)

[7] Y. Kosmann-Schwarzbach and F. Magri, *Poisson-Lie groups and complete integrability*, Ann. Inst. Henri Poincarè 49, 443-460 (1988)

[8] B. Kostant, *Toda lattice and representation theory*, Adv. Math. 34, 195-338 (1979)

[9] B.A. Kupershmidt, *Mathematics of Dispersive Water Waves*, Commun. Math. Phys. 99, 51-73 (1985) ; *Discrete Lax Equation and Differential-Difference Calculus*, Astérisque 123, Societé Mathematique de France, Paris, (1985)

[10] F. Magri, *A simple model of the integrable Hamiltonian equation*, J. Math. Phys. 19, 1156-1162 (1978)

[11] W. Oevel: *R-structures, Yang-Baxter Equations and Related Involution Theorems*, to appear in J. Math. Phys. (1989)

[12] W. Oevel, B. Fuchssteiner, H. Zhang, O. Ragnisco : *Mastersymmetries, Angle Variables and Recursion Operator of the Relativistic Toda Lattice*, to appear in J. Math. Phys. (1989)

[13] W. Oevel and O. Ragnisco, *R-Matrices and Higher Poisson Brackets for Integrable Systems*, to appear in J. Math. Phys. (1989)

[14] T. Ratiu, *Involution Theorems*, Lect. Notes Math. **775**, 219-257 (1980)

[15] M.A. Semenov-Tian-Shansky, *What is a classical R-matrix ?*, Funct. Anal. Appl. **17**, 259-272 (1983) ; *Dressing Transformations and Poisson Group Actions*, Publ. RIMS, Kyoto Univ. **21**, 1237-1260 (1985)

[16] W.W. Symes, *Systems of Toda Type, Inverse Spectral Problems and Representation Theory*, Invent. Math. **59**, 13-51 (1980)

Classical R-Matrix and Semi-Simple Lie Algebras

Liu Zhangju and Qian Min

Department of Mathematics, Beijing University, Beijing 100871,
People's Rep. of China

INTRODUCTION. Let g be a Lie algebra. A linear operator $R \in \text{End}(g)$ is called a r-matrix (see [5]) if the bracket given by

$$[X, Y]_R = [RX, Y] + [X, RY] \qquad (1)$$

is also a Lie bracket on g. Such a pair (g, R) is called a double Lie algebra. Moreover, if there is a nondegenerate invariant bilinear form on g and R is skew-symmetric (g, R) becomes a Lie bialgebra ([1], [5]). It is known that (g, R) is a double Lie algebra iff the following bilinear map $B_R: g \times g \to g$ given by

$$B_R(X, Y) = [RX, RY] - R([X, Y]_R) \qquad (2)$$

is ad-invariant, i.e., the equation

$$[X, B_R(Y, Z)] + [Y, B_R(Z, X)] + [Z, B_R(X, Y)] = 0 \qquad (3)$$

holds for all $X, Y, Z \in g$. Particularly, the equation

$$B_R(X, Y) = 0, \qquad \forall\ X, Y \in g$$

is the Yang-Baxter equation. The modified Yang-Baxter equation was defined in [5] as follows:

$$B_R(X, Y) = -[X, Y] \qquad (4)$$

Obviously, the equation (4) means (3) hold.

In this paper we give a skew-symmetric linear operator J (the Koszul operator) for every compact semi-simple Lie algebra satisfying the equation as follows

$$B_J(X, Y) = [X, Y] \qquad (5)$$

Obviously, such a J satisfies the equation (3). In fact the equation (5) is nothing but the condition of integrability for a almost complex stucture on a homogeneous space. It is also proved that the Lie bialgebra structure with respect to J is equivalent to that given in [4] . In §2, we point out that $\sqrt{-1}$ J is a r-matrix for a real split semi-simple Lie algebra and also skew-symmetric. In §3, we give a involution theorem which may be considered as a variant of the Kostant-Symmes lemma for double Lie algebras. Finally, it is seen that some results about the Lax equation given by Goodman and Wallach ([2]) become very clear from the viewpoint of r-matrix.

§1. In this section we assume that g is a compact semi-simple Lie algebra. Let G be the connected compact semi-simple Lie group with the Lie algebra g, H a connected Lie subgroup of G with a subalgebra h of g. The Koszul's theorem say that the coset space G/H has a G-invariant complex structure if and only if there is a linear endomorphism J of g (the Koszul operator) satisfying

(i) $J|_h = 0$, $J^2|_m = -id$, where $m = h^\perp$ with respect to the Killing form of g.

(ii) $ad(x) \circ J = J \circ ad(x)$, $\forall x \in h$

(iii) $[JX, JY] - J([JX, Y] + [X, JY]) = [X, Y]$ (mod h)

Notice that the left side of equation (iii) is just the bilinear map B_J defined by (2). It is also known that the coset space G/H is a homogeneous Kaehler manifold if H is a maximal toral subgroup of G Combining these facts we can get the following theorem

Theorem 1 Let g be a compact semi-simple Lie algebra, h a Cartan subalgebra of g. Then the corresponding Koszul operator J is skew-symmetric with respect to the Killing form of g and satisfies the equation

$[JX, JY] - J([JX, Y] + [X, JY]) = [X, Y]$, $\forall X, Y \in g$ (6)

Consequently, J is a classical r-matrix and g is a Lie bialgebra.

In [4] , Lu and Weinstein give a Lie bialgebra structure for compact semi-simple Lie algebras by means of the Iwasawa's decomposition and Manin's triple. The following proposition shows that it is equivalent to ours given by the Koszul operator J. Notice that g is a compact real form of g^c and can be expressed as

$$g = h + \sum_{\alpha \in \Delta^+} R(X_\alpha - X_{-\alpha}) + \sum_{\alpha \in \Delta^+} Ri(X_\alpha + X_{-\alpha}) \qquad (7)$$

Write

$$t = \sum_{\alpha \in \Delta^+} R(X_\alpha - X_{-\alpha}), \quad p = h + \sum_{\alpha \in \Delta^+} Ri(X_\alpha + X_{-\alpha}) \qquad (8)$$

Thus, $g = t + p$ is a Cartan decomposition of g. Let

$$n_\pm = \sum_{\alpha \in \Delta^\pm} (RX_\alpha + RiX_\alpha), \quad s_\pm = ih + n_\pm \qquad (9)$$

considered as real nilpotent Lie algebras and set

$$\tilde{n}_\pm = \sum_{\alpha \in \Delta^\pm} RX_\alpha, \quad \tilde{h} = ih \quad \text{and} \quad \tilde{s}_\pm = \tilde{h} + \tilde{n}_\pm \qquad (10)$$

Proposition 2 With the notation as above and let the symbol g_J denote the Lie algebra g equipped with the bracket

$$[X,Y]_J = [JX,Y] + [X,JY] \qquad (11)$$

Then we have

(a) g_J is isomorphic to $s_\pm$.

(b) p is a subalgebra of g_J and isomorphic to $\tilde{s}_\pm$.

§2. Let $g = t + p$ be the Cartan decomposition of a compact semi-simple Lie algebra given by (8) in the last section. Write $\tilde{p} = ip$, then

$$\tilde{g} = t + \tilde{p} \qquad (12)$$

is a real split semi-simple Lie algebra with the Cartan subalgebra $\tilde{h} \subset \tilde{p}$ and (12) gives a Cartan decomposition of $\tilde{g}$. It is easy to see that the operator $\tilde{J} = iJ$ ($i = \sqrt{-1}$) is a linear endomorphism on $\tilde{g}$ and satisfies the modified Yang-Baxter equation

$$B_{\tilde{J}} = -[X,Y], \quad \forall X,Y \in \tilde{g} \qquad (13)$$

It is easy to see that $\tilde{J}$ is also skew-symmetric with respect to the Killing form of $\tilde{g}$. Thus we have

Theorem 3 Every real split semi-simple Lie algebra is a Lie bialgebra with the r-matrix $\tilde{J}$ above which satisfies the modified Yang-Baxter equation.

§3. In order to apply r-matrix approach to integrable systems we give a involution theorem for a spliting double Lie algebra. Suppose (g,R) have a decompposition

$$g = a+b, \quad [a, a] \subset a, \quad [b,b]_R \subset b \quad (14)$$

That is, $a(b)$ is a subalgebra of $g(g_R)$ respectively.

The decomposition $g = a + b$ gives a corresponding decomposition $g^* = b^\circ + a^\circ$ which allows an identification

$$b^\circ \sim a^*, \quad a^\circ \sim b^*$$

Let $\pi_a : g^* \longrightarrow a^*$, $\pi_b : g^* \longrightarrow b^*$ be the projections and $f_a = f \circ \pi_a$, $f_b = f \circ \pi_b$ for all $f \in C^\infty(g^*)$. The following theorem gives a variant of the Kostant-Symmes lemma for double Lie algebras.

Theorem 4 Let (g,R) be a double Lie algebra and R satisfy (4) or (5). Suppose that (g,R) have the decomposition (14) and satisfy the condition $b \subset a$. Then

(i) For ad* - invariant functions $f, h \in C^\infty(g^*)$, one has

$$\{f_b, h_b\}_R (\xi) = 0, \quad \forall \xi \in a^\circ \sim b^* \quad (15)$$

(ii) For a ad* -invariant function f, the corresponding Euler equation of f_b on b^* with respect to the bracket $\{\cdot,\cdot\}_R$ can be written in the form

$$\dot{\xi} = ad^* (Rdf_b(\xi)) \xi, \quad \xi \in b^* \quad (16)$$

Now we return to our special case. With the same notation as used in §1, §2, we notice that, for a function $f \in C^\infty(p)$ there is ad-invariant function $F \in C^\infty(g)$ such that $F\big|_p = f$ iff f is ad(t)-invariant. In the orther hand, $(g = t+p, \tilde{J})$ satifies the condition of the theorem 4. Consequently, we get

Corollary 5 (i) All ad(t) -invariant functions on p are in involution with respect to the bracket $\{\cdot, \cdot\}_{\tilde{J}}\big|_p$. (ii) For a ad(t)-invariant function $f \in C^\infty(p)$, the corresponding Euler equation (16) can be written in the form

$$\dot{X} = [Jdf(X), X] \qquad X \in p \qquad (17)$$

Remark. The same conclusion is true for the double Lie algebra $(\tilde{g}, \tilde{J})$ with the decomposition $\tilde{g} = t + \tilde{p}$.

Example (the non-periodic Toda lattice). Let $g = su(n)$. Thus, $t = o(n)$. $p = i\, symm(n)$ and $J|_p : p \to t$ is given by $iX \mapsto X^- - X^+$ where $X^\pm$ are the upper (lower) triangular parts of X repectively. Take

$$f(iX) = tr\, X^2, \qquad X \in symm(n)$$

Then the corresponding Euler equation (17) is

$$\dot{X} = [X^- - X^+,\ X] \qquad (18)$$

The equation (18) is the standard Lax form for the non-periodic Toda lattice (see [3]). In this time, the property (i) of theorem 4 is the same with the Kostant-Symmes lemma.

To consider the periodic Toda lattice we should extend J to on loop algebras (see [5]). Here we won't write down it in detail because the following proposition means that all results on the periodic Toda lattice given in [2] may be described by the langurage of r-matrix. Notice we have the decomposition

$$g = t + p, \qquad g_R^C = t + p + s_+ \qquad (19)$$

and

$$\tilde{g} = t + \tilde{s}_+, \qquad \tilde{g} = t + \tilde{p} \qquad (20)$$

where g_R^C represents g^C considered as a real Lie algebra. Let π_t ($\tilde{\pi}_t$) be the projection from g ($\tilde{g}$) to t along $p + s_+$ ($\tilde{s}_+$) respectively. Then we have

Proposition 6 $\qquad J\big|_p = -i\pi_t\ ,\quad \tilde{J}\big|_{\tilde{p}} = \tilde{\pi}_t$

This proposition means that the Lax equation on Riemann symmetric spaces given in [2] is the same with the Euler equation given in the corollary 5 for the case that the Lie algebra is spltied over R.

References

[1] V.G. Drinfel'd, Quantum groups, Proc. ICM, Berkeley 1986, Vol. I, 789-820.

[2] R. Goodman and N.R. Wallach, Commun. Math. Phys. 94, 177-217 (1984).

[3] V. Guillemin and S. Sternberg, Symplectic techiniques in physics, Cambridge University press, (1984).

[4] J.H. Lu and A. Weinstein, Poisson Lie group, dressing transformation and Bruhat decompositions, Preprint.

[5] Semenov-Tyan-Shansky, What is a classical r-matrix? Funct. Anal. Appl. 17 (4) (1983) 259-272.

Witten's Approach, Braid Group Representations and X-Deformations

M.L. Ge, F. Piao, L.Y. Wang, and K. Xue

Theoretical Physics Division, Nankai Institute of Mathematics,
Tianjin 300071, People's Rep. of China

The Witten's approach of link polynomials based on (2+1) Chern–Simons Lagrangian is used to simplify the calculations of braid group representations(BGR) for SU(2) algebra. On the basis of the direct derivations of BGR the "x–deformation" scheme is presented to generate explicitly the quantum R(x)–matrix for given BGR.

The braid group representations and the related topics are intensively studied because of their importance for many subjects such as Statistical Mechanics, Quantum Group, 1+1 Conformal Field Theory (CFF) and so on (1–16,22). Some derivations of braid group representations(BGR) are based on reducing the known quantum R(X)–matrix satisfying the Yang–Baxter equations (YBE) with spectral parameter x–dependence to yeild BGR. For instance by taking the symmetry breaking limit of R(x)–matrix for the spin models AW(9) derived the link polynomials other than Jones the cases of groups of B_n, C_n, and D_n in their fundamental representations were discussed by Turaev in terms of reduction of Jimbo's R(x)–matrix (10). Both of them are not direct derivations of the BGR. As we know that the derivations of R(x) is much more difficult than those of R(0) which is just BGR in Jimbo's "gauge". Therefore from point of view of Physics we prefer to the inverse, namely, to determine the R(x) from known BGR corresponding to certain representation of Lie algebra. Speakig in another way, we devide the Jimbo's Q.G. version into two steps by different approach. The first one is to establish the BGR for a representation of Lie algebra with the help of Witten's version and x–independent YBE. Next we then give the x–dependent R(x)–matrix by "x–deformation". In comparison with the Q.G. calculations of R(x) by Jimbo (17–19) our approach is more simple and powerful. Under such a backgroud we first should set up a direct approach to calculate BGR by a standard device. In this respect in the Refs(6,7,8,11) Kauffman and Reshetikhin gave the answers in general forms that in principle can be used to construct BGR for given representations of simple Lie algebras.

However the general statement are not easy to be carried out because the state expansions are complicated and the calculations of C–G coefficients of Q.G. are difficult in certain cases. Recently in his paper Witten(14) presented a new version to find the eigenvalues of BGR, framing factor of SU(N) based on 2+1 Chern–Simons action and CFT(14,20). It has been shown that the Witten's version is universal for any simple Lie algebras (without the multiplicity)(11.23).In this note we shall show how the calculations of BGR can be simplified with the Witten's version which also make the connection between the theory with the CFF.

The general scheme of extended Yang–Baxter state models has been discussed in detail in Refs(15,22,8).Now we would like to take two examples of the spin model as the representatives to show a "minimum scheme" in computing BGR that can be used to deal with the much more complicated cases which are hard to make in terms of the methods shown in Refs(7,11).

Let us consider the spin models .For the cases spin less than 5/2 there had been direct calculations (15,16).They are more and more tedious as the dimensions of the matrix increase .Now let us calculate the BGR for spin 5/2 and 3 (n=2s+1=6 and 7) in terms of the present new method.

(I) The Relationship Between State Models and Witten's Approach

1.Witten's Approach

A decomposition of the direct product of representations R for simple Lie algebra gives rise to the eigenvalues of BGR(14,20,21)

$$\lambda_i = \pm q^{(\Delta_R - \frac{1}{2}\Delta_{E_i})} \tag{1.1}$$

where

$$q = \exp(2i\pi/(K+C_v))$$

and Δ stands for the corresponding Casimir eigenvalues.

The reduction relation for the BGR is

$$\prod_{i=1}^{n}(S-\lambda_i) = 0 \tag{1.2}$$

The framing factor f is given by

$$< \chi S^m \psi > = f^m P(L_{m-1}) \tag{1.3}$$

where $P(L_{m-1})$ denotes (m–1) crossing polynomials .The meaning of $< \chi ...... \psi >$ and other conventions are refered to Refs(21).The skein relation has the form

$$P(L_{n-1}) - f^{-1}(\sum_{i=1}^{n}\lambda_i)P(L_{n-2}) + f^{-2}(\prod_{i<j}^{n}(\lambda_i\lambda_j))P(L_{n-3})$$

$$-... + f^{-n}(-1)^n(\prod_{i=1}^{n}\lambda_i)P(L_{-1}) = 0 \tag{1.4}$$

153

2. The Markov trace is defind as

$$\phi(A)=\mathrm{tr}(AH) \tag{1.5}$$

where the matrix A can be any crossing block and H is given by

$$H = h \times h \times \ldots \times h.$$

The diagonal matrix h is given by (9.11)

$$h_{ab} = \delta_{ab} t^{-2\langle \delta, W_a \rangle} = \delta_{ab} t^{-L(a)}$$

where δ is the half sum of simple roots and W_a denotes the weight labelled by index a. We note that the sum

$$\sum_a S^{ab}_{ab} h_{bb} = \tau \qquad \sum_a (S^{-1})^{ab}_{ab} h_{bb} = \bar{\tau} \tag{1.7}$$

is independent of index a. The nomarlization of S^{aa}_{aa} can be made in an arbitrary way, for example

$$S^{aa}_{aa} = 1 \tag{1.8}$$

or

$$S^{aa}_{aa} = t^{-(W_R)^2} \tag{1.9}$$

which gives rise to

$$\sum_b S^{ab}_{ab} h_{bb} = t^{-\Delta_R} = \tau \tag{1.10}$$

$$\sum_b (S^{-1})^{ab}_{ab} h_{bb} = t^{\Delta_R} = \bar{\tau} = \tau^{-1} \tag{1.11}$$

The W_R denotes the highest weight of representation R.
The factor

$$\alpha^{1/2} = (\bar{\tau}/\tau)^{1/2} = t^{\Delta_R}$$

plays the important role in construction of polynomials just as f in eq.(1.3). Obviously, in comparison with the Witten's approach we have

$$f = \alpha^{1/2} \tag{1.12}$$

if it holds the condition

$$t = q \tag{1.13}$$

which is know as the trace—cross channel unitrarity(7,15)

3. Kauffman's State Model

The link polynomial is defined by

$$P_K = (A)^{-W(K)} \sum_S [K|S] \, t^{\|S\|} \qquad (1.14)$$

where

$$[\mathcal{C\!\!\!\!\!\!\!\!\!\!}] = A\,[O]$$
$$[\mathcal{C\!\!\!\!\!\!\!\!\!\!}] = A^{-1}[O] \qquad (1.15)$$

The notations adopted here refer to Refs(7,15). For example the norm $\|S\|$ is given by

$$S = \sum_{\text{comp}(s)} \text{rot}(a).<2\delta, W_a> \qquad (1.16)$$

where a suitable label set a is understood (12,22).

Because the calculations show

$$A = \tau$$

and the writhe

$$W(\underbrace{\times \cdots \cdots \times}_{m}) = -m$$
$$W(\underbrace{\times \cdots \cdots \times}_{m}) = m \qquad (1.17)$$

the eq.(1.14) is equivalent to the defination of link polynomials in Refs(7).

There are two possibilities to closure the graph

One of them is trivial

The other provides the trace cross–channel unitarity

$$\begin{aligned}
\text{\Large$\infty$\!\!\!\!\!\!\!\!\!\!\!\!\!\!\!\!} &= \sum_{a,b,c} S^{ca}_{ca}(S^{-1})^{cb}_{cb}\, t^{-L(a)}\, t^{-L(b)}\, t^{-L(c)} \\
&= \sum_c (\sum_a S^{ca}_{ca}\, t^{-L(a)})(\sum_b (S^{-1})^{cb}_{cb}\, t^{-L(c)}) \\
&= \sum_c t^{-L(c)} = [O]
\end{aligned} \qquad (1.18)$$

Thus the trace cross–channel unitarity is automaticly satisfied provided t = q does .

Now through the above discussions are closely related and sometimes they equivalent.The Witrten's approach is more powerful not only in the explicit forms of the eigenvalues but also in their connections with the CFF.

In the following we shall employ anyone of the three approaches when it is able to simplify the calculations .As we see later that this "combining operation" is effective in deriving BGR and is a general method.

(II).The Direct Calculations for the Cases of Spin 5/2

The Casimir eigenvalues for SU(2) is simply j(j+1) so that the eigenvalues of BGR under the case are

$$1, -t^5, t^9, -t^{12}, t^{14}, -t^{15}, \tag{2.1}$$

where

$$t = \exp(2i\pi/(K+2))$$

The general considerations lead to the block–diagonal form for the BGR S(7,16)

$$S = (\ A^{(5)},...,A^{(1)},A^{(0)},...,A^{(5)}) \tag{2.2}$$

where

$$A^{(1)}=1, \quad A^{(2)}=\begin{bmatrix} 0 & -t^{5/2} \\ -t^{5/2} & 1-t^5 \end{bmatrix}$$

$$A^{(3)}=\begin{bmatrix} 0 & 0 & t^5 \\ 0 & t^4 & q_1 \\ t^5 & q_1 & (1-t^4)(1-t^5) \end{bmatrix}$$

$$A^{(2)}=\begin{bmatrix} 0 & 0 & 0 & -t^{15/2} \\ 0 & 0 & -t^{-11/2} & q_2 \\ 0 & -t^{-11/2} & (1-t^4)(t^3+t^4) & q_3 \\ t^{15/2} & q_2 & q_3 & (1-t^3)(1-t^4)(1-t^5) \end{bmatrix}$$

where q_1, q_2 and q_3 are functions of t and to be determined later.

$$A^{(1)}=\begin{bmatrix} 0 & 0 & 0 & 0 & t^{10} \\ 0 & 0 & 0 & t^7 & q_4 \\ 0 & 0 & t^6 & q_5 & q_6 \\ 0 & t^7 & q_5 & p_1 & q_7 \\ t^{10} & q_4 & q_6 & q_7 & (1-t^2)(1-t^3)(1-t^4)(1-t^5) \end{bmatrix}$$

$$A^{(0)} = \begin{bmatrix} 0 & 0 & 0 & 0 & 0 & -t^{25/2}q_8 \\ 0 & 0 & 0 & 0 & -t^{17/2}q_9 & q_{10} \\ 0 & 0 & 0 & -t^{13/2} & q_9 & q_{11} & q_{12} \\ 0 & 0 & -t^{13/2} & p_2 & q_{11} & q_{12} \\ 0 & -t^{17/2} & q_9 & q_{11} & p_3 & q_{13} \\ -t^{25/2} & q_8 & q_{10} & q_{12} & q_{13} & \prod_{i=1}^{5}(1-t^m) \end{bmatrix}$$

where again the parameters $q_6,...,q_{13}, p_1,...,p_3$ will be determined in the following.

1. By taking the trace of $A^{(1)}$ and using the eigenvalues $1,...,t^4$ we have
$$p_1 = t^2 + t^3 + t^4 - t^5 - 2t^6 - 2t^7 - t^8 + t^9 + t^{10} + t^{11}$$

2. By taking the Markov trace we get
$$(1-t^3)(1-t^4)(1-t^5)(t^{-5/2}) + p_1 t^{-3/2} + p_2 t^{-1/2} + t^6 t^{1/2} = t^{-5/2}$$
$$p_2 = t^4 + t^5 + t^6 - t^7 - t^8 - t^9$$

3. With eq.(2.1) the trace of $A^{(0)}$ leads to
$$p_3 = t + t^2 + 2t^8 + 3t^9 + t^{10} - t^4 - 3t^5 - 2t^6 - t^{12} - t^{13}$$

The off-diagonal parameters q's should be determined in terms of the YBE that can be performed by the extended YBE-state model. The standard device is refered to Refs(15).

The results are as following.

$$(q_1)^2 = t^{14} - t^{10} - 2t^9 - t^8 + t^{13} + t^5 + t^4$$

$$(q_2)^2 = (1-t^5)(t^{10} - t^{11}) + t^4 q_1^2$$

$$(q_3)^2 = (1-t^5)^2(1-t^4)(t^3 + t^4) + t^5(1-t^4)(1-t^5)$$
$$-(1-t^4)^2(t^3 + t^4)^2(1-t^5) - t^{11}(1-t^4)(1-t^5)$$

$$(q_4)^2 = (1-t^5)(t^{15} - t^{14}) + (q_2)^2 t^4$$

$$(q_5)^2 t^4 = (1-t^4)^2(1+t)^2 t^{12} + (1-t^4)(t^3 + t^4)(t^{11} - t^{12})$$

$$(q_6)^2 = (1-t^4)^2(1-t^5)^2 t^6 + (1-t^4)(1-t^5)(t^{10} - t^{12})$$
$$+ (q_1)^2(1-t^4)(t^3 + t^4) - q_5^2(1-t^5)$$

$$(q_7)^2 = p_1(1-t^5)^2 - p_1^2(1-t^5) + (1-t^3)(1-t^4)(1-t^5)(t^5 - t^{14})$$
$$-(q_5)^2(1-t^4)(1-t^5)$$

$$(q_8)^2 = (1-t^5)(t^{20}-t^{17}) + q_4^2 t^4$$

$$t^4(q_9)^2 = (1-t^4)(t^3+t^4)(t^{14}-t^{13}) + t^6 q_5^2$$

$$(q_{10})^2 = (1-t^4)(1-t^5)(t^{15}-t^{13}) + t^6(q_3)^2$$
$$+ (q_2)^2(1-t^4)(t^3+t^4)(q_9)^2(1-t^5)$$

$$(q_{11})^2 t^4 = p_2(1-t^4)(t^3+t^4)^2 + p_1 t^{11} + t^6 q_3^2$$
$$- t^{13} p_1 - (p_2)^2(1-t^4)(t^3+t^4)$$

$$(q_{13})^2 = p_2(1-t^4)^2(1-t^5)^2 + p_1 q_1^2 + (1-t^3)(1-t^4)(1-t^5)(t^{10}-t^{13})$$
$$- p_2^2(1-t^5)(1-t^4) - q_{11}(1-t^5)$$

$$(q_{14})^2 = p_3(1-t^5)^2 + (1-t^2)(1-t^3)(1-t^4)(1-t^5)(t^5-t^{17})$$
$$- p_3^2(1-t^5) - q_9^2(1-t^3)(1-t^4)(1-t^5) - q_{11}^2(1-t^4)(1-t^5).$$

Thus all of the unknown parameters have been expressed by t, e.g. we obtain the explicit form of BGR for spin 5/2. Substituting these expressions into the defination of link polynomial or by Witten's version the polynomial can only be determined by the eigenvalues and the framing factor. No explicit BGR is needed. However for constructing the R(x)—matrix it is necessary.

(III). The case of spin 3.

Under the case the eigenvalues are

$$1, -t^6, t^{11}, -t^{15}, t^{18}, -t^{20}, t^{21}. \qquad (3.1)$$

The block diagonal sub—matrix are as following.

$$A^{(6)} = 1, \qquad A^{(5)} = \begin{bmatrix} 0 & -t^3 \\ -t^3 & 1-t^6 \end{bmatrix}$$

$$A^{(4)} = \begin{bmatrix} 0 & 0 & t^6 \\ 0 & t^5 & \bar{q}_1 \\ t^6 & \bar{q}_1 & (1-t^5)(1-t^6) \end{bmatrix}$$

$$A^{(3)} = \begin{bmatrix} 0 & 0 & 0 & -t^9 \\ 0 & 0 & -t^7 & \bar{q}_2 \\ 0 & -t^7 & (1-t^5)(1+t)t^4 & \bar{q}_3 \\ -t^9 & \bar{q}_2 & \bar{q}_3 & (1-t^4)(1-t^5)(1-t^6) \end{bmatrix}$$

$$A^{(2)} = \begin{bmatrix} 0 & 0 & 0 & 0 & t^{12} \\ 0 & 0 & 0 & t^9 & \bar{q}_4 \\ 0 & 0 & t^8 & \bar{P}_5 & \bar{q}_6 \\ 0 & t^9 & \bar{q}_5 & \bar{P}_1 & \bar{q}_7 \\ t^{12} & \bar{q}_4 & \bar{q}_6 & \bar{q}_7 & (1-t^3)(1-t^4)(1-t^5)(1-t^6) \end{bmatrix}$$

$$A^{(1)} = \begin{bmatrix} 0 & 0 & 0 & 0 & 0 & -t^{15} \\ 0 & 0 & 0 & 0 & t^{11} & \bar{q}_8 \\ 0 & 0 & 0 & -t^9 & \bar{q}_9 & \bar{q}_{10} \\ 0 & 0 & -t^9 & \bar{P}_2 & \bar{q}_{11} & \bar{q}_{12} \\ 0 & -t^{11} & \bar{q}_9 & \bar{q}_{11} & \bar{P}_3 & \bar{q}_{13} \\ -t^{15} & \bar{q}_8 & \bar{q}_{10} & \bar{q}_{12} & \bar{q}_{13} & (1-t^2)(1-t^3)(1-t^4)(1-t^5)(1-t^6) \end{bmatrix}$$

$$A^{(0)} = \begin{bmatrix} 0 & 0 & 0 & 0 & 0 & 0 & t^{18} \\ 0 & 0 & 0 & 0 & 0 & t^{13} & \bar{q}_{14} \\ 0 & 0 & 0 & 0 & t^{10} & \bar{q}_{15} & \bar{q}_{16} \\ 0 & 0 & 0 & t^9 & \bar{q}_{17} & \bar{q}_{18} & \bar{q}_{19} \\ 0 & 0 & t^{10} & \bar{q}_{17} & W & \bar{q}_{20} & \bar{q}_{21} \\ 0 & t^{13} & \bar{q}_{15} & \bar{q}_{18} & \bar{q}_{20} & V & \bar{q}_{22} \\ t^{18} & \bar{q}_{14} & \bar{q}_{16} & \bar{q}_{19} & \bar{q}_{21} & \bar{q}_{22} & \prod_{1}^{6}(1-t^m) \end{bmatrix}$$

Following the same strategy as that in the section(II) and making use of the traces, Markov trace and extended Yang–Baxter state expansions we derive the following results.

$$\bar{P}_1 = 1 - t^6 - t^8 + t^{11} - t^{15} + t^{18} - \prod_{m=3}^{6}(1-t^m)$$

$$\bar{P}_2 = t^{-2} - t^{10} - \bar{P}_1 t^{-1} - t^{-2}(1-t^4)(1-t^5)(1-t^6)$$

$$\bar{P}_3 = 1 - t^6 + t^{11} - t^{15} + t^{18} - t^{20} - \bar{P}_2 - \prod_{m=3}^{6}(1-t^m)$$

$$w = t^{-2} - t^{10} - t\bar{P}_2 - t^{-1}\bar{P}_3 - t^{-2}\prod_{m=1}^{6}(1-t^m)$$

$$v = 1 - t^6 - t^9 + t^{11} - t^{15} + t^{18} - t^{20} + t^{21} - w - \prod_{m=1}^{6}(1-t^m)$$

$$(\bar{q}_1)^2 = t^5(1-t^6)^2 + (1-t^6)(t^6 - t^{10})$$

$$(\bar{q}_2)^2 = t^5(\bar{q}_1)^2 + (1-t^6)(t^{12} - t^{14})$$

$$(\bar{q}_3)^2 = (1-t^5)(t^4+t^5)(1-t^6)^2 + (1-t^5)(1-t^6)(t^6-t^{14})$$
$$-(1-t^5)^2(t^4+t^5)^2(1-t^6)$$

$$(\bar{q}_4)^2 = t^5(\bar{q}_2)^2$$

$$(\bar{q}_5)^2 = t^3(1-t^5)^2(t^4+t^5)^2 + (1-t^5)(t^4+t^5)(t^9-t^{11})$$

$$(\bar{q}_6)^2 = t^8(1-t^5)^2(1-t^6)^2 + (1-t^5)(t^4+t^5)\bar{q}_1^2$$
$$+(1-t^5)(1-t^6)(t^{12}-t^{16}) - (\bar{q}_5)^2(1-t^6)$$

$$(\bar{q}_7)^2 = \bar{p}_1(1-t^6)^2(\bar{p}_1)^2(1-t^6) + (t^6-t^{18})(1-t^4)(1-t^5)(1-t^6)$$
$$-(\bar{q}_5)^2(1-t^5)(1-t^6)$$

$$(\bar{q}_8)^2 = t^5(\bar{q}_4)^2 + (1-t^6)(t^{24}-t^{22})$$

$$(\bar{q}_9)^2 = t^3(\bar{q}_5)^2$$

$$(\bar{q}_{10})^2 = t^8(\bar{q}_3)^2 + (\bar{q}_2)^2(1-t^5)(t^4+t^5) - (\bar{q}_9)^2(1-t^6)$$

$$t^5(\bar{q}_{11})^2 = \bar{p}_2(1-t^5)^2(t^4+t^5)^2 + \bar{p}_1 t^{14} + t^9(\bar{q}_3)^2 - \bar{p}_1 t^{18}$$
$$-(\bar{p}_2)^2(1-t^5)(t^4+t^5)$$

$$(\bar{q}_{12})^2 = \bar{p}_2(1-t^5)^2(1-t^6)^2 + (\bar{q}_1)^2 \bar{p}_1 + (1-t^4)(1-t^5)(1-t^6)(t^{12}-t^{18})$$
$$-(\bar{p}_2)(1-t^5)(1-t^6) - (\bar{q}_{11})^2(1-t^6)$$

$$(\bar{q}_{13})^2 = (\bar{p}_3)(1-t^6)^2 + (t^6-t^{22})\prod_{m=3}^{6}(1-t^m) - (\bar{q}_9)^2 \prod_{m=4}^{6}(1-t^m)$$
$$-(\bar{q}_{11})^2(1-t^5)(1-t^6) - (\bar{p}_3)^2(1-t^6)^2$$

$$(\bar{q}_{14})^2 = t^5(\bar{q}_8)^2 + (1-t^6)(t^{30}-t^{26})$$

$$(\bar{q}_{15})^2 t^5 = (\bar{q}_9)^2 t^8 + (1-t^5)(t^4+t^5)(t^{22}-t^{20})$$

$$(\bar{q}_{16})^2 = t^8(\bar{q}_6)^2(1-t^5)(1+t)t^4(\bar{q}_4)^2 + (1-t^5)(1-t^6)(t^{24}-t^{20})$$
$$-(\bar{q}_{15})^2(1-t^6)$$

$$(\bar{q}_{17})^2 = t(\bar{p}_2)^2$$

$$(\bar{q}_{18})^2 t^5 = t^9 (\bar{p}_1)^2 (\bar{q}_5)^2 (\bar{p}_2) - (\bar{q}_{17})^2 (1-t^5)(t^4+t^5)$$

$$(\bar{q}_{19})^2 = t^9 \prod_{m=4}^{6}(1-t^m)^2 + (\bar{q}_3)^2 \bar{p}_2 + (\bar{q}_2)^2 \bar{p}_1$$
$$-(\bar{q}_{17})^2(1-t^5)(1-t^6)-(\bar{q}_{18})^2(1-t^6)$$

$$(\bar{q}_{20})^2 t^5 = (\bar{q}_3)^2 \bar{p}_2 + (1-t^5)^2(t^4+t^5)^2 w + \bar{p}_3(t^{14}-t^{20})$$
$$-\bar{p}_1(\bar{q}_{17})^2 - w^2(1-t^5)(t^4+t^5)$$

$$(\bar{q}_{21})^2 = (1-t^5)^2(1-t^6)^2 w + (\bar{q}_1)^2 \bar{p}_3 + (t^{12}-t^{20})\prod_{m=3}^{6}(1-t^m)$$
$$-(\bar{q}_{20})^2(1-t^6)-w^2(1-t^5)(1-t^6)-(\bar{q}_{17})^2 \prod_{m=4}^{6}(1-t^m)$$

$$(\bar{q}_{22})^2 = (1-t^6)^2 v + (t^6-t^{26})\prod_{m=2}^{6}(1-t^m) - v^2(1-t^6)$$
$$-(\bar{q}_{20})^2(1-t^5)(1-t^6)-(\bar{q}_{18})^2 \prod_{m=4}^{6}(1-t^m)$$
$$-(\bar{q}_{15})^2 \prod_{m=3}^{6}(1-t^m)$$

(IV). The x–deformation for SU(2) Algebras

As was shown by Jimbo that a q–difference analogue of the universal enveloping algebra U(g) can be introduced.Its structure and representations are studied in the case of SL(2).Now let us see how to make the x–deformations for a given BGR in the case of SU(2) spin models so that the resultant ones give the same results as the same as those obtain by Jimbo based on Q.G. in Ref(17).

For given BGR with eigenvalues λ_i (i=0,1,...,N) the constructions of the quantum R(x)–matrix can be made by the simple substitutions

$$\lambda_i \rightarrow P_i = \prod_{m=1}^{i}(\frac{\lambda_{m-1}}{\lambda_m}x+1)\prod_{m=i+1}^{N}(x+\frac{\lambda_{m-1}}{\lambda_m}) \qquad (4.1)$$

(i = 0,1,2,...,N)

A BGR expressed by S has the very simple eigenvalues denoted by $\lambda_0, \lambda_1, ..., \lambda_N$. Introducing the simular thansformation matrix L we are able to express S in the form

$$S = L \begin{bmatrix} \lambda_0 & & & \\ & \lambda_1 & & \\ & & \lambda_2 & \\ & & & \ddots \\ & & & & \lambda_N \end{bmatrix} L^{-1} = \sum_{i=1}^{N} \lambda_i P_i \qquad (4.2)$$

where P_i is the projector onto the irreducible component R_{2j-i} of $R_j \times R_j$. Where

$$R_j \otimes R_j = \bigoplus_{i=0}^{2j} R_{2j-i} \qquad (4.3)$$

The eigenvalues relarted to the Hilbert space of R_{2j-i} are easy to find by Witten's version (14,21). We know that λ_i are related to the Casimir C_i, i.e.

$$C_i = (1/4)((D_i+1)^2 - 1) \qquad (4.4)$$

where
$$D_i = 2(2j-i)$$

It follows from Witten's approach that

$$\lambda_{i+1}/\lambda_i = -t^{2(2j-i)} \qquad (4.5)$$

Now in order to make the x-deformation we perform the substitution bases on eq.(4.2)

$$\lambda_i \longrightarrow P_i(x)$$

$$S \longrightarrow R(x)$$

by

$$R(x) = \sum_{i=0}^{N} P_i(x) P_i \qquad (4.6)$$

It is easy to find from eq.(4.1) that

$$P_{i+1}(x)/P_i(x) = (\frac{\lambda_i}{\lambda_{i+1}} x + 1)/(x + \frac{\lambda_i}{\lambda_{i+1}})$$
$$= \frac{x + \lambda_{i+1}/\lambda_i}{1 + x \cdot \lambda_{i+1}/\lambda_i} \qquad (4.7)$$

When $i=2\nu$ eq.(4.7) gives the results of Jimbo's paper (17). We thus conclude that our x-deformatoin scheme is correct. Eq.(4.6) provides a pratical method to generate quantum R(x)-matrix. Some particular examples have been checked by the x-deformation scheme, for examples, for j=1 the R(x) derived by using this methods is the same as the exjpression given by Jimbo for the fundamental representation of B_1 in Ref(18).

We are grateful to Prof. C.H.Gu, Prof. H.S.Hu, Prof. M.Wadati, Prof. L. Takhtajan and Prof. F.Smirnov for helpful discussions.

References

(1). Birman J.S.   Ann Math Studies No.82,(1976) Princeton University Press.
(2). Birman J.S. and Wenzel,H. "Braids ,Link Polynomials and a new algebra",to appear in Trans A.M.S.
(3). Frohlich,J.   "Statistics of fields ,the Yang–Baxter equatoins and the theory of Knots and links,ETH–Honggerberg preprint.
(4). Jones,V.F.R.   Bull Amer Math Soc.12,103(1985)   Ann of Math 126,335(1987)
(5). Jones,V.F.R.   "On knot invariants related to some statistical mechanics models ",jpreprint 1988.
(6). Kauffman L.H.   "State model for knot polynomials— An introduction",preprint. Ann of Math Studies # 115.
(7). Kauffman,L.H.   "Knot theory and Applications",talk at UT Austin,March 1988
(8). Kauffman,L.H.   "Braid Group,Knot Theory and Statistical Mechanics" P.27,edited C.N.Yang,World Scientific,1989.
(9). Akutsu,Y. and Wadati,M.
    J.Phys soc Jan,56,839,3039(1987)
    "Braid Group,Knot Theory and Statistical Mechanics",p 151, edited by C.N.Yang,World Scientific,1989.
(10). Turaev,V.G.   Invent Math 92,527(1988)
(11). Reshetikhin,N.   "Quantized universal enveloping algebras,the Yang–Baxter equation and invariants of links I,II" LOMT preprint E–4–87
(12). kohno,T.   "Quantum University enveloping algebras and monodromy of braid group" Nagoya preprint 1989.
    Nankai lectures on Math Physics,Tianjin ,1988.
(13). Takhtajan,L.   Nankai lectures on Quantum Group,Tianjin,1989.
(14). Witten,E.   Comm Math Phys,121,135(1989)
(15). Ge,M.L.,Wang,L.Y.,Xue,K.,and Wu,Y.S.,
    Inter J.Mod Phys 4,3351(1989).
(16). Lee,H.C.,Ge,M.L.,Couture,M.and Wu,Y.S.
    Inter J.Mod Phys 4,2333(1989)
(17). Jimbo,M.   Lett Math Phys 10,63(1985).
(18). Jimbo,M.   Comm Math Phys 102,537(1986).
(19). Jimbo,M.   Lett Math Phys 11,253(1986)
(20). Moore,G. and Seiberg,N.
    Phys Lett B212,451(1988),IAS preprint,HEP–88/39.
(21). Kniznik,V.G. and Zamolodchikov,A.B.
    Nucl Phys B2147,83 (1984).
(22). Ge,M.L.Wang L.Y. and Xue K,
    "Extended State Expandions and the University of Witten's

Version of Link Polynomial Theory"

Nankai preprint 1989,to appear in Inter.Mod Phys.

(23).Yamagishi,K.,Ge M.L., and Wu Y.S.

UU—HEP—89—1.

"New Hierachies of Knot Polynomials from Topological Chern—Simons gauge Theory".

Part IV

Physical Phenomena

Nonlinear Evolution Equations, Solitons, Chaos and Cellular Automata

M.J. Ablowitz, B.M. Herbst*, and J.M. Keiser

Program in Applied Mathematics, University of Colorado,
Boulder, CO 80309, USA
*Permanent address: Department of Applied Mathematics, University of the Orange Free State, Bloemfontein 9300, South Africa

1 Introduction

Among the remarkable discoveries in applied mathematics during the past twenty or so years, have been the concept of the soliton and completely integrable nonlinear partial differential equations, chaotic phenomena associated with nonintegrable equations, and the understanding of cellular automata. Certainly, these developments have been largely responsible for the current widespread interest in nonlinear phenomena. Although these concepts are rather disparate, nevertheless, there are certain links between the different areas and one typically finds that ideas and techniques developed for one field can be applied to other areas as well. In this article we wish to touch upon two of these – chaos in numerically perturbed versions of soliton equations, and soliton solutions of cellular automata. First some background material is provided.

The soliton is a remarkably stable solution satisfying certain nonlinear evolution equations. In fact, it is well-known that when a faster moving soliton catches up with a slower moving one, the two solitons emerge virtually unchanged from the interaction (apart from a phase shift), with the faster soliton ahead of the slower one.

The paradigm equation for solitons is the Korteweg-de Vries (KdV) equation,

$$u_t + 6uu_x + u_{xxx} = 0. \tag{1}$$

Historically it was the first equation discovered with the soliton solution and many other important properties. There is an extensive literature on this subject and we refer the reader to the monograph and survey papers [1,2,3] for a review of this field. It should be noted that in 1965 Zabusky and Kruskal [4] discovered the soliton property of KdV, from numerical simulations. Shortly thereafter Gardner, Greene, Kruskal and Muira in 1967 [5] found that the Cauchy problem for the KdV equation corresponding to initial values $u(x,0) = f(x)$, vanishing sufficiently rapidly as $|x| \to \infty$, could be linearized by employing methods of direct and inverse scattering. Indeed, Lax showed in 1968 [6] that there is a rather general formulation by which the KdV equation could be viewed as a compatibility condition between two linear operators. He also investigated in considerable detail the

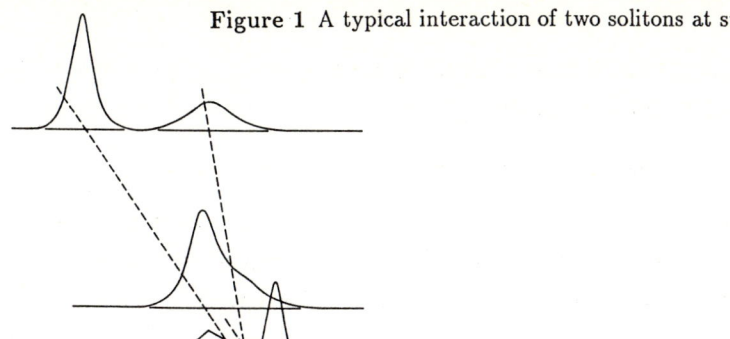

Figure 1 A typical interaction of two solitons at succeeding times.

phase shift

interaction properties of the KdV solitons. In Figure 1 we show a typical two-soliton interaction. Note that the fast soliton is pushed forward and the slower one retarded during the interaction.

Subsequent studies have established the wide ranging significance and occurrence of soliton solutions. In 1972 Zakharov and Shabat [7] found that the physically significant cubic nonlinear Schrödinger equation,

$$iu_t + u_{xx} + Qu^2 u^* = 0, \qquad (2)$$

(here $i^2 = -1$, Q is a real parameter and u^* is the complex conjugate of u), admitted soliton solutions and that the Cauchy problem (for decaying data on $|x| \to \infty$) could be linearized by inverse scattering methods. In fact, Ablowitz, Kaup, Newell and Segur [8] demonstrated that these ideas apply to a class of nonlinear evolution equations, including the physically interesting sine-Gordon,

$$u_{xt} = \sin u \qquad (3)$$

and modified KdV (mKdV) equation,

$$u_t + 6u^2 u_x + u_{xxx} = 0. \qquad (4)$$

They termed the method of solution the Inverse Scattering Transform (IST) in analogy with the linear technique of Fourier Transforms.

Not only does the method apply to nonlinear partial differential equations, but it also applies to discrete equations. Examples include (a) the Toda lattice [9]

$$\ddot{u}_n = e^{-(u_n - u_{n-1})} - e^{-(u_{n+1} - u_n)} \tag{5}$$

which in a suitable continuous limit tends to the KdV equation, and (b) a differential-difference nonlinear Schrödinger equation [10]

$$i\dot{U}_n + \frac{1}{h^2}(U_{n+1} + U_{n-1} - 2U_n) + \tfrac{1}{2}QU_n U_n^*(U_{n+1} + U_{n-1}) = 0, \tag{6}$$

which tends to the cubic NLS in the continuous limit, $h \to 0$. This system has been studied extensively from a theoretical point of view, see for example, [10,11]. It is also a viable scheme for numerical computations. In fact, in section 3 we show how its inherent properties induce superior numerical properties as compared with a standard discretization of (2). There are nonlinear partial difference equations (discrete in space and time) which are in the IST class as well [12].

An extensive literature on this subject exists and we refer the reader to the monograph [1] for a review of some of the work in this field.

It should also be mentioned that the ideas of IST can be suitably extended in order to study certain nonlinear singular integro-differential equations such as the Benjamin-Ono equation,

$$u_t + 6uu_x + H(u_{xx}) = 0, \tag{7}$$

where Hu is the Hilbert transform of u:

$$Hu(x) = \frac{1}{\pi} \mathcal{P} \int_{-\infty}^{\infty} \frac{u(\xi)}{\xi - x} d\xi, \tag{8}$$

and $\mathcal{P}$ denotes the Cauchy principal value, and various multidimensional equations, e.g. the Kadomtsev-Petviashvili (KP) equation (a 2+1 dimensional generalization of KdV)

$$(u_t + 6uu_x + u_{xxx})_x = -3\sigma^2 u_{yy}, \tag{9}$$

and the so-called Davey-Stewartson (DS) equations (a 2+1 dimensional generalization of the NLS equation),

$$\begin{aligned} iu_t + \sigma^2 u_{xx} + u_{yy} &= (\phi - |u|^2)u \\ \phi_{xx} - \sigma^2 \phi_{yy} &= \pm 2(|u|^2)_{xx}. \end{aligned} \tag{10}$$

where $\sigma^2 = \pm 1$ in both cases. There is a more general form of (10) given by

$$\begin{aligned} iu_t + \sigma_1 u_{xx} + u_{yy} &= (\phi - |u|^2) \\ \phi_{xx} + \frac{1}{a}\phi_{yy} &= -\frac{b}{a}(|u|^2)_{xx}, \end{aligned} \tag{11}$$

(σ_1, a, b constant) which reduces to (10) by taking $\sigma_1 = -\frac{1}{a} = \sigma^2$, $b = 2/\sigma_2$, $\sigma_1, \sigma_2 = \pm 1$. This generalized DS system is of physical interest (see [13]) and, depending on the choices of σ_1, a, and b, many different types of behavior are exhibited, for example, self focusing singularities, instability of plane solitons, etc. It should be noted that appropriate boundary conditions on $\phi(x,y,t)$ must be supplied, and the choice of the sign of a is critical. This issue is addressed as well as obtaining the general solution to (10) in the strong coupling limit in [14].

Soliton solutions in multi-dimensions have been obtained in isolated cases and is a topic of current research. For instance, weakly decaying solitons have been found for the KP and DS equations (2+1 dimensions), see for example [15,16]. Subsequently, strongly decaying soliton solutions have been found for the DSI equation ($\sigma^2 = 1$) [17] and these have been related to the IST in 2+1 dimensions ([18]). We are not aware of any significant results in 3+1 dimensions.

Whereas the soliton is a remarkable stable entity, chaos, in a sense, represents the opposite end of the spectrum – solutions that are extremely sensitive to small disturbances. The history of chaos began with the seminal paper of Lorenz in 1963 [19] who investigated a three mode Galerkin truncation of the Oberbeck-Boussinesq equations for two dimensional Rayleigh-Bénard convection. Subsequently, chaos has been observed in many different physical situations and nonlinear dynamical systems with chaotic solutions have been studied extensively, see for example [20,21]. One of the few situations where a relatively complete description of the mechanism responsible for the chaos has been given, is in those cases where the chaos results from the homoclinic structure of an underlying unperturbed equation. More recently, it was found that perturbations of some of the soliton equations, most notably, the sine-Gordon (3) and NLS (2) equations, may also lead to temporal chaos, see for example, [22,23,24] and the references therein. In these studies, a homoclinic structure of the underlying unperturbed partial differential equation has been identified as being responsible for the temporal chaos, as well.

It has been observed that certain discretizations of the NLS equation such as the Fourier spectral scheme, may induce irregular temporal behavior at intermediate levels of grid refinement, see, for example, [25,26] and in fact, Shen and Nicholson [26] showed that the irregular behavior in the Fourier spectral scheme is associated with a positive Lyapunov exponent, an indication of chaos. In section 3 we argue, concentrating on a finite difference scheme, that the homoclinic structure associated with the NLS equation is responsible for the destabilization of the standard discretizations of the NLS equation, concentrating on a finite difference scheme. For this reason, the instability is refered to as Numerical Homoclinic Instability (NHI), see [27,28,29]. In marked contrast with the standard discretizations we find that the *integrable* discretization of Ablowitz and Ladik [10] has quasi- periodic solutions at all levels of discretization.

Both solitons and chaos were at first associated with solutions of partial or ordinary differential equations. However they also exist for discrete systems – a cellular automaton may be viewed as an extreme form of discretization where the solution is not only defined at discrete locations in space and time, but the solution itself exists in a small number of states, often only two. In some cases has it been shown that the discretization can be associated with a differential equation, see for example [31,32]. Chaotic solutions have been associated with certain cellular automata (see [33] for a collection of papers discussing various aspects of cellular automata), and it was only recently that Park, Steiglitz and Thurston [34] introduced their Parity Rule Filter Automata (PRFA) admitting *soliton* solutions. Considering the history of soliton solutions of partial differential equations in multi-dimensions, it is remarkable that the PRFA rule may easily be generalized to yield solitary waves and sometimes even soliton interactions in 2+1 *and* 3+1 dimensions. The PRFA rule will be discussed in more detail in section 4 and soliton interactions in 1+1, 2+1 and 3+1 dimensions will be illustrated.

2 Homoclinic structure of the NLS equation

In order to describe the homoclinic structure associated with the NLS equation, it is convenient to study (2) with periodic boundary conditions, $u(x + L, t) = u(x, t)$. The starting point for developing the homoclinic structure is to find a suitable 'fixed point', in this case given by

$$\tilde{u}(x,t) = a \exp(iQ|a|^2 t), \tag{12}$$

where a is a complex constant. Note that it is possible to get rid of the time dependence by a simple transformation, justifying our use of the term 'fixed point'. Next we investigate the stability of the fixed point by considering small perturbations of the form (see, for example, [35])

$$u(x,t) = \tilde{u}(x,t)(1 + \epsilon(x,t)) \tag{13}$$

where $|\epsilon| \ll 1$. Substituting (13) into (2) and keeping linear terms in ϵ leads to

$$\epsilon_t = i\epsilon_{xx} + iQ|a|^2(\epsilon + \epsilon^*).$$

Assuming

$$\begin{aligned}\epsilon(x,t) &= \hat{\epsilon}_{-n}(t)\exp(-i\mu_n x) + \hat{\epsilon}_n(t)\exp(i\mu_n x) \\ &= (\hat{\epsilon}_{-n}(t) + \hat{\epsilon}_n(t))\cos\mu_n x + i(\hat{\epsilon}_{-n}(t) - \hat{\epsilon}_n(t))\sin\mu_n x \end{aligned} \tag{14}$$

where $\mu_n = 2\pi n/L$, it follows easily that the growth rate, σ_n, of the n-th mode, $\hat{\epsilon}_n$, is given by

$$\sigma_{n\pm} = \pm\mu_n\sqrt{2Q|a|^2 - \mu_n^2}. \qquad (15)$$

It follows that the 'fixed point' is hyperbolic, provided, $2\pi^2 < QL^2|a|^2$. The dimension of the (linear, complex) unstable eigenspace is determined by the number of unstable modes and is given by $2n$ where,

$$0 < \mu_n^2 < 2Q|a|^2, \qquad (16)$$

and forms a lower bound for the dimension of the solution space, see, e.g., [36]. Note that the dimension increases with increasing values of $Q|a|^2$.

These results mean that all Fourier modes of arbitrary small pertubations, satisfying (16), will grow away from the 'fixed point' at an exponential rate. It is well-known that these perturbations will eventually decay and return to their initial configuration – a phenomenon known as recurrence [37,38]. More recently numerical and analytical evidence have been given [29,39] that if the perturbations are taken in exactly the right 'direction', the solution may only return to its initial configuration after an infinitely long time, indicating the existence of a homoclinic orbit. In fact, a homoclinic orbit is associated with each unstable mode allowed by (16).

The unstable direction, hence the homoclinic orbit, associated with σ_{n+}, translates into an initial condition,

$$u(x,0) = a + \epsilon(\mu_n^2 + i\sigma_{n+})\sin(\mu_n x + \phi), \qquad (17)$$

in the limit as $\epsilon \to 0$. Note that ϕ is an arbitrary real parameter.

As in the case of finite dimensional dynamical systems, the homoclinic orbit in this case also forms a separatrix between different dynamical behavior. Figure 2a,b shows the different trajectories followed by orbits starting from initial conditions on different sides of the homoclinic orbit. Figure 2a shows $\cos(\mu x)$ appearing periodically whereas interchanges between $\cos(\mu x)$ and $\cos(\mu(x + \frac{1}{2}L))$ take place in Figure 2b.

As alluded to in the introduction, the homoclinic structure is 'broken' under suitable perturbations of the NLS equation and it was shown, see for example [22], that this may lead to a weak form of temporal chaos. In the next section we briefly discuss how a destabilization of a standard difference approximation of the NLS, is related to the homoclinic structure.

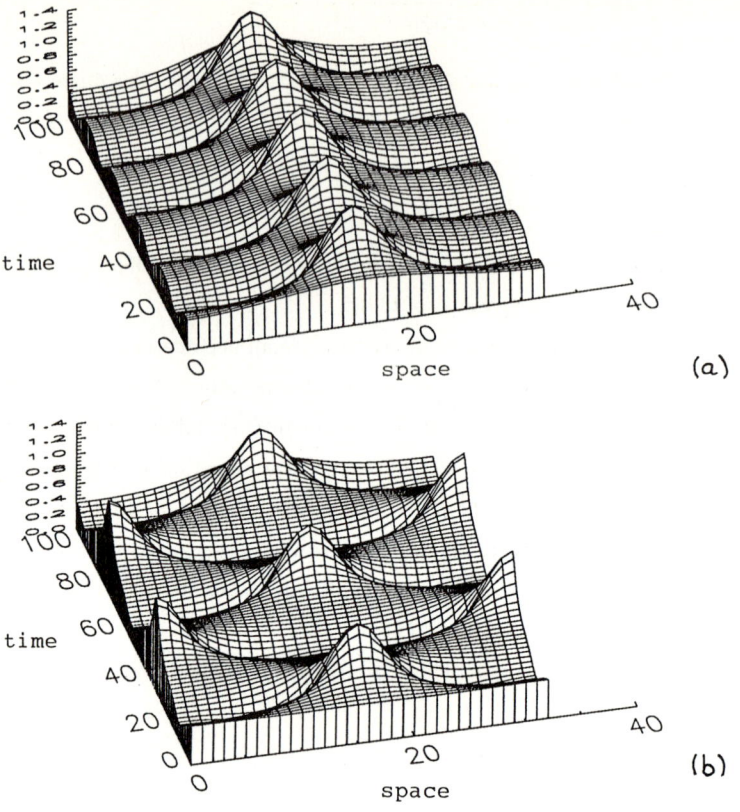

Figure 2 The solution of (2) with $Q = 2$ and initial conditions (a) $u(x,0) = \frac{1}{2} + 0.1 \cos \mu x$, (b) $u(x,0) = \frac{1}{2} + 0.1i \cos \mu x$.

3 Numerical homoclinic instability (NHI).

In this section we wish to point out some of the *qualitative* differences between the integrable discretization, (6), of the NLS equation derived by Ablowitz and Ladik [10] and the standard difference approximation given by

$$i\dot{U}_j + \frac{1}{h^2}(U_{j-1} - 2U_j + U_{j+1}) + Q|U_j|^2 U_j = 0. \tag{18}$$

Periodic boundary conditions are used throughout; given by $U_{j+N} = U_j$.

The two schemes, (6) and (18), are both of second order accuracy and differ only on $O(h^2)$. It is therefore quite remarkable that the two schemes display completely different behavior at intermediate levels of discretizations – *the standard difference scheme, (18), suffers from NHI*, whereas the integrable scheme (6) has quasi-periodic solutions at all levels of discretization, by virtue of its integrability. In order to illustrate some of the dif-

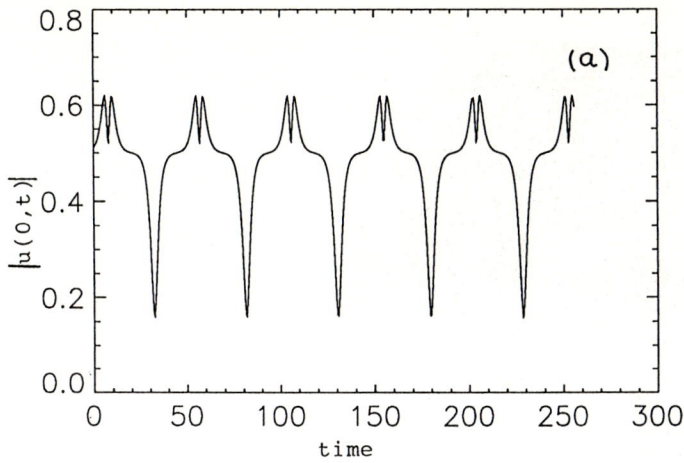

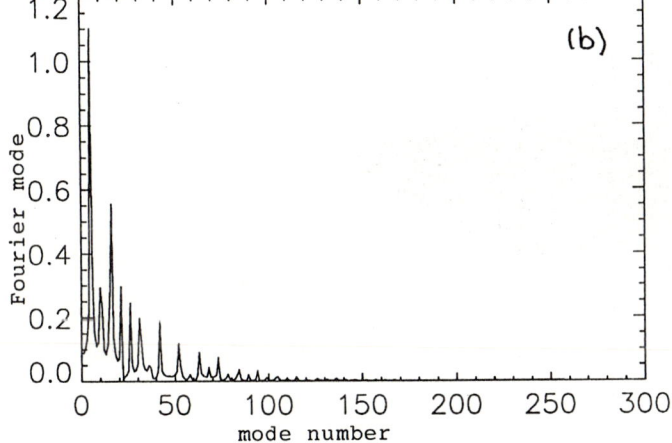

Figure 3 The integrable solution, (6), $Q = 2$, $N = 26$. (a) $|u(0,t)|, 0 \leq t \leq 256$. (b) The Fourier spectrum of the solution shown in (a).

ferences between the two schemes, we attempt to calculate the homoclinic orbit obtained from (17) with $\phi = -\pi/4$, i.e., we solve (6) and (18), using the initial condition,

$$u(x,0) = \tfrac{1}{2} + 0.01(1+i)(\cos\mu x - \sin\mu x),$$

where $\mu = 2\pi/L$, $L = 2\sqrt{2}\pi$ together with $Q = 2$ and $N = 26$. According to (16), these parameter choices imply that only the first mode ($n = 1$) is unstable, showing maximal instability for the given parameter values. The solution obtained from (6) is shown in Figure 3a,b. In Figure 3a, the time evolution of the modulus of the solution at $x = 0$ is shown for 256 time units. The solution is quasi-periodic – a fact that is confirmed by its Fourier spectrum shown in Figure 3b (but note the large number of Fourier frequencies

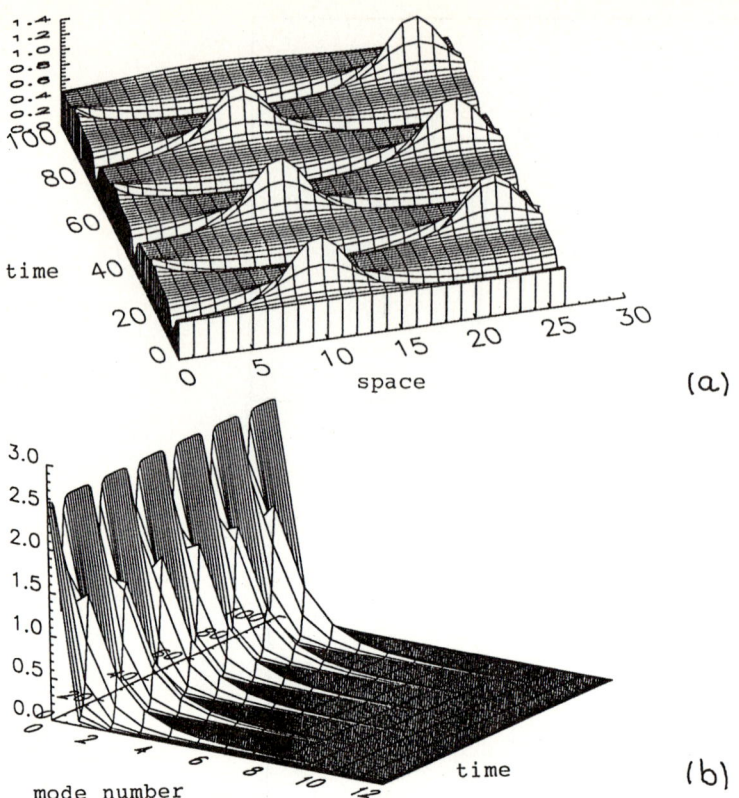

Figure 4 The integrable soltution, (6), $Q = 2$, $N = 26$. (a) $|u(x,t)|$, $-\sqrt{2}\pi \leq x \leq \sqrt{2}\pi$, $0 \leq t \leq 150$. (b) The time evolution of the Fourier decomposition of the spatial structure.

– the time evolution is not simple!). The time evolution of the full spatial structure over the first 150 time units is shown in Figure 4a and we note the recurrent behavior of the solution. Figure 4b show the time evolution of the Fourier decomposition of the spatial structure. Note that the spatial structure is dominated by a small number of modes, i.e. we do not anticipate any particular difficulty in obtaining a good approximation. However, the solution obtained from (18) for the same parameter values, is in sharp contrast, as shown in Figures 5a,b and 6a,b. Figure 5a indicates that the quasi-periodic time evolution of $|u(0,t)|$, $0 \leq t \leq 256$, is lost and this is supported by the Fourier spectrum shown in Figure 5b. Figure 6a shows that this is a consequence of a loss of spatial structure. Although some regularity in the time evolution is clearly retained, the Fourier spectrum, Figure 5b, is quite different from the quasi-periodic spectrum shown in Figure 3b. It is therefore not clear whether the time evolution on a sufficiently long time scale, may tend to be chaotic, for these parameter values.

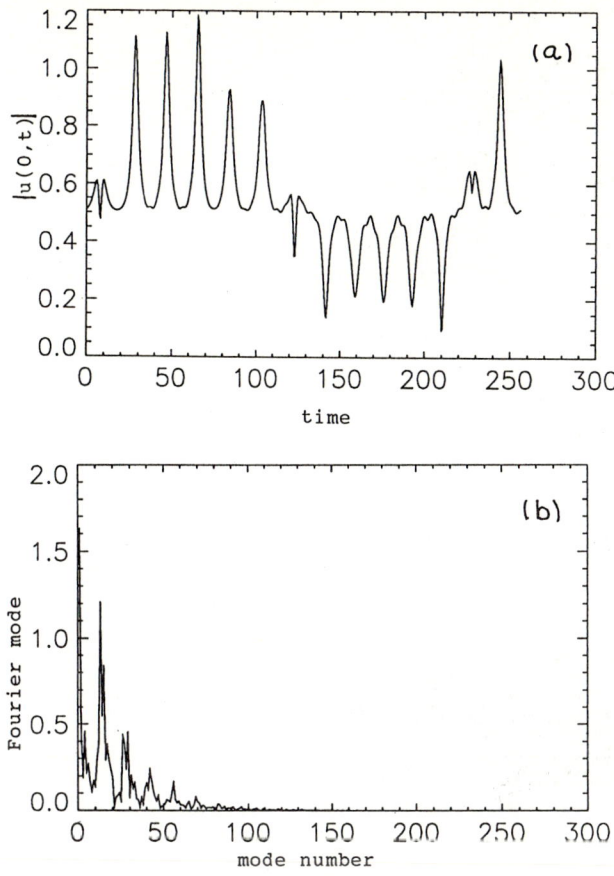

Figure 5 The finite difference solution, (18), $Q = 2$, $N = 26$. (a) and (b) as in Figure 3.

Additional information is provided by Figure 6b, showing the time evolution of the Fourier decomposition of the spatial structure shown in Figure 6a. Again the spatial structure is dominated by the first two modes, the constant mode and $\cos(\mu_1 x + \phi)$, where $\mu_1 = 2\pi/L$ and ϕ denotes the phase. In fact, a comparison of Figures 4b and 6b shows little difference, indicating that it is the same basic spatial structure appearing each time in Figure 6a, with a small phase shift at each appearance. The phase shift becomes more irregular as the spatial resolution (i.e., N) is reduced. Also note that the highest Fourier modes are already, apparently, negligible small. Judging from the Fourier spectrum, Figure 6, one might therefore be tempted to conclude that an adequate spatial resolution has been obtained. This is clearly not the case. In fact, even for $N = 64$, marked differences with the integrable scheme, (6), is still observed. (There is no *qualitative* difference between solutions of (6) with $N = 26$ and $N = 64$.)

175

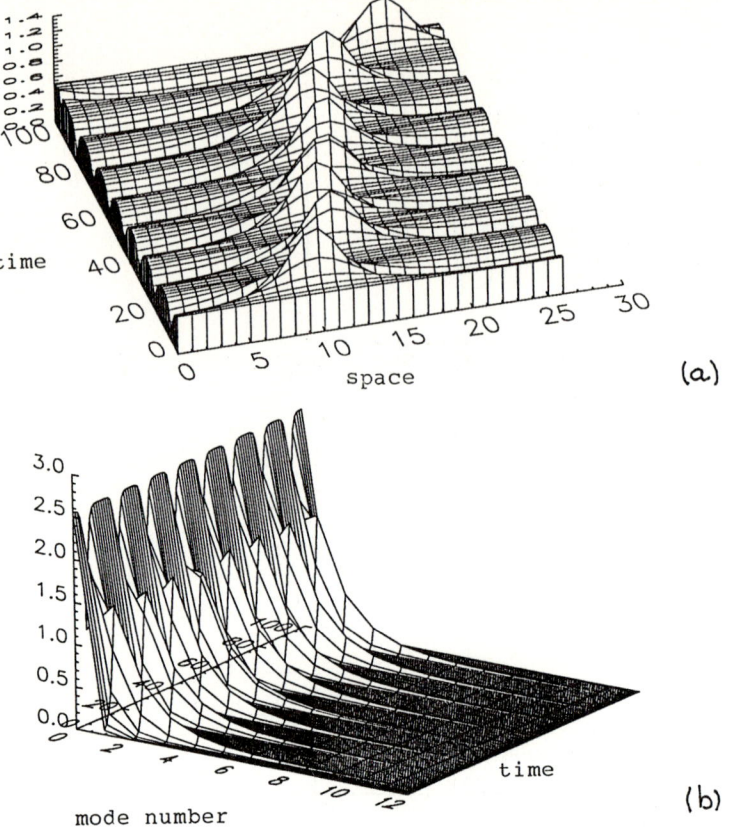

Figure 6 The finite diference solution, (18), $Q = 2$, $N = 26$. (a) and (b) as in Figure 4.

Stronger evidence of a weak form of chaos is obtained by changing the parameter values (see also [27,28,29]). For $Q = 4$ and the rest of the parameter values as before, it follows from (16) that the second mode ($n = 2$) is on the edge of the instability region. It is known, see for example [30], that the finite difference discretization, (18), moves the second mode into the instability region. In addition, the growth rate of the first mode is increased from $\sigma_1 = \frac{1}{2}$ for $Q = 2$ to $\sigma_1 = \frac{1}{2}\sqrt{3}$ for $Q = 4$. Therefore, the problem becomes significantly more unstable when Q is increased from 2 to 4. For our calculations with $Q = 4$ we use the initial condition,

$$u(x,0) = \tfrac{1}{2}(1 + 0.1 \cos \mu x)$$

and first force spatial symmetry on the numerical solution by solving the problem over only half the spatial interval, extending the solution symmetrically over the other half. (Otherwise, rounding error may destroy the symmetry, as we show in a moment; see

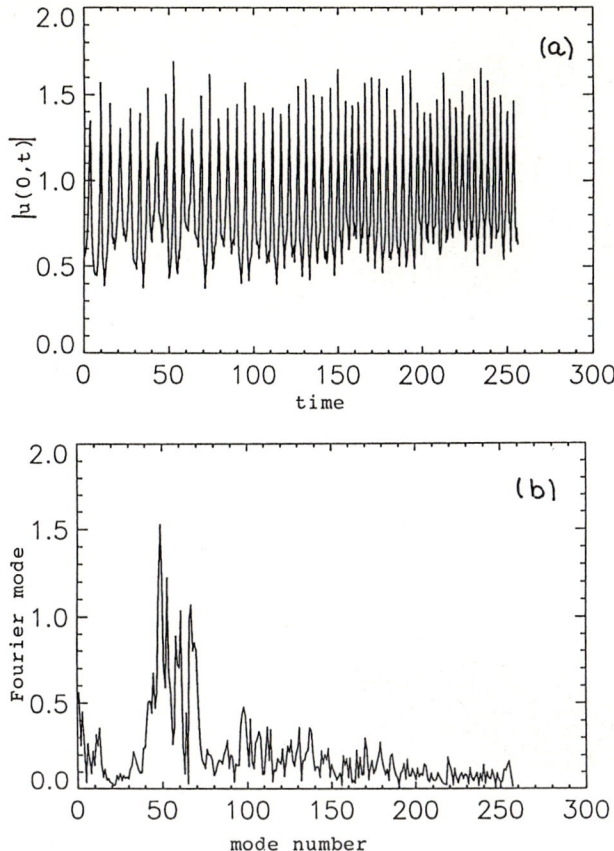

Figure 7 The finite difference solution, (18), $Q = 4$, $N = 32$. (a) and (b) as in Figure 3.

also [29]). This means that the phase is not allowed to vary freely as in the previous experiments. Nevertheless, the time evolution for (18) is again irregular as shown in Figure 7a,b for $N = 32$. Note the broad band Fourier spectrum of the time evolution in Figure 7b, indicating a weak form of chaos.

In order to illustrate some of the differences between the irregular behavior shown in Figures 5 and 7, we investigate in Figure 8a,b the time evolution of the spatial structure of the solution of Figure 7a. (Note that the temporal resolution in Figure 8a is coarser than in Figure 7a.) There is no indication of the phase shift observed in Figure 6. Apparently, the irregular time evolution is caused by an irregular interchange of energy between the two unstable modes ($n = 1$ and $n = 2$). There exist numerical evidence that this may involve an interchange between the homoclinic orbits associated with each of the unstable modes.

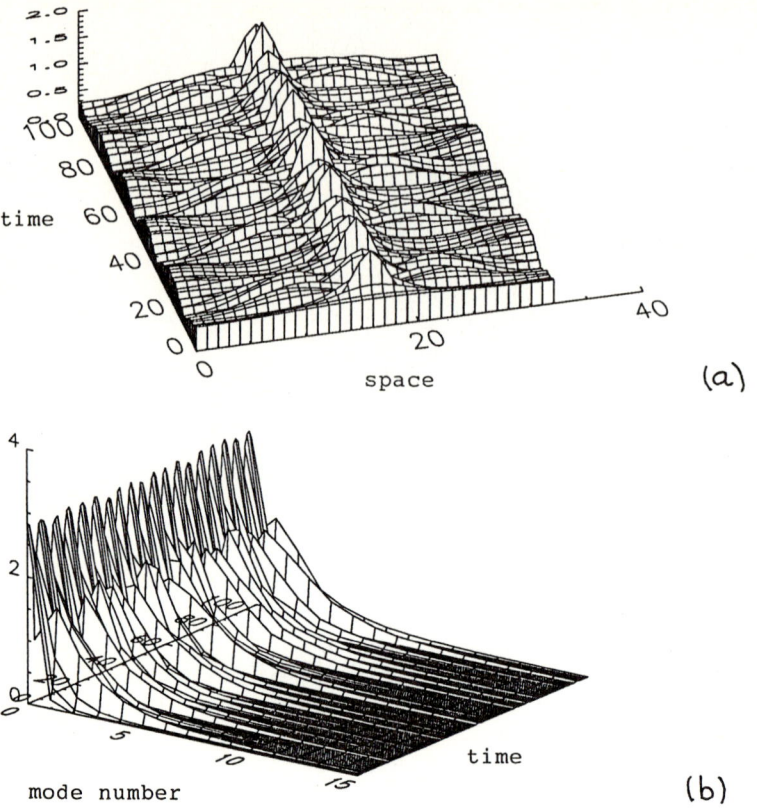

Figure 8 The finite difference solution, (18), $Q = 4$, $N = 32$. (a) $|u(x,t)|$, $-\sqrt{2}\pi \leq x \leq \sqrt{2}\pi$, $0 \leq t \leq 100$. (b) The time evolution of the Fourier decomposition of the spatial structure shown in (a).

Before proceeding to the integrable scheme, we again solve the problem with the standard finite difference method, (18). However, this time we do not impose spatial symmetry on the problem as before. Theoretically this should not give results any different from our previous calculations (Figures 7 and 8), since the standard finite difference scheme naturally respects the spatial symmetry. However, Figure 9 shows that the time evolution differs completely from that of Figure 7. The reason is that rounding error quickly destroys the spatial symmetry, as explained in [29] and observed, for instance, in Figure 6a. In addition to the mechanisms responsible for the instability shown in Figures 7 and 8, in Figure 9 the instability is enhanced by a loss spatial symmetry. It is quite remarkable that this effect of rounding error disappears as the grid is refined and the standard finite difference solution converges to the analytical solution.

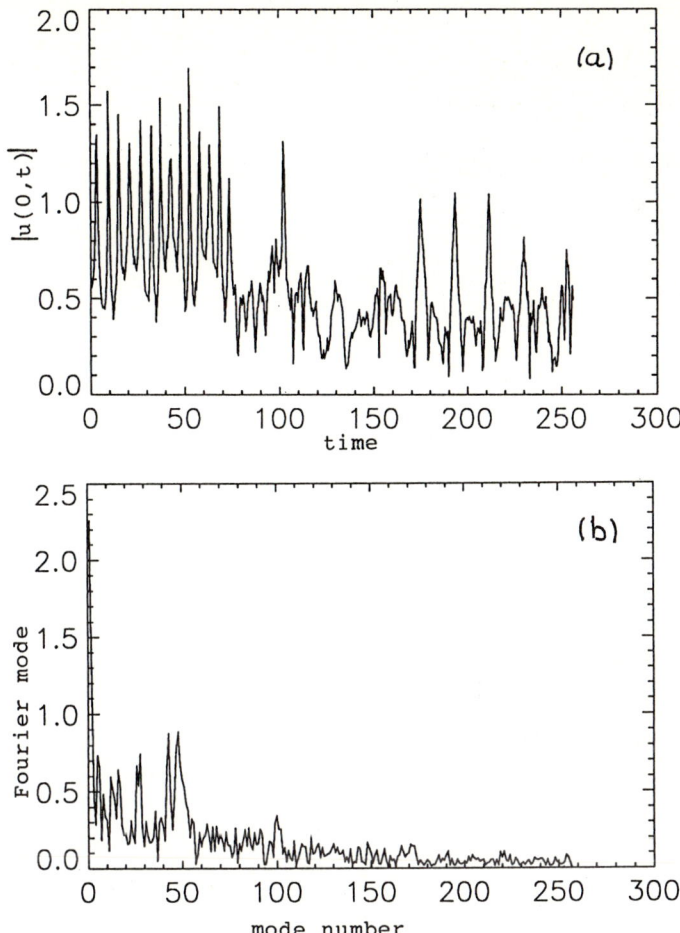

Figure 9 Calculating the full problem. Parameters as in Figure 8.

The solution of the same problem ($N = 32$) obtained from the integrable scheme, (6), is shown in Figure 10a,b. No instabilities have been encountered with this scheme. The time evolution is quasi-periodic at all levels of grid refinement and no loss of spatial symmetry has been observed over the time intervals investigated.

The instability described above has also been observed in connection with Fourier spectral and pseudo-spectral methods. As we have observed, it can also be triggered by rounding error and in all cases investigated, the instability disappears as the spatial resolution is refined [27,28,29], as it should. A detailed analytical description of this numerical homoclinic instability has not yet been given, but preliminary analysis indicates that it is closely related to the homoclinic structure of the NLS equation. The standard difference

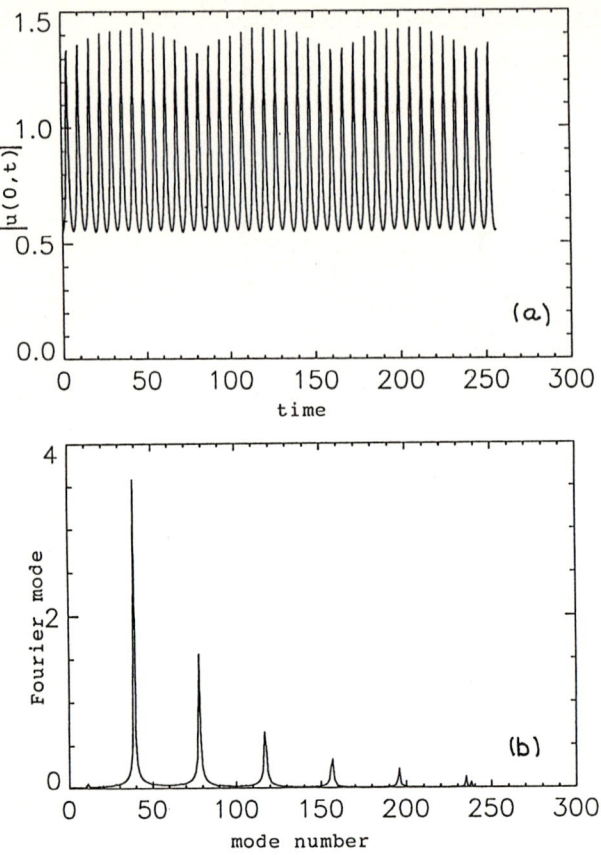

Figure 10 The integrable solution, (6), $Q = 4$, $N = 32$. (a) and (b) as in Figure 3.

approximation, (18), is not integrable, but may be viewed as a perturbation of the integrable problem. The perturbation apparently allows the phase space to be 'opened up' in the sense that the solution is free to cross the homoclinic orbits of the underlying integrable equation. As the spatial resolution is refined and the numerical scheme converges, the homoclinic structure is restored and the anomalous behavior disappears.

4 Cellular automata and solitons

In this section we turn to the question of finding soliton interactions in Cellular Automata (CA). It turns out that there is a class of CA that admits soliton interactions, bearing a rather close resemblence to KdV solitons, cf., Figure 1.

The prototype model which shall be discussed here is called the Parity Rule Filter Automata (PRFA), recently introduced by Park, Steiglitz and Thurston [34] in 1986. The PRFA is given by the following implicit rule. Define the sum

$$S(a_i^{t+1}) = \sum_{j=1}^{r} a_{i-j}^{t+1} + \sum_{j=0}^{r} a_{i+j}^{t} \tag{19}$$

on the interval $-\infty < i < \infty$ where a_i^t denotes values at site i, time t with a_i^t taking on values 0,1 only. The new state at level $t+1$ is given by

$$a_i^{t+1} = \begin{cases} 0 & \text{if } S(a_i^{t+1}) \text{ is odd or zero} \\ 1 & \text{if } S(a_i^{t+1}) \text{ is even, nonzero} \end{cases} \tag{20}$$

The computation is carried out by sweeping from left to right, assuming that at the initial time there is a finite number of nonzero sites, a_i^t, and that to the left we always have an infinite number of zeros.

An alternative formulation of the PRFA, referred to as the Fast Rule Theorem (FRT), is given by [40],

$$a_{i-r}^{t+1} = \begin{cases} a_i^t & \text{if } i \notin B(t) \\ \bar{a}_i^t & \text{if } i \in B(t) \end{cases}, \tag{21}$$

where $\bar{a}_i^t = 1 - a_i^t$, i.e., the complement of a_i^t, and $B(t)$ is a certain set described as follows:

(a) The site of the first nonzero value is an element of $B(t)$.

(b) We place all subsequent sites in steps of $r+1$ bits in $B(t)$ so long as there is at least one nonzero value in any of the intervening $r+1$ bits.

(c) If there is at least $r+1$ zeros after a site in $B(t)$, then we go to the next available nonzero value and repeat step (b).

Steps (a) to (c) are repeated until all nonzero values are exhausted. The FRT theorem has been used to study many of the analytical features of the aforementioned CA, including, stability, soliton interactions and periodic particles, see for example, [40,41,42]. Here we restrict ourselves to examples of soliton interactions in 1+1, 2+1 and 3+1 dimensions.

Figure 11a shows two particles, initially completely separated by a large number of zeros, emerging intact after an interaction. Note that a '1' is represented by a black box and a '0' by a blank space. The resemblance with the KdV soliton interaction of Figure 1 is striking. Note, for instance, a similar phase shift. Changing the representation of the CA to plot the 'sum profile' allows the resemblance to be even more pronounced. In Figure 11b the particles of Figure 11a are represented by their sum profiles (note that the CA rule is not changed). The sum profile at level $t+1$ is simply given by $S(a_i^{t+1})$, $-\infty < i < \infty$, in (19). It should be noted that the value of $S(a_i^{t+1})$ is always in the range zero to $2r+1$.

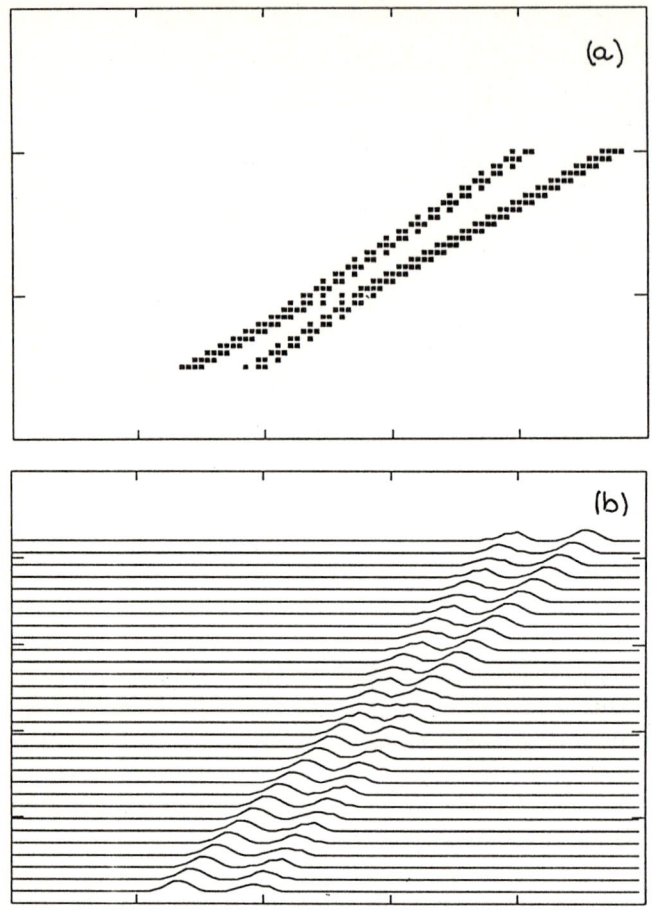

Figure 11 Solitonic interaction of two particles ($r = 3$). (a) The two particles. (b) Representation by their sum profiles.

These ideas can be easily extended to higher dimensions. Consider a two-dimensional grid and let a_{ij}^t denotes the value at site (i,j) at time level t. In order to advance to the next time level the one-dimensional PRFA is applied to each of the successive horizontal levels, sweeping from left to right. Next the PRFA is successively applied to each of the vertical levels, sweeping from top to bottom (these sweeping directions are clearly not important). The two-dimensional automaton is represented by its sum profile by adding the horizontal and vertical sums. The main observation is that we find that in multi-dimensions it is easy to construct localized solitary waves. Indeed, interactions of these waves can be solitonic. Figures 12 and 13 show the interaction of two different pairs of particles in 2+1 dimensions ($r = 3$). We start initially with two localized particles, at the

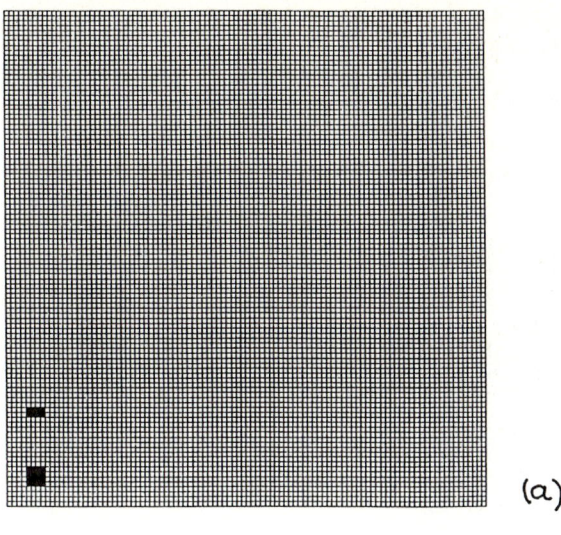

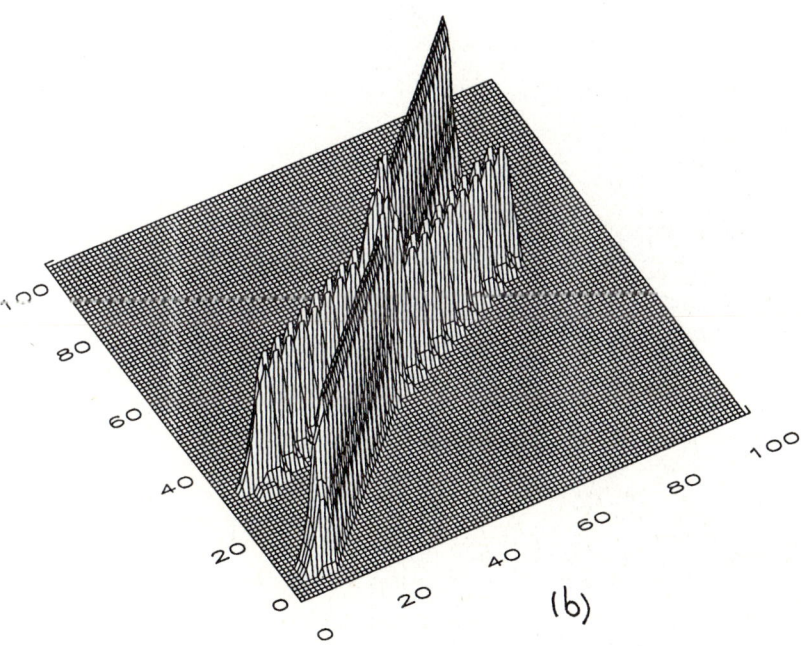

Figure 12 Solitonic interaction in 2+1 dimensions ($r = 3$). (a) The initial positions of the particles. (b) The interaction.

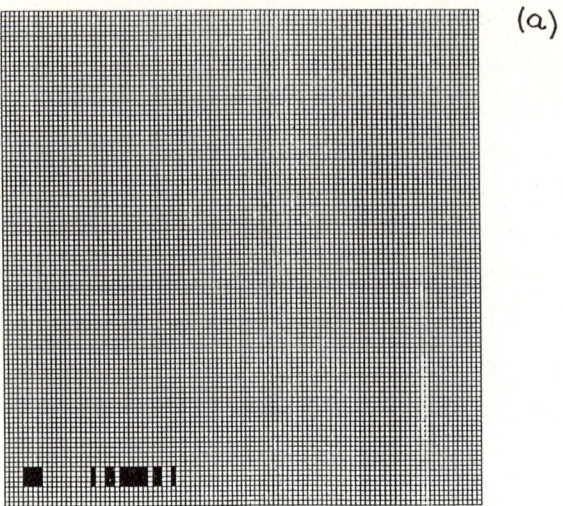

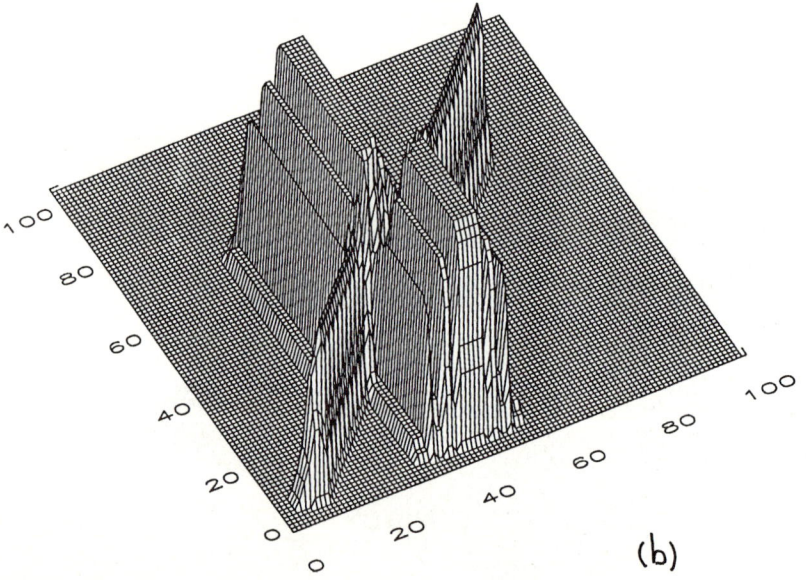

Figure 13 Solitonic interaction in 2+1 dimensions ($r = 3$). Note that the particles differ from Figure 12. (a) and (b) as in Figure 12.

positions shown in Figures 12a and 13a. As the time evolves, the images of the particles at the old positions are frozen in Figures 12b and 13b. Again the particles emerge virtually intact from the interaction, apart from a phase shift; the particles behave like solitons.

These ideas are generalized in a straightforward manner to three spatial dimensions. Now we apply the one dimensional PRFA successively in each of the three spatial directions in order to advance to the next time level. Again the order in which the rule is applied is not important. Figure 14 shows the positions of two particles at different times, in different frames. Now we again use the usual representation of the particle where a dot indicates a '1' and a blank space a '0'.

Finally we note that the way in which the PRFA is generalized here, increases dissipation. By this we mean that the 'energy' decrease is greater in higher dimensions and that a larger number of particles will loose their character under propagation by this rule. Nevertheless, the fact that localized solitary waves exist and the fact that some particles maintain their character, even under interactions as we have seen, merit our interest in these straightforward generalizations. Also, these generalizations are by no means the only possibilities. In fact, 2+1 dimensional soliton interactions have been observed for

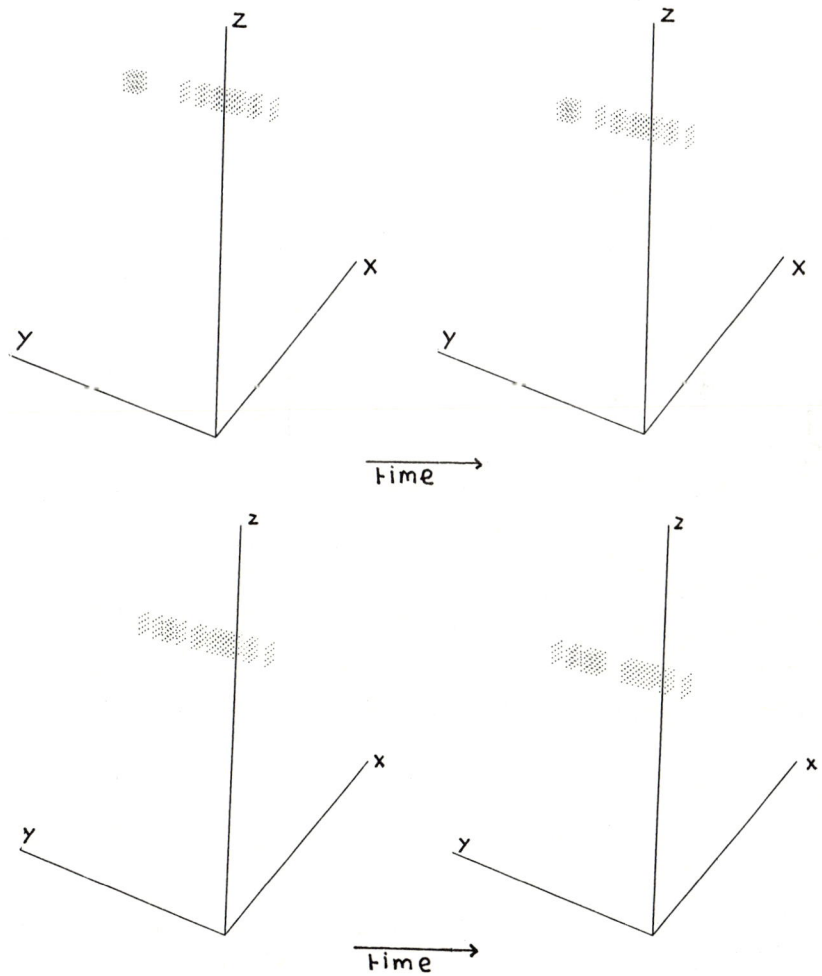

Figure 14 Solitonic interaction in 3+1 dimensions ($r = 3$).

185

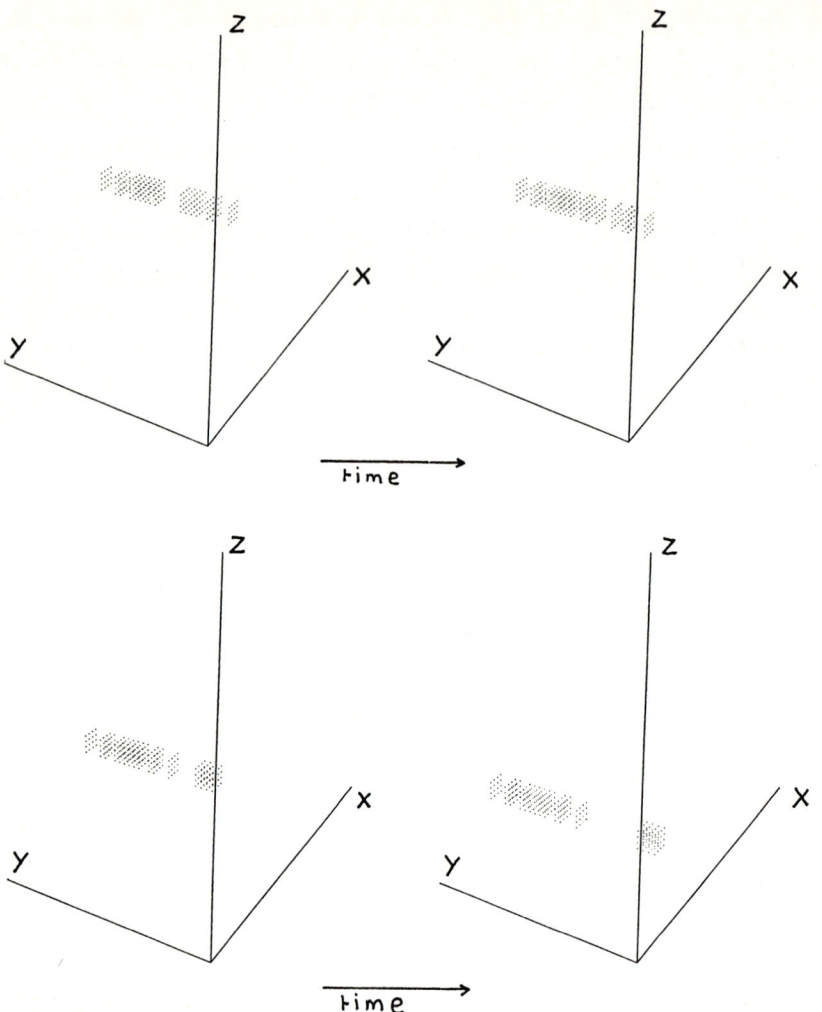

Figure 14 (Continued)

rules based on the FRT mentioned above. Although not quite as straightforward as the ones described above, preliminary studies show that it may be less dissipative.

Acknowledgements This work (MJA) is partially supported by the NSF, Grants No. DMS-8803471, the Office of Naval Research, Grant No. N00014-88-K-0447 and the Air Force Office of Scientific Research, Grant No. AFOSR-88-0073. One of us (BMH) would like to express his appreciation for support of the University of the Orange Free State and colleagues in the department of Applied Mathematics.

†Permanent address: Department of Applied Mathematics, University of the Orange Free State, Bloemfontein 9300, South Africa.

References

[1] M.J. Ablowitz and H. Segur. Solitons and the Inverse Scattering Transform. SIAM, Philadelphia, 1981.

[2] M.J.Ablowitz and A.S. Fokas. In *Nonlinear Phenomena*, ed. K.B. Wolf, p3, Springer (1983).

[3] A.S. Fokas and M.J. Ablowitz. In *Nonlinear Phenomena*, ed. K.B. Wolf, p137, Springer (1983).

[4] N.J. Zabusky and M.D. Kruskal. Phys. Rev. Lett., **15**, p240 (1965).

[5] C.S. Gardner, J.M. Greene, M.D. Kruskal and R.M Muira. Phys. Rev. Lett., **19**, p1095 (1967).

[6] P.D. Lax. Comm. Pure Appl. Math., **21**, p467 (1968).

[7] V.E. Zakharov and A.B. Shabat. Sov. Phys. JETP, **34**, p62 (1972).

[8] M.J. Ablowitz, D.J. Kaup, A.C. Newell and H. Segur. Stud. Appl. Math., **53**, p249 (1974).

[9] H. Flaschka. Phys. Rev. B, **9**, p1924 (1974); H. Flaschka. Prog. Theoret. Phys., **51**, p703 (1974).

[10] M.J. Ablowitz and J.F. Ladik. Stud. Appl. Math., **55**, p213 (1976).

[11] N.N. Bogolyubov and A.K. Prikarpat-skii. Sov. Phys. Dokl., **27**, p113 (1982).

[12] Thiab R. Taha and Mark J. Ablowitz. J. Comput. Phys., **55**, p192 (1984); J. Comput. Phys., **55**, p203 (1984); J. Comput. Phys., **55**, p231 (1984); J. Comput. Phys., **77**, p540 (1988).

[13] M.J. Ablowitz and H. Segur. J. Fluid Mech., **92**, p691 (1979).

[14] M.J. Ablowitz and C. Schultz. Strong coupling limit of certain multi dimensional nonlinear equations. Stud. Appl. Math., **80**, p229 (1989).

[15] A.S. Fokas and M.J. Ablowitz. Stud. Appl. Math., **69**, p211 (1983).

[16] A.S. Fokas and M.J. Ablowitz. J. Math. Phys., **25**, p2494 (1984).

[17] M. Boiti, J.JP. Leon, L. Martina and F. Pempinelli. Solitons in two dimensions. To be published in *Integrable systems and applications*, Eds. M. Balabane, P. Lochak, D.W. McLaughlin, C. Sulem. Lecture Notes, Springer.

[18] A.S. Fokas and P.M. Santini. Solitons in multidimensions. INS report # 106, Clarkson University (1988).

[19] E.N. Lorenz. J. Atmos. Sci., **20**, p130 (1963).

[20] J. Guckenheimer and P. Holmes. Nonlinear oscillations, dynamical systems, and bifurcations of vector fields. Springer, New York (1983).

[21] S. Wiggins. Global bifurcations and chaos. Springer, New York (1988).

[22] A.R. Bishop, D.W. McLaughlin, M.G. Forest and E.A. Overman II. Phys. Lett., **127A**, p335 (1988).

[23] N. Ercolani, M.G. Forest and D.W. McLaughlin. Geometry of the modulational instability Part III: Homoclinic orbits for the periodic sine-Gordon equation. To appear Physica D (1989).

[24] G. Terrones, D.W. McLaughlin E.A. Overman II and A.J. Pearlstein. Stability and bifurcation of spatially coherent solutions of the damped-driven NLS equation. Preprint (1989).

[25] J.A.C. Weideman. Computation of instability and recurrence phenomena in the nonlinear Schrödinger equation. Ph.D. Thesis, University of the O.F.S. (1986).

[26] Mei-Mei Shen and D.R. Nicholson. Phys. Fluids, **30**, p3150 (1987).

[27] B.M. Herbst and M.J. Ablowitz. Phys. Rev. Lett., **62**, p2065 (1989).

[28] B.M. Herbst and M.J. Ablowitz. On numerical chaos in the nonlinear Schrödinger equation in Integrable Systems and Application, M. Balabane, P. Lochak and C Sulem (eds.). Lecture Notes in Physics 342, Springer- Verlag, Berlin (1989).

[29] B.M. Herbst and M.J. Ablowitz. Rounding error and the loss of spatial symmetry associated with discretizations of the nonlinear Schrödinger equation. INS Report # 121, Clarkson University (1989).

[30] J.A.C. Weideman and B.M. Herbst. SIAM J. Sci. Stat. Comput., **8**, p988 (1987).

[31] B.M. Boghosian and C.D. Levermore. Complex Systems, **1**, p17 (1987).

[32] J.L. Lebowitz. E. Orlandi and E. Presutti. Physica, **33D**, p 165 (1988).

[33] S. Wolfram. Theory and applications of cellular automata. World Scientific, Singapore (1986).

[34] J. Park, K. Steiglitz and W. Thurston. Physica, **19D**, p423 (1986).

[35] J.T. Stuart and R.C. DiPrima. Proc. R. Soc. London, **A362**, p27 (1978).

[36] C.R. Doering, J.D. Gibbon, D.D. Holm and B.Nicolaenko. Phys. Rev. Lett., **59**, p2911 (1987) and Nonlinearity, **1**, p279 (1988).

[37] H.C. Yuen and B.M. Lake. Phys. Fluids, **18**, p956 (1975).

[38] H.C. Yuen and W.E. Ferguson. Phys, Fluids, **21**, p1275 (1978).

[39] M.G. Forest. Private communication.

[40] T.S. Papatheodorou, M.J. Ablowitz and Y.G. Saridakis. Stud. Appl. Math., **79**, p173 (1988).

[41] A.S. Fokas, E. Papadopoulou, Y. G. Saridakis and M.J. Ablowitz. Interaction of simple particles in soliton cellular automata. INS report # 97, Clarkson University (1988), to appear Stud. Appl. Math.

[42] J. Keiser. On the computation of periodic particles for cellular automata. Masters Thesis, Clarkson University (1989).

Kadomtsev-Petviashvili Equations in the Description of Water Waves

D. Levi

Dipartimento di Fisica, Università di Roma "La Sapienza",
Piazzale A. Moro 2, I-00185 Roma, Italy
and Istituto Nazionale di Fisica Nucleare, Sezione di Roma, Italy

Here we discuss the nonlinear evolution equations which describe the evolution of the surface of an infinite shallow basin filled with a fluid subject just to the force of gravity. In the approximation of long waves at least two possible situations can arise: i) almost one dimensional waves and ii) waves with equal characteristic length in any direction of the plane.
As is well known in case (i) the surface amplitude evolves according to the Kadomtsev-Petviashvili equation. In case (ii) we find that, at the lowest order in the perturbation parameter, the surface amplitude evolves according to a Korteweg-de Vries equation which depends parametrically on an extra variable and, at the next order, to the second equation in the hierarchy of nonlinear partial differential equations associated to the "time" dependent Schrödinger spectral problem.

I. Introduction

Water covers about two thirds of the Earth surface and understanding its behaviour is very important for mankind. A problem of fundamental interest is to provide an accurate description of the waves on the surface of the water. Naturally there are many kinds of waves; they can be characterized by their wavelength. We are interested mainly to consider the case of long waves as they represent a coherent phenomena on the water surface, generated by the periodic forcing of the tide, in contrast with the short waves, generated by wind stresses, which are chaotic in behaviour. Moreover we shall limit ourselves to waves of small amplitude to exclude the problem of breaking.

The motion of a Newtonian fluid in an external field is well described by Euler equation of motion [1] when, as is the case for most of the physical applications, we can neglect the viscosity of the fluid. Moreover we can very well consider water as an incompressible fluid. Under these hypothesis the equations of motion for the fluid under the action of gravity read:

$$\frac{D\rho}{Dt} = 0 \qquad (1.1a)$$

$$\vec{\nabla} \cdot \vec{v} = 0 \qquad (1.1b)$$

$$\rho \frac{D\vec{v}}{Dt} = \rho \vec{g} - \vec{\nabla} p \qquad (1.1c)$$

where by $D/Dt = \partial/\partial t + \vec{v} \cdot \vec{\nabla}$ we denote the convective derivative, i.e. the time derivative along the fluid trajectory, $\vec{\nabla}$ is the vector derivative defined as $\vec{\nabla} = (\partial/\partial x, \partial/\partial y, \partial/\partial z)$, $\rho(x,y,z,t)$ is the density field, $\vec{v} = (u,v,w)(x,y,z,t)$ the velocity field, $\vec{g} = (0, 0, g)$ the gravity force per mass and $p(x,y,z,t)$ is the pressure field, assumed isotropic at any point of the fluid.

To the Euler equation we have to add the boundary conditions at the free surface $z = \eta(x,y,t)$ and at the bottom $z = -H$ (H being the stationary level of the fluid over the bottom surface, here assumed to be constant; some results in the case $H = H(x,y)$ are considered in refs. [2] and [3]):

$$w = \left.\frac{D\eta}{Dt}\right|_{z=\eta} \qquad (1.2a)$$

$$p = 0 \left.\right|_{z=\eta} \qquad (1.2b)$$

$$w = 0 \left.\right|_{z=-H} \qquad (1.2c)$$

where eq.(1.2a) is obtained by requiring that the surface moves with the fluid, eq.(1.2b) that the pressure is constant on the surface due to the small variations of the atmospheric pressure compared with those involved in the waves on the water's surface and eq.(1.2c) tells us that the bottom is impenetrable.

Density variations appear when one is considering a large depth of fluid; however, as we limit ourselves to the case of shallow water, at first approximation, we can take the density to be constant. The extension to a variable density, corresponding to a stratified fluid have been partly done [4] but, here, for the sake of clarity, we limit ourselves to just one stratum of fluid of constant density. In such a situation eq.(1.1a) is identically satisfied.

Euler equation is a system of first order partial differential equations and thus, to solve it, we must give the appropriate initial conditions; here we consider the case when we look for the behaviour of our system in all space once given the dependent variables on a given surface for all time. This Cauchy problem is often more near to the experimental situation one can encounter in nature or create in a laboratory [5]. A discussion on the results one can obtain considering instead of this Cauchy problem the case when we give initially all dependent variables at a fixed instant of time is given in ref.[6].

It is well known that Euler equation (1.1), together with its boundary conditions (1.2) sustains a great variety of possible flows. The description of a particular type of waves is carried out by the introduction of appropriate rescalings of the dependent and independent variables characteristic of the system at study, which reduces the system (1.1,1.2) to dimensionless form and introduces a small parameter into the theory.

Let us introduce L_x, L_y, the characteristic lengths of the waves in the directions x, y of the plane, H, the characteristic depth of the fluid and N_0 the characteristic amplitude of the waves. Then, as our main interest is to treat waves propagating in one direction, which we can always take to be x, we can introduce the following dimensionless primed variables:

$$x' = x/L_x; \quad y' = y/L_y; \quad z' = z/H; \quad t' = t\, c_0/L_x \qquad (1.3)$$

$$u' = u/c_0; \quad v' = v/c_0; \quad w' = w/c_0; \quad p' = p/(\rho g H); \quad \eta' = \eta/N_0$$

where $c_0 = \sqrt{gH}$ is the characteristic velocity of gravity waves in a fluid of depth H [1].

When rewriting Euler equation (1.1) and its boundary conditions (1.2) in terms of the dimensionless variables (1.3) we find out that they just depend on three physical parameters, which, for later convenience, we write as:

$$\alpha = \left[\frac{H}{L_x}\right]^2 \quad \beta = \left[\frac{L_x}{L_y}\right]^2 \quad \gamma = \left[\frac{N_0}{H}\right] \qquad (1.4)$$

If we want to write the Euler equation in the long-wave approximation then we have to require that α is small, i.e. $\alpha = \alpha_0 \epsilon$ with ϵ a small parameter and α_0 a constant of order 1. If we want to consider the small amplitude approximation then we must require γ to be small, for example $\gamma = \gamma_0 \epsilon^s$ with s a positive constant of order 1. In principle a physical model for our system can be obtained for any value of s. We shall choose s = 1 to get results compatible with the Korteweg de Vries (KdV) equation [7].

For what concerns β we can make two different hypothesis:

(a) L_x and L_y are of a different order of magnitude.

In this case we can always choose $L_y \gg L_x$ and thus set, for example, $L_x \simeq \epsilon^{1/2} L_y$ so that $\beta = \beta_0 \epsilon$. This implies that, if x'-ay'=0 is a line in the normalized (x',y')-plane, then the corresponding line in the physical (x,y)-plane is x-a$\beta^{1/2}$y=0, which, as β is of order ϵ, makes an angle $\epsilon^{1/2}$ times smaller with the y axis then the one the original line did with the y' axis.

This is what is usually denoted as an almost one dimensional approximation as we can treat only interactions between waves which are at a very small angle between themselves. This choice gives rise to the Kadomtsev - Petviashvili (KP) equation [8] or its generalizations in the case of variable bottom [3] i.e for H different from a constant.

(b) L_x and L_y are of the same order of magnitude.

In such a case we can set $L_x \cong L_y$ so that β is of order of unity and, for the sake of simplicity, we can choose it to be equal to one. In such a situation $y'/x' = y/x$ so we can take into consideration interactions between waves propagating at any angle. In this case our system is better represented in cylindrical coordinates $r = (x^2 + y^2)^{1/2}$, $\theta = \arctan(y/x)$. In terms of the radial velocity $v_r = u \cos\theta + v \sin\theta$ and angular velocity $v_\theta = (v \cos\theta - u \sin\theta)/r$ (replacing, for notational convenience, in the final result u for v_r and v for v_θ) the dimensionless variables now read:

$$r' = r / L; \quad z' = z / H; \quad t' = c_0 t / L \quad (1.5)$$

$$u' = u/c_0; \quad v' = Lv/c_0; \quad w' = w/c_0; \quad \eta' = \eta/N_0; \quad p' = p/(\rho g H)$$

where $L_x = L_y = L$. This choice has been studied in a series of papers [4,6] for one fluid or a stratified fluid. In Section 2 we briefly discuss the results one gets in this case at the lowest order in the perturbation expansion and then go to higher order, obtaining an equation which contains corrections to the 2+1 dimensional KdV equation previously obtained. This last equations is genuinely 2+1 dimensional and reduce to the KP equation in the almost one dimensional approximation. Section 3 is devoted to the study this equation.

II. 2+1 dimensional KdV equations

Euler equation (1.1) and the boundary conditions (1.2), in the dimensionless cylindrical variables (1.5) read (dropping the primes on the dimensionless variables):

$$\alpha^{1/2} [u_{,r} + v_{,\theta} + u/r] + w_{,z} = 0 \quad (2.1a)$$

$$\alpha^{1/2} [p_{,r} + u_{,t} + u u_{,r} + v u_{,\theta} - r v^2] + w u_{,z} = 0 \quad (2.1b)$$

$$\alpha^{1/2} [p_{,\theta}/r^2 + v_{,t} + uv_{,r} + v v_{,\theta} + 2uv/r] + w v_{,z} = 0 \quad (2.1c)$$

$$\alpha^{1/2} [w_{,t} + u w_{,r} + v w_{,\theta}] + w w_{,z} + 1 + p_{,z} = 0 \quad (2.1d)$$

$$p = 0 \Big|_{z=\gamma\eta} \quad (2.2a)$$

$$w = \gamma \alpha^{1/2} [\eta_{,t} + u \eta_{,r} + v \eta_{,\theta}] \Big|_{z=\gamma\eta} \quad (2.2b)$$

$$w = 0 \Big|_{z=-1} \quad (2.2c)$$

As eqs.(2.1,2.2) depend explicitly on a small parameter ϵ through α and γ then we can assume that all dependent variables will have a power expansion in ϵ. A consistent way of doing it is through the following ϵ-expansion:

$$u = \sum_{j=0}^{\infty} \epsilon^{j+1} u_j \ ; \quad v = \sum_{j=0}^{\infty} \epsilon^{j+1} v_j \ ; \quad w = \alpha^{1/2} \sum_{j=0}^{\infty} \epsilon^{j+1} w_j \quad (2.3)$$

$$p = \sum_{j=0}^{\infty} \epsilon^{j} p_j \ ; \quad \eta = \sum_{j=0}^{\infty} \epsilon^{j} \eta_j$$

Introducing the expansion (2.3) into (2.1,2.2) and equating terms with the same power of ϵ, we get that, the lowest order approximation of the surface amplitude, η_0, satisfies a wave equation

$$\eta_{0,tt} = \eta_{0,rr} + \eta_{0,r}/r + \eta_{0,\theta\theta}/r^2 \quad (2.4)$$

So, at the lowest level of the ϵ-expansion, the wave amplitude sustain a wave motion in an arbitrary direction ϕ. This result is valid for such interval of time ($t \ll 1/\epsilon$) in which the nonlinear terms of the Euler equation do not play any role. For a longer period of time ($t \simeq 1/\epsilon$) we have to go over to the coordinate system of the wave, so as to be able to take into account also the higher order terms of the ϵ-expansion of (2.1,2.2).

To do so we introduce a new (ϵ-dependent) coordinate system, whose form is dictated by the characteristic variables of the solution of eq.(2.4)

$$R = r \sin(\theta + \phi) \pm t \quad (2.5a)$$

$$\Theta = \theta \quad (2.5b)$$

$$Z = z \quad (2.5c)$$

$$T = \epsilon r \quad (2.5d)$$

In the following we shall choose the sign (+) in eq.(2.5a) as the results in the other case can be obtained just by changing the sign of the velocity field in all formulas.

The solution of the Cauchy problem in T of the resulting equations corresponds to a Cauchy problem in r in the original Euler equation. This corresponds to the case, frequently met in observations in the seas or in experiments in a basin, when we fix the initial condition in a point for a long period of time and then ask ourselves what happens far away from it.

Let us transform eqs.(2.1,2.2) into the new coordinate system (2.5) and carry out the expansion (2.3) of the dependent fields, then the lowest order terms of the ϵ-expansion of eqs.(2.1,2.2) are satisfied provided $\eta_0(R,\Theta,T)$ satisfies the following NEE:

$$\sin(\Theta+\phi)\eta_{0,T} + \frac{1}{T}\cos(\Theta+\phi)\eta_{0,\Theta} + \frac{1}{6}\alpha_0\eta_{0,RRR} + \frac{3}{2}\gamma_0\eta_0\eta_{0,R} = 0 \qquad (2.6)$$

Equation (2.6) is a KdV type equation in 2+1 dimensions; it is easy to see that by introducing the following new variables:

$$\tau = -T\sin(\Theta+\phi)$$

$$\sigma = T\cos(\Theta+\phi) \qquad (2.7)$$

$$\psi = R$$

we reduce eq.(2.6) to the pure KdV equation for $\eta_0(\psi,\sigma,\tau)$:

$$\eta_{0,\tau} - \frac{1}{6}\alpha_0\,\eta_{0,\psi\psi\psi} - \frac{3}{2}\gamma_0\,\eta_0\eta_{0,\psi} = 0 \qquad (2.8)$$

Eq.(2.8), being a pure KdV equation, is completely integrable [9] and its solutions will depend parametrically on σ. This does not imply, however, that it has just trivial solutions in 2+1 dimensions. If the initial condition for eq.(2.8) is σ independent than no angular dependence can appear. However if this is not the case then we can provide interesting solutions in 2+1 dimensions using the Inverse Scattering transform for the KdV equation. As an example of a possible solution of eq.(2.6), let us consider the one soliton solution of eq.(2.8):

$$\eta_0(\psi,\sigma,\tau) = \frac{4\alpha_0}{3\gamma_0}p^2(\sigma)\,\mathrm{sech}^2\left[p(\sigma)\,[\psi+\frac{2}{3}\alpha_0 p^2(\sigma)\tau+\psi_0(\sigma)]\right] \qquad (2.9)$$

where p and ψ_0 are arbitrary functions of σ, to be determined in terms of the initial datum. We shall consider for simplicity just the case of $\phi=\pi/2$, i.e. main propagation along the x-axis; then in terms of the physical variables we have $\tau = -\epsilon x/L$, $\sigma = -\epsilon y/L$ and $\psi = (x + c_0 t)/L$. This implies that, unless for $x = x_0$ we choose an initial condition $\eta(x_0,y,t)$ which is independent from y, i.e. independent from σ, $\eta(x,y,t)$ will depend on all three variables. Let us notice, first of all, that the hidden variable σ is just proportional to y; so we can introduce any y dependence we want into the solutions of the KdV equation and, in particular, we can recover all soliton, periodic or almost periodic solutions of the KP equation. In such a way we have an explanation of the fact that the experimental observations made in basins seemed to be coherent with the solution of KP also in the case when the almost one dimensional approximation was no more valid [5]. As we shall show in the following we can also allow for much more general dependence from y than those consistent with the KP theory.

So here we consider examples of one and two soliton solutions which possess y-dependence which can be physically interesting.

In Figs.(1-3) we plot the one soliton solution (2.9) for (i)$p(\sigma)=1$, $\psi_0(\sigma)=\sin(\sigma)$, (ii) $p(\sigma)=1$, $\psi_0(\sigma)$ is a random function of σ uniformly distributed in the interval (0,1) and (iii) $p(\sigma)=\text{sech}(\sigma)$, $\psi_0(\sigma)=0$.

Fig.1; Plotting of the one soliton solution (2.9) with σ given by eq.(2.11), $\epsilon = 0.1$, $\alpha_0 = 6$, $\gamma_0 = 8$ for $p(\sigma)=1$ and $\psi_0(\sigma)=\sin(\sigma)$. The wave amplitude η is plotted against the x and y dimensionless physical variables at t=0; x varies in the interval [-6,6] while y varies in the interval [-40,40].

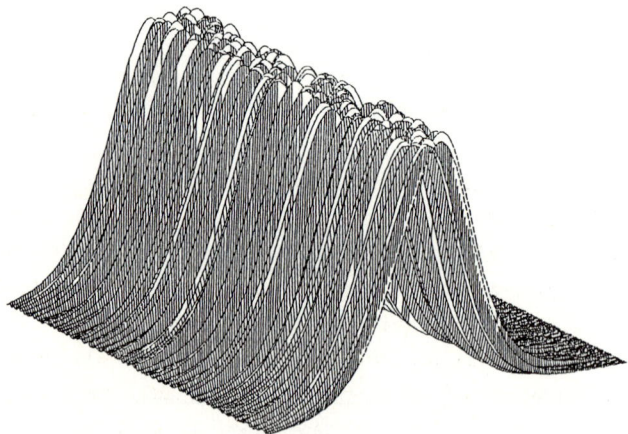

Fig.2; Plotting of the one soliton solution (2.9) with σ given by eq.(2.11), $\epsilon = 0.1$, $\alpha_0 = 6$, $\gamma_0 = 8$ for $p(\sigma)=1$ and $\psi_0(\sigma)$ is a random function of σ uniformly distributed in the interval (0,1). The wave amplitude η is plotted against the x and y dimensionless physical variables at t=0; x varies in the interval [-6,6] while y varies in the interval [-40,40].

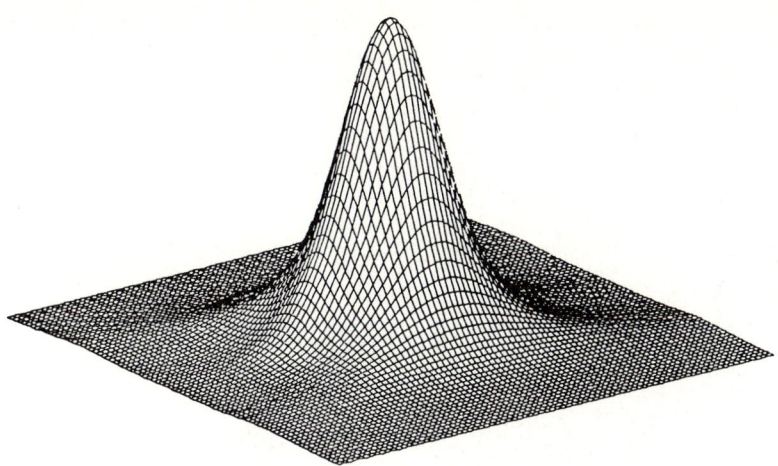

Fig.3; Plotting of the one soliton solution (2.9) with σ given by eq.(2.11), $\epsilon = 0.1$, $\alpha_0 = 6$, $\gamma_0 = 8$ for $p(\sigma)=\text{sech}(\sigma)$ and $\psi_0(\sigma)=0$. The wave amplitude η is plotted against the x and y dimensionless physical variables at t=0; x varies in the interval [-6,6] while y varies in the interval [-40,40].

The two soliton solution for the KdV equation (2.8) is given by [10]:

$$\eta_0(\psi,\sigma,\tau)=\frac{4\alpha_0}{3\gamma_0}(p_2^2(\sigma)-p_1^2(\sigma))\frac{p_2^2(\sigma)\text{csch}^2(\chi_2)+p_1^2(\sigma)\text{sech}^2(\chi_1)}{[p_2(\sigma)\coth(\chi_2)-p_1(\sigma)\tanh(\chi_1)]^2} \quad (2.10)$$

where χ_j (j=1,2) is given by:

$$\chi_j = p_j(\sigma) [\psi + \frac{2}{3}\alpha_0 p_j^2(\sigma) \tau + \psi_{0j}(\sigma)]$$

If we want our solution to be bounded, we have to require that for any σ $p_2(\sigma) > p_1(\sigma)$. This implies that not all one soliton solutions can interact giving rise to a bounded solution; for example two bounded solitons of the type (iii) will always be singular somewhere. A soliton solution of the type (iii) can, however, interact with a soliton of constant amplitude p_2 which is greater than 1. Solitons of the kind (i) or (ii), at difference with the case of the Kadomtsev Petviashvili equation, can interact at any angle in the physical coordinate system. We present in Fig.4 a time sequence of the interaction of two solitons characterized by $p_1=1$, $\psi_{01}=0$, $p_2=1.58$, $\psi_{02}=2\sigma$ which are at $90°$ one with the other.

These plotting clearly show the richness of solutions we can present for the description of water waves on the surface of a two dimensional sea. Their interaction, however, is just a local effect and, at this level of approximation, the phase shift appear only in the direction of propagation of the waves.

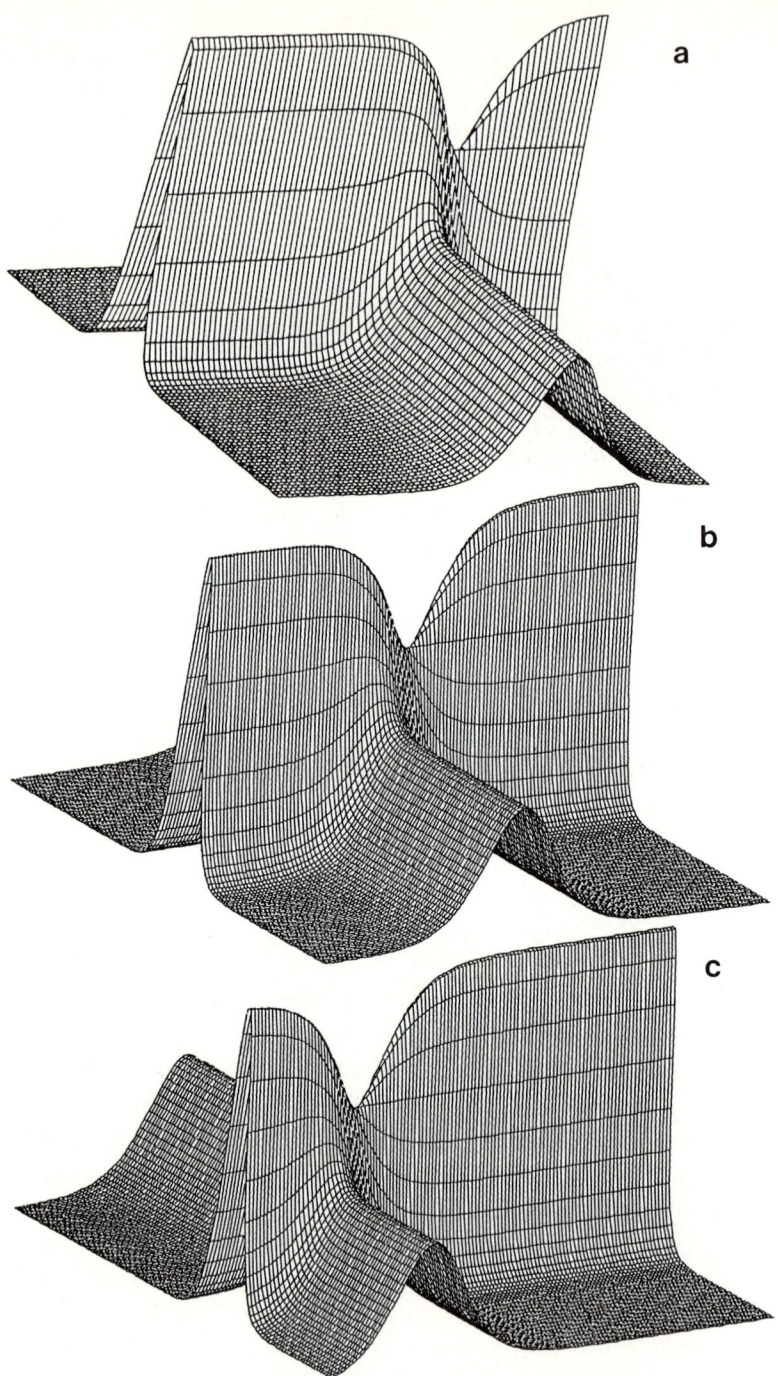

Fig.4; Plotting of the two soliton solution (2.10) with σ given by eq.(2.11), $\epsilon = 0.1$, $\alpha_0 = 6$, $\gamma_0 = 8$ for $p_1=1$, $\psi_{01}=0$, $p_2=1.58$, $\psi_{02}=2\sigma$. The wave amplitude η is plotted against the x and y dimensionless physical variables at (a) t=-2, (b) t=0 and (c) t=2; x varies in the interval [-8,8] while y varies in the interval [-40,40].

To show the stability of these solutions and see what happens to them in the long time scale we go over to the next order in the perturbation parameter ϵ. Using the symbolic malipulation program MACSYMA we get the following linear equation for $\eta_1(R,\Theta,T)$:

$$\eta_{1,T} + \frac{1}{T}\cot(\Theta+\phi)\,\eta_{1,\Theta} + \csc(\Theta+\phi)\left[\frac{1}{6}\alpha_0\,\eta_{1,RRR} + \frac{3}{2}\gamma_0(\eta_0\eta_1)_{,R}\right] + \csc(\Theta+\phi)\left[\frac{1}{9}\alpha_0^2\eta_{0,RRRRR}(\frac{\csc^2(\Theta+\phi)}{8} - \frac{2}{5}) + \right.$$

$$\alpha_0\gamma_0\,\eta_{0,R}\eta_{0,RR}\,(\frac{3\csc^2(\Theta+\phi)}{8} - \frac{5}{3}) + \alpha_0\gamma_0\eta_0\eta_{0,RRR}\,(\frac{\csc^2(\Theta+\phi)}{4}$$

$$\left. - \frac{5}{6}) + \gamma_0^2\,\eta_0^2\eta_{0,R}\,(\frac{9\csc^2(\Theta+\phi)}{8} - \frac{15}{4})\right] +$$

$$\csc^3(\Theta+\phi)\left[(\frac{3}{2}\gamma_0\eta_0\eta_{0,\Theta} + \frac{1}{6}\alpha_0\eta_{0,RR\Theta})\frac{\cos(\Theta+\phi)}{T} + \frac{1}{2T^2}\int ds\eta_{0,\Theta\Theta}\right]$$

$$- \frac{\csc^4(\Theta+\phi)}{2T}\left[\frac{1}{T}\cos(\Theta+\phi)\int ds\eta_{0,\Theta} + \frac{3}{4}\gamma_0\eta_0^2 + \frac{1}{6}\alpha_0\eta_{0,RR}\right] = 0 \quad (2.11)$$

Let's define the field $\zeta(R,\Theta,T)$ as a linear combination of η_0 and η_1:

$$\zeta(R,\Theta,T) = \eta_0(R,\Theta,T) + \epsilon\,\eta_1(R,\Theta,T) \quad (2.12)$$

Taking into account eq.(2.12), (2.6) and (2.11) and then by performing the linear transformation (2.7) we get for $\zeta(\psi,\sigma,\tau)$ the following equation (valid up to order ϵ):

$$\zeta_{,\tau} + \frac{1}{6}\alpha_0\,\zeta_{,\psi\psi\psi} + \frac{3}{2}\gamma_0\,\zeta\zeta_{,\psi} - \epsilon\left[\frac{11}{360}\alpha_0^2\zeta_{,\psi\psi\psi\psi\psi} + \right. \quad (2.13)$$

$$\left. \frac{7}{12}\alpha_0\gamma_0\zeta\zeta_{,\psi\psi\psi} + \frac{31}{24}\alpha_0\gamma_0\zeta_{,\psi}\zeta_{,\psi\psi} + \frac{21}{8}\gamma_0^2\zeta^2\zeta_{,\psi} - \frac{1}{2}\int ds\zeta_{,\sigma\sigma}\right] = 0$$

III. Resolution of the nonlinear evolution equation (2.13).

Let's here analyze eq.(2.13). First of all we notice that this equation, in the form in which it is written, is not integrable; if one construct its symmetry group, using a well defined algorithm contained for example in Olver's book [11], one finds that it has just a five parameter group, not bigger than that of the KdV equation and much smaller than that we should expect for an integrable equation in 2+1 dimensions [12].

For $\epsilon = 0$, eq.(2.13) is a KdV equation but at order ϵ it contains the term $\zeta_{,\sigma\sigma}$, proper of the KP equation. Let's notice then that, if we make the change of variable (corresponding to the almost unidirectional approximation)

$$\sigma' = \epsilon^{1/2} \sigma$$

then eq.(2.13) reduces to the usual KP equation for $\zeta(\psi,\sigma',\tau)$ at the lowest order in ϵ.

Moreover, at order ϵ, we have the same kind of terms which appear in the higher order KdV equation, but with different coefficients.

If we take into account that eq.(2.13) has been obtained leaving out terms of order ϵ^2, we can try, by applying the following transformation:

$$\zeta(\psi,\sigma,\tau) = \frac{2\alpha_0}{3\gamma_0} \chi(\psi,\sigma,\tau) + \epsilon \, F(\frac{2\alpha_0}{3\gamma_0} \chi) \tag{3.1}$$

to get for χ an integrable equation. This approach has been considered by Kodama [13] and applyed with success by him to get solutions of a perturbed Nonlinear Schrödinger equation [14]. It is clear that the result we obtain by applying the standard perturbation technique [15] to eq.(2.13) and those obtained by introducing into (3.1) the values of χ obtained by solving its nonlinear integrable evolution equation, are to be the same up to the first order in ϵ. However, while the perturbative approach allows one to get a solution of the <u>perturbed equation</u> for a <u>given</u> solution of the unperturbed one, going over to an equivalent <u>integrable equation</u> we can obtain <u>any</u> solution of our perturbed equation.

Introducing the ansatz (3.1) into eq.(2.13) and keeping just terms up to order ϵ, we can choose χ in such a way that it satisfies the nonlinear partial differential equation:

$$\chi_{,\tau} = -\frac{1}{6} \alpha_0 \left[\chi_{,\psi\psi\psi} + 6\chi\chi_{,\psi} + \frac{3\epsilon}{\alpha_0} \int ds \chi_{,\sigma\sigma} \right] +$$

$$\frac{11\alpha_0^2 \epsilon}{360} \left[\chi_{,\psi\psi\psi\psi\psi} + 10 \, \chi\chi_{,\psi\psi\psi} + 20 \, \chi_{,\psi} \chi_{,\psi\psi} + 30 \, \chi^2 \chi_{,\psi} \right] \tag{3.2}$$

if F satisfies the following linear equation:

$$F_{,\tau} + \frac{1}{6} \alpha_0 F_{,\psi\psi\psi} + \alpha_0 \, (F \chi)_{,\psi} =$$

$$-\frac{\alpha_0^3}{18\gamma_0} \left[\chi\chi_{,\psi\psi\psi} + 3 \, \chi_{,\psi} \chi_{,\psi\psi} + 3 \, \chi^2 \chi_{,\psi} \right] \tag{3.3}$$

Eq.(3.3) is a nonhomogeneous linear partial differential equation which, by the Fredholm alternative theorem can have a solution only if its r.h.s. is orthogonal to χ, the generic solution of the adjoint of the l.h.s. of (3.3) with respect to a scalar product in $\mathbb{R}^2$. This is verified[1] and then we can find a bounded solution (no secular terms involved) of eq.(3.3) of the form:

$$F = \frac{\alpha_0^2}{9\gamma_0} \left[z^2 + z_{,\psi} \int^\psi ds\, z(s,\sigma,\tau) \right] \tag{3.4}$$

Eq.(3.2) is just the truncation at order ϵ of the higher equation in the hierarchy with respect to the KP equation [16], given by:

$$z_{,\tau} = -\frac{1}{6}\alpha_0 \left[z_{,\psi\psi\psi} + 6zz_{,\psi} + \frac{3\epsilon}{\alpha_0}\int ds\, z_{,\sigma\sigma} \right] +$$

$$\frac{11\alpha_0^2 \epsilon}{360} \left[z_{,\psi\psi\psi\psi\psi} + 10\, zz_{,\psi\psi\psi} + 20\, z_{,\psi} z_{,\psi\psi} + 30\, z^2 z_{,\psi} \right. +$$

$$\frac{5\epsilon}{\alpha_0}\left[2 z_{,\sigma\sigma\psi} + \int ds\, (z^2)_{,\sigma\sigma} + 2 z_{,\psi} \int ds \int ds'\, z_{,\sigma\sigma} \right. +$$

$$\left. 4z_{,\sigma}\int ds z_{,\sigma\sigma} + 4z\int ds z_{,\sigma\sigma} \right] + \left. \frac{5\epsilon^2}{\alpha_0^2}\int ds \int ds' \int ds''\, z_{,\sigma\sigma\sigma\sigma} \right] \tag{3.5}$$

So, given any solution of the higher order KP equation (3.5), we can obtain a solution of eq.(2.13) by taking into account eqs.(3.1) and (3.4). As eq.(2.13) is valid up to order ϵ, so the obtained solution is to be considered at this same order of approximation in ϵ.

For the sake of simplicity let's consider here just the one soliton solution. From eqs.(3.1) and (3.4) we get:

$$\zeta(\psi,\sigma,\tau) = \frac{4\alpha_0}{3\gamma_0} p^2 \left(1 - \frac{2}{3}\alpha_0 p^2 \epsilon\right) \text{sech}^2(\Phi) +$$

$$\frac{4\alpha_0^2}{9\gamma_0}\epsilon p^4 \text{sech}^4(\Phi) - \frac{8\alpha_0^2}{9\gamma_0}\epsilon p^4 \text{sech}^4(\Phi) \tanh(\Phi) \tag{3.6}$$

with $\Phi = p(\psi + q\sigma + c\tau) = p[R + qT\cos(\Theta+\phi) + cT\sin(\Theta+\phi)]$, p and q are arbitrary positive constants and $c = c_0 + \epsilon c_1 + O(\epsilon^2)$ with $c_0 = -\frac{2}{3}\alpha_0 p^2$ and $c_1 = \frac{22}{45}\alpha_0^2 p^4 - \frac{1}{2}q^2$.

---

[1] We write down here the following proposition:
given a linear nonhomogeneous partial differential equation for a field F, if the Fredholm alternative theorem is satisfied by the generic solution g of the corresponding adjoint homogeneous equation then there may exist an explicit solution F = $\mathcal{F}(g)$ where $\mathcal{F}$ is well defined functional.
An analysis of a possible proof of such a proposition and of the applications of such a proposition in other physically interesting situations is under study.

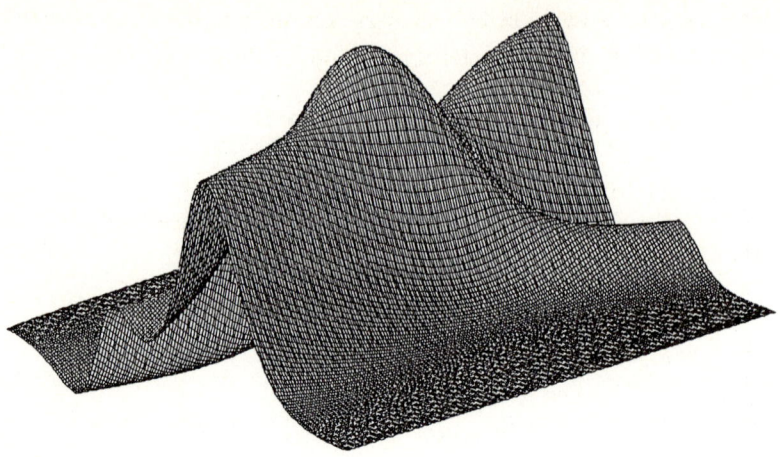

Fig.5; Plotting of the one soliton solution (3.6) with $\epsilon = 0.1$, $\alpha_0 = 6$ and $\gamma_0 = 8$, for $p = 1$ and $q = 15$. The wave amplitude $\eta = \zeta$ is plotted against x and y, the dimensionless physical variables, at $t = 0$; both x and y vary in the interval $[-3,3]$.

If we choose $\phi = \pi/2$ then $\tau = \epsilon x$, $\sigma = \epsilon y$ and $\psi = x + t$ and then

$$\Phi = p [x (1 + \epsilon c) + t + \epsilon q y] \qquad (3.7)$$

Eq.(3.7) shows that p and q are related to the phase velocity in the x and y directions respectively, but from eq.(3.6) we derive that the soliton amplitude depends just on the phase speed in the x direction.

Fig.5 presents a plotting of the solution (3.6) of eq.(2.13), limited up to order ϵ, for a somehow exaggerated value of q in such a way to be able to show its corrections with respect to the solution of the pure KP equation. As we can see from Fig.5 the extra terms in eq.(3.6) make the solution no more positive definite, they have a kinklike form and propagate at an angle with respect to the soliton's propagation direction.

Acknowledgements.

We thank P.M. Santini, O. Ragnisco and M. Bruschi for numerous discussions. This research was supported by the Italian Ministry of Public Education.

V.References.

1. A.R. Paterson, *A first course in fluid dynamics*, Cambridge U.P., Cambridge, England, 1983

2. R.S. Johnson, On the development of a solitary wave moving over an uneven bottom, *Proc. Cambridge Philos. Soc.* 73(1973)183-203

R.H.J. Grimshaw, Long nonlinear internal waves in channels of arbitrary cross section, *J. Fluid Mech.* 86(1978)415-431

3. D. David, D. Levi and P. Winternitz, Integrable nonlinear equations for water waves in straits of varying depth and width, *Stud. Appl. Math.* 75(1987)133-168

4. D. Levi, Alternatives to the Kadomtsev-Petviashvili equation in the description of water waves in more dimensions, *Phys. Lett.* A135(1989)453-457; Alternatives to the Kadomtsev-Petviashvili equation in the description of water waves in more dimensions, submitted to *Phys. Lett.* for the publication.

5. See for example the experimental setting of the experiment done by Hammack, Scheffner and Segur (J. Hammack, N. Scheffner and H. Segur, Periodic waves in shallow water, to be published in the Proceeding of the CIX Summer Course of the International School of Physics Enrico Fermi , edited by A.R. Osborne , Elsevier, 1989) or take into consideration that most sea observations are done by mooring a chain of thermistors in one position.

6. D. Levi, Integrable nonlinear evolution equations in the description of waves in the shallow water long wave approximation, to be published in the Proceedings of the I.M.A. workshop on "Mathematical theory of solitons" edited by P.J. Olver and D. Sattinger, IMA Volumes in Mathematics and its Applications, Springer and Verlag, 1989

7. D.J. Korteweg and G. de Vries, On the change of form of long waves advancing in a rectangular canal, and on a new type of long stationary waves, *Phil. Mag.* 39(1895) 422-443.

8. B.B. Kadomtsev and V.I. Petviashvili, On the stability of solitary waves in weakly dispersing media, *Soviet Phys. Dokl.* 15(1970) 539-541

9. F. Calogero and A. Degasperis, *Spectral transform and solitons*, Vol.1, North-Holland, Amsterdam, 1982.

10. G.L. Lamb, jr., *Elements of soliton theory*, Wiley, New York, 1980.

11. P.J. Olver, *Applications of Lie groups to differential equations* Springer Verlag, New York 1986.

12. P. Winternitz, Kac-Moody-Virasoro symmetries of integrable nonlinear partial differential equations, in *Symmetries and Nonlinear Phenomena* edited by D. Levi and P. Winternitz, World Scientific, Singapore 1988

13. Y. Kodama, On integrable systems with higher order corrections, *Phys. Letts.* 107A(1985) 245-249; Nearly integrable systems, *Physica* 16D(1985) 14-26

14. Y. Kodama and A. Hasegawa, Nonlinear pulse propagation in a monomode dielectric guide, *IEEE Quantum Elect.* QE-23(1987) 510-524.

15. Y. Kodama and M.J. Ablowitz, Perturbations of solitons and solitary waves, *Stud. Appl. Math.* 64(1981) 225-245 M. Tanaka, Perturbations on the K-dV solitons - An approach based on the multiple time scale expansion, *J. Phys. Soc. Japan* 49(1980) 807-812

16. P.M. Santini and A.S. Fokas, Recursion operators and bi-Hamiltonian structures in multidimensions. I, *Commun. Math. Phys.* 115(1988) 375-419

Three-Dimensional Lattice Model Based on Soliton Theory

N. Saitoh

Department of Applied Mathematics, Yokohama National University,
Hodogaya-ku, Yokohama 240, Japan

A three dimensional lattice model which leads to Hirota's bilinear *difference* equation is studied. The corresponding Lagrangean is shown to be equivalent to the hopping term of the Hubbard model for antiferromagnetism which is related to high T_c superconductivity under certain conditions.

1. Introduction

We would like to report in this article some properties of Hirota's bilinear *difference* equation (HBDE) and its application to antiferromagnetic systems.

Hitota's bilinear difference equation[1] reads

$$\alpha f(l+1,m,n)f(l,m+1,n) + \beta f(l,m,n)f(l+1,m+1,n) \\ + \gamma f(l+1,m,n+1)f(l,m+1,n-1) = 0, \quad (1)$$

where l, m, n are independent variables and α, β and γ are parameters satisfying $\alpha + \beta + \gamma = 0$. It is a discrete version of two dimensioal Toda lattice equation and a completely integrable lattice model in three dimensions. It was shown[1] to reduce to various soliton equations, such as the two and one dimensional Toda lattice equations, the KP(Kadomtsev-Petviashvili) equation, the KdV equation, the sine-Gordon equation, the nonlinear Klein Gordon equation and so on. Moreover it was shown by Miwa[2] that it is satisfied by all of the solutions to the equations in the KP-hierarchy. The KP-hierarchy is an infinite set of equations whose solutions are known completely through the theory developed by M.Sato and Y.Sato[3].

As is known generally Bäcklund transformations (BT) create solutions successively if an initial solution is given. The BT for the HBDE was found by Hirota in the form of pair equations[1]. In the preceding works[4,5], we have expressed Hirota's BT in the gauge covariant form to linearize the HBDE and obtained general solutions to it[6]. We can regard this set of equations as equations for wave functions defined on two dimensional square lattice interacting between two adjacent layers. HBDE is derived as a compatibility condition to the BT and determines behaviour of gauge potential under an appropriate gauge condition. When the gauge field is governed by HBDE, the wave function is obtained solving a single linear equation. The local gauge invariance of the BT follows to the fact that the BT appears in a bilinear form of the wave function and the gauge potential. The gauge freedom comes from the ambiguity of the factorization of two fields.

One of the most remarkable feature of this scheme is the *duality* relation between the gauge potential and the wave function. Because the theory is symmetric under the exchange of the two fields, the wave function itself not only solves the single linear equation but also solves the HBDE. This enables us to employ the solution as a gauge potential which governs the single linear equation. In this manner we can generate an infinite series of solutions of HBDE. For instance starting from a constant we can generate a series of multi-solitons successively. This example suggests a strong correlation between behaviour of the gauge potential and the wave function.

On the other hand, we know that in the gauge theory of high T_c superconductivity, there appears a gauge symmetry arose from the ambiguity of factorization of electron fields into spin part and non-spin part[7,8]. Furthermore, the superconducting materials such as La_2CuO_4 based compound consist of two dimensional square lattice of $Cu-O$ planes weakly interacting along the third axis. The situation resembles to the equations of the BT above mentioned. Finally we mention that some kind of correlation between electrons and deformation of lattice plays important roles in the ordinary superconductivity. Especially in one dimensional polyacetylene soliton like lattice deformation traps electrons and move together. This is a behaviour expected from the duality relation discussed above. All these features of the superconductivity make us try to analyze the features of high T_c superconductivity using HBDE.

Based on these motivations, we will study in this report a model Lagrangean of three dimensional lattice with two sublattice structure and having right moving and left moving electrons. It reduces to the Hubbard Lagrangean in the case that right moving and left moving waves are degenerate. Under appropriate physical assumptions for the materials, the equation of motion derived from this Lagrangean turns out to be the BT of HBDE written in the gauge covariant form.

The contents of this report are arranged in the following manner; In section 2, we will introduce HBDE and its BT in the gauge covariant formulation. The next section is devoted to the derivation of the model Lagrangean of Hubbard system and to the discussion of physical behaviour of electrons, holes, charges and spins of the La_2CuO_4 based compounds. In the final section some discussions are given.

2. Gauge Covariant Bäcklund Transformation for HBDE

We start the discussion of this section with the following gauge covariant coupled linear equations

$$\nabla_+ \phi(l,m,n) = c_+ E_+ \phi(l,m,n), \qquad (2a)$$

$$\nabla_- \phi(l,m,n) = c_- E_- \phi(l,m,n), \qquad (2b)$$

where we define the covariant difference operators $\nabla_\pm$ by

$$\nabla_\pm \equiv U_\pm D_\pm U_\pm^{-1}$$

together with the difference operators $D_+ \equiv e^{\partial_l} - 1$, $D_- \equiv e^{\partial_m} - 1$, and $U_\pm = U_\pm(l,m,n)$ are the gauge fields defined on the lattice site (l,m,n). $c_\pm$ are coupling constants and $E_\pm$ are the operators defined by

$$E_+ \equiv U_- e^{\partial_l + \partial_n} U_-^{-1}, \qquad E_- \equiv U_+ e^{\partial_m - \partial_n} U_+^{-1}.$$

Under the local gauge transformations

$$\phi(l,m,n) \to V(l,m,n)\phi(l,m,n) \quad \text{and} \quad U_\pm(l,m,n) \to V(l,m,n)U_\pm(l,m,n)$$

the linear equations (2) are invariant. The set of equations (2) are essentially the same as Hirota's BT for HBDE [1] under an appropriate gauge condition. We emphasize that eqs.(2) are the generalization of BT for two dimensional Toda lattice equation when $\alpha + \beta + \gamma = 0$, which comes from the generalized recurrence formulae[9] to this equation.

From the expression of the operators, it is easy to show that the commutation relations $[\nabla_+, E_-]$, $[\nabla_-, E_+]$ are vanishing; namely $[\nabla_\pm, E_\mp] = 0$ hold. Owing to these relations the compatibility condition for the equations (2) is realized as the commutation relation

$$[\nabla_+, \nabla_-] = c_+ c_- [E_-, E_+]. \tag{3}$$

Since eq.(3) is written by only the gauge fields, the gauge potential $U_\pm$ must obey eq.(3). When eq.(3) holds for $U_\pm$, ϕ is governed by a single linear equation

$$\{\nabla_+, \nabla_-\}\phi = c_+ c_- \{E_-, E_+\}\phi, \tag{4}$$

where we denote by $\{\cdot, \cdot\}$ an anticommutaion relation. The explicit form of eq.(3) is

$$\frac{U_+(l,m,n)U_-(l,m,n)}{U_+(l+1,m,n)U_-(l+1,m+1,n)} - \frac{U_-(l,m,n)U_+(l,m+1,n)}{U_-(l,m+1,n+1)U_+(l+1,m+1,n)}$$
$$= c_+ c_- \left(\frac{U_+(l,m,n)U_-(l,m+1,n-1)}{U_+(l,m+1,n-1)U_-(l+1,m+1,n)} \right.$$
$$\left. - \frac{U_-(l,m,n)U_+(l+1,m,n+1)}{U_-(l+1,m,n+1)U_+(l+1,m+1,n)} \right). \tag{5}$$

If we fix, at this stage, the gauge condition as

$$U_-(l,m,n+1) = U_+(l,m,n) \equiv U(l,m,n), \tag{6}$$

the compatibility condition becomes

$$\frac{U(l+1,m,n)U(l,m+1,n)}{U(l,m,n)U(l+1,m+1,n)} - c_+ c_- \frac{U(l+1,m,n)U(l,m+1,n-1)}{U(l,m,n)U(l+1,m+1,n)} = -\frac{\beta}{\alpha}, \tag{7}$$

with $\frac{\beta}{\alpha}$ an n-independent parameter. Or writing $c_+ c_- \equiv -\frac{\gamma}{\alpha}$, eq.(7) turns out to be HBDE (1) itself for the gauge potential $f(l,m,n) = U(l,m,n)$. At the same time, the single linear equation (4) for the wave function $\phi(l,m,n)$ is given by

$$\frac{\beta}{\alpha} \frac{U(l,m,n)U(l,m,n-1)}{U(l+1,m,n)U(l,m+1,n-1)}\phi(l+1,m+1,n)$$
$$+ \frac{U(l,m,n)}{U(l+1,m,n)}\phi(l+1,m,n) + \frac{U(l,m,n-1)}{U(l,m+1,n-1)}\phi(l,m+1,n) - \phi(l,m,n) = 0. \tag{8}$$

In summary we have shown that the pair of eqs.(2a) and (2b) are equivalent to the set of HBDE (7) and the linear difference equation (8).

We now notice that we can rewrite eqs.(2) in the dual form

$$\tilde{\nabla}_\pm \tilde{\phi} = c_\pm \tilde{E}_\pm \tilde{\phi}. \tag{9}$$

through the exchange of the gauge potential and the wave function

$$U(l,m,n) \to \tilde{\phi}(l,m,n), \qquad \phi(l,m,n) \to \tilde{U}(l,m,n). \tag{10}$$

In this expression, the operators are defined by

$$\tilde{V}_+ \equiv \tilde{U}_+(1 - e^{-\partial_l})\tilde{U}_+^{-1}, \qquad \tilde{V}_- \equiv \tilde{U}_-(1 - e^{-\partial_m})\tilde{U}_-^{-1}$$
$$\tilde{E}_+ \equiv -\tilde{U}_- e^{-\partial_n - \partial_l}\tilde{U}_-^{-1} \qquad \tilde{E}_- \equiv -\tilde{U}_+ e^{\partial_n - \partial_m}\tilde{U}_+^{-1}, \tag{11}$$

with
$$\tilde{U}_-(l, m, n-1) = \tilde{U}_+(l, m, n) \equiv \tilde{U}(l, m, n).$$

The equivalence of eq.(2) and eq.(9) is most easily seen if we write eq.(2) explicitly under the gauge condition (6) and compare with eq.(9). In fact we will see that they are equivalent to Hirota's BT[1].

What do we learn from this? Since the dual expression (9) shares all the features possessed by (2), we obtain HBDE (1) for the wave function $\phi(l, m, n) = \tilde{U}(l, m, n)$ as a compatibility condition. At the same time the gauge potential $U(l, m, n) = \tilde{\phi}(l, m, n)$ is given as a solution of the single linear equation when the $\phi(l, m, n)$ is a solution of HBDE. In this manner we can generate new solutions of HBDE successively. This is an auto-BT and we call it *duality* between the wave function and the gauge potential.

The general solution of HBDE has been known explicitly and is given in terms of Reimann theta function[10,11]. This, however, does not mean that we know everything about HBDE. For example the soliton solutions must be included in the general solution in certain limit of parameters. Using our scheme of the BT we obtain a chain of soliton solutions[11] if we start from the simplest solution $U(l, m, n) = 1$.

3. Model Lagrangean of Antiferromagnetism and Application of HBDE

We would like to study in this section some properties of high T_c superconducting materials,*i.e.*, La_2CuO_4 based compounds, from the gauge theoretical point of view. We are not going to clarify the mechanism of high T_c superconductivity. Instead we show that the behaviour of electrons in the systems are described by HBDE under appropriate assumptions of physical conditions. Layers of two-dimensional square lattice of copper ions constitute the La_2CuO_4 compound. Between two layers of copper ion lattice two layers of lanthanum are sandwitched. At the ground state the $d_{x^2-y^2}$ orbit of electrons on each Cu^{2+} site is filled by only one electron whose spin is aligned either parallel or anti-parallel to the a-axis *i.e.*, along the direction of the diagonal of the lattice. In fact it is in an antiferromagnetic state and has 2-sublattice structure which we call the A-sublattice and B-sublattice, respectively.

Another characteristic feature of the compound is the so-called perovskite structure
Every copper ion on a lattice site is surrounded by six oxygen ions which constitute an octahedron. Four of them are on the Cu-lattice plane and each coupled to two adjacent Cu ions. Two other oxygens are located along the vertical z-axis to the plane and are directly connected to the d_{z^2} orbit of Cu^{2+}. The octahedrons are known to deform or rotate rather easily which cause various types of phase transitions. In particular there exists a deformation of the square of four O^{2-} ions on the x-y-plane in which two of them along the x (resp. y)-axis come close to the Cu^{2+} at A (resp. B)-sublattice or vice-versa. Two such possible configurations are degenerate if there is no doping of holes. In this case these two states will undergo an oscillation. We assume that, when holes are doped, the degeneracy is solved and either one of two states, say the first possibility, is chosen. This leads to a spontaneous breaking of chiral symmetry in the sense that a hole at A (resp. B)-sublattice tends to attracts electrons selectively along the x (resp. y)-axis and distinguish the left and right moving electrons.

We consider the Lagrangean for interacting electrons on layers of square lattice:

$$L = L_0 + L_1, \tag{12a}$$

$$L_0 = 2ia^3 \sum_j (\psi_{jR}^\dagger \partial_t \psi_{jR} + \psi_{jL}^\dagger \partial_t \psi_{jL})$$
$$+ \sum_{j,\hat{n}} \{\psi_{jR}^\dagger h_{\hat{n}} (\psi_{j+\hat{n},R} - \kappa \psi_{jR}) + \psi_{jL}^\dagger h_{\hat{n}} (\psi_{j-\hat{n},L} - \kappa \psi_{jL})\}, \tag{12b}$$

$$L_1 = 16ga^3 \sum_j (\psi_{jR}^\dagger \psi_{jL})(\psi_{jL}^\dagger \psi_{jR}), \tag{12c}$$

where we distinguish the right and left moving fields ψ_R and ψ_L. $\hat{n}$ denotes a unit vector along the spacial axis with the lattice constant a. Note that (12) is invariant under the global chiral transformation

$$\psi_R \to e^{-i\alpha} \psi_R, \qquad \psi_L \to e^{i\alpha} \psi_L. \tag{13}$$

When

$$h_{\hat{n}} = 2ia^2 \sigma^n, \quad \text{and} \quad \kappa = 1,$$

with σ^n, ($n = 1, 2, 3$) the Pauli matrices, this reduces, in the continuum limit $a \to 0$, to the Nambu Jona-Lasinio Lagrangean[12]:

$$\int d^3x \{-\bar{\psi}\gamma_\mu \partial^\mu \psi + g[(\bar{\psi}\psi)^2 - (\bar{\psi}\gamma_5 \psi)^2]\}$$

where

$$\psi = \begin{pmatrix} \psi_R + \psi_L \\ \psi_R - \psi_L \end{pmatrix}.$$

If ψ_R and ψ_L are identical and $h_{\hat{n}}$ are constant h, the Lagrangean (12) is the standard Hubbard model[14]:

$$L_0 = 4ia^3 \sum_j \psi_j^\dagger \partial_t \psi_j + h \sum_{<j,k>} \psi_j^\dagger \psi_k + \mu \sum_j n_j, \tag{14a}$$

$$L_1 = -U \sum_j (n_{j\uparrow} n_{j\downarrow}), \tag{14b}$$

where $n_{j\uparrow} = \psi_{j\uparrow}^\dagger \psi_{j\uparrow}$ (resp. $n_{j\downarrow} = \psi_{j\downarrow}^\dagger \psi_{j\downarrow}$) are number density operators of up (resp. down) spin electrons and $n_j = n_{j\uparrow} + n_{j\downarrow}$, $\sum_{<j,k>}$ denotes the summation over pairs of nearest neighbour sites and $\mu = -2\kappa h$. In eq.(14b), reflecting the features of La_2CuO_4, $U = -32ga^3$ represents the strength of Coulomb repulsive force acting between two $d_{x^2-y^2}$ orbit electrons of copper with opposite spins at one site. When U is large there is no hopping of electrons and the $d_{x^2-y^2}$ orbit is half filled unless holes are doped from outside. We will consider, in what follows, the case that $h_{\hat{n}}$ are constant but ψ_R and ψ_L are independent.

Actual material is not as simple as described by eqs.(12). The gauge theory of antiferromagnetism has been developed in terms of the slave boson technique. In this theory, an electron on the lattice site j is expressed as

$$\psi_{j\sigma} = \zeta_{j\sigma} \phi_j^\dagger + \sigma d_j \zeta_{j,-\sigma}^\dagger \qquad (\sigma = +, -). \tag{15}$$

Here $\zeta_{j\sigma}$ denotes the spin configuration of the electron and ϕ_j represents the state of the

j-th lattice site in which the electron is missing while d_j corresponds to the state in which the $d_{x^2-y^2}$ orbit of copper is filled by an extra electron. Owing to the requirement[14] for the algebras obeyed by electron creation and annihilation operators, ϕ_j, d_j and $\zeta_{j\sigma}$ are subject to the completeness condition:

$$\phi_j^\dagger \phi_j + d_j^\dagger d_j + \sum_\sigma \zeta_{j\sigma}^\dagger \zeta_{j\sigma} = 1. \tag{16}$$

The expression of (15) admits a local gauge symmetry:

$$\phi_j \to \phi_j V_j, \qquad d_j \to d_j V_j, \qquad \zeta_{j\sigma} \to \zeta_{j\sigma} V_j, \tag{17}$$

under which ψ_j, hence the Lagrangean itself, is invariant.

The antiferromagnetic Néel state and/or the RVB state[13] are realized in the strong limit of the Coulomb repulsive force $U \to \infty$ in the Hubbard model. After integration over the d_j field one obtains an effective Lagrangean:

$$L_{eff} = L_0 - J \sum_{<j,k>} (\vec{S}_j \cdot \vec{S}_k - \frac{1}{4} n_j n_k), \tag{18a}$$

where $J = 4h^2/U$ and $\vec{S}_j = \frac{1}{2}\zeta_j^\dagger \vec{\sigma} \zeta_j$. The second term of eq.(18a) describes the Néel state in which spins are frozen antiferromagnetically. This term can be also expressed as

$$-J \sum_{<j,k>} b_{jk}^\dagger b_{jk}, \qquad ; \quad b_{jk} = \frac{1}{\sqrt{2}}(\zeta_{j+}\zeta_{k-} - \zeta_{j-}\zeta_{k+}), \tag{18b}$$

which represents the RVB state[14].

When a small number of holes are doped to the ground state, a spin can move from a site to another without going through a higher energy excitation due to the strong Coulomb repulsive force. In other words the motion of electrons is a real hopping effect. This is one of the point of view on which our investigation is standing. Now taking into account this effect and the assumption that the deformation of the octahedra due to the doping are such that holes at A (resp. B)-sublattice attract electrons along the x (resp. y) direction, the L_0 part of the Lagrangean (18a) can be written as

$$\begin{aligned} L_0 = 2ia^3 \sum_{\chi=R,L} &\left(\sum_{j \in A} \phi_{j\chi}^A \zeta_j^\dagger \partial_t \zeta_j \phi_{j\chi}^{A\dagger} + \sum_{j \in B} \phi_{j\chi}^B \zeta_j^\dagger \partial_t \zeta_j \phi_{j\chi}^{B\dagger} \right) \\ &+ h \sum_{j \in A} \left(\phi_{jL}^A W_{j,j+\hat{x}} \phi_{j+\hat{x},L}^{B\dagger} - c_+ \phi_{jL}^A W_{j,j+\hat{x}+\hat{z}} \phi_{j+\hat{x}+\hat{z},L}^{B\dagger} \right. \\ &\qquad \left. + \phi_{jR}^A W_{j,j-\hat{x}} \phi_{j-\hat{x},R}^{B\dagger} - c_+ \phi_{jR}^A W_{j,j-\hat{x}-\hat{z}} \phi_{j-\hat{x}-\hat{z},R}^{B\dagger} \right) \\ &+ h \sum_{j \in B} \left(\phi_{jL}^B W_{j,j+\hat{y}} \phi_{j+\hat{y},L}^{A\dagger} - c_- \phi_{jL}^B W_{j,j+\hat{y}-\hat{z}} \phi_{j+\hat{y}-\hat{z},L}^{A\dagger} \right. \\ &\qquad \left. + \phi_{jR}^B W_{j,j-\hat{y}} \phi_{j-\hat{y},R}^{A\dagger} - c_- \phi_{jR}^B W_{j,j-\hat{y}+\hat{z}} \phi_{j-\hat{y}+\hat{z},R}^{A\dagger} \right) \\ &+ \frac{\mu}{2} \sum_{\chi=R,L} \left(\sum_{j \in A} \phi_{j\chi}^A \phi_{j\chi}^{A\dagger} + \sum_{j \in B} \phi_{j\chi}^B \phi_{j\chi}^{B\dagger} \right). \end{aligned} \tag{19}$$

In this expression W_{jk} denote

$$W_{jk} = \zeta_j^\dagger \zeta_k,$$

and $c_\pm$ represent strength of the interaction between two layers. By ϕ_j^A (resp. ϕ_j^B)

we denote the state of A (resp. B)-sublattice at the lattice site j which has a hole of electron. In the derivation of above expression we used the fact that, when the number of holes

$$\delta = \sum_{\chi=R,L} \left(\sum_{j \in A} \phi_{j\chi}^{A\dagger} \phi_{j\chi}^{A} + \sum_{j \in B} \phi_{j\chi}^{B\dagger} \phi_{j\chi}^{B} \right),$$

is sufficiently small, the completeness condition (16) is approximately replaced by

$$\zeta_j^\dagger \zeta_j = 1 \quad for \ all \ j.$$

The Lagrangean (19) is invariant under the gauge transformation (17) and W_{jk} play the role of link operators. The link operator W_{jk} represents a flow of spin from the k-site to the j-site. Since the spins of $d_{x^2-y^2}$ electrons at Cu-site are directed either 45° or 225° from the x-axis, there are two types of flow of spins. One is the flow of a spin polarized 45° from the direction of the flow and the other of 225°[15] . We denote them as $W_{jk}^{(+)}$ and $W_{jk}^{(-)}$, respectively. Corresponding to these flows we consider two independent spin operators $\zeta_j^{(\pm)}$ which constitute the link operators in the following manner:

$$W_{jk}^{(\pm)} = \zeta_j^{(\pm)\dagger} \zeta_k^{(\pm)}.$$

They are subject to the conditions

$$\zeta_j^{(\pm)\dagger} \zeta_j^{(\pm)} = 1 \quad for \ all \ j,$$

which we can solve as

$$\zeta_j^{(\pm)} = \zeta_0^{(\pm)} U_{j\pm}^{-1},$$

where $\zeta_0^{(\pm)}$ are constant spinors independent of j satisfying $\zeta_0^{(\pm)\dagger} \zeta_0^{(\pm)} = 1$ and $U_{j\pm}$ are functions of j satisfying $U_{j\pm}^\dagger U_{j\pm} = 1$.

In order to proceed further we have to take into account the actual processes of the propagation of holes discussed so far. The holes can exist either on the $d_{x^2-y^2}$ or d_{z^2} orbit of copper. Let us look at the behaviour of holes at a particular site j of A-sublattice. Under the assumption that the octahedron deforms such that two O^{2-} ions on the x-axis approach to the center of the j-site of A-sublattice when a hole is created there, up spins will be brought from the d_{z^2} orbit of the neighbour sites along the x-axis if the hole is on the $d_{x^2-y^2}$ orbit, whereas down spins can be supplied from the $d_{x^2-y^2}$ orbit of the nearest sites of the adjacent layers if the hole is on the d_{z^2} orbit. Exactly the same processes undergo at the B-sublattice if we change the role of up and down spins and replace the x-axis by the y-axis. The corresponding link operators are then given as follows:

When W links ϕ_L's,

$$W_{jk} = U_{j+} U_{k+}^{-1}$$

if $(j,k) \in (A,B)$ on the same $Cu-O$ plane or $(j,k) \in (B,A)$ on the adjacent planes,

$$W_{jk} = U_{j-} U_{k-}^{-1}$$

if $(j,k) \in (B,A)$ on the same $Cu-O$ plane or $(j,k) \in (A,B)$ on the adjacent planes, whereas when W links ϕ_R's, U_+ and U_- are reversed.

We are interested in the behaviour of holes. From the effective Lagrangean (19),after substituting explicit forms of W's, we derive the Euler-Lagrange equations for $\phi^\dagger$'s:

$$2ia^3 U_{j+}\partial_t U_{j+}^{-1}\phi_{jL}^{A\dagger} + h(U_{j+}e^{\theta_i}U_{j-}^{-1} - c_+ U_{j-}e^{\theta_i+\theta_i}U_{j-}^{-1})\phi_{jL}^{B\dagger} + \frac{\mu}{2}\phi_{jL}^{A\dagger} = 0$$
$$2ia^3 U_{j+}\partial_t U_{j+}^{-1}\phi_{jR}^{A\dagger} + h(U_{j-}e^{-\theta_i}U_{j-}^{-1} - c_+ U_{j+}e^{-\theta_i-\theta_i}U_{j+}^{-1})\phi_{jR}^{B\dagger} + \frac{\mu}{2}\phi_{jR}^{A\dagger} = 0$$
$$2ia^3 U_{j-}\partial_t U_{j-}^{-1}\phi_{jL}^{B\dagger} + h(U_{j-}e^{\theta_j}U_{j-}^{-1} - c_- U_{j+}e^{\theta_j-\theta_i}U_{j+}^{-1})\phi_{jL}^{A\dagger} + \frac{\mu}{2}\phi_{jL}^{B\dagger} = 0$$
$$2ia^3 U_{j-}\partial_t U_{j-}^{-1}\phi_{jR}^{B\dagger} + h(U_{j+}e^{-\theta_j}U_{j+}^{-1} - c_- U_{j-}e^{-\theta_j+\theta_i}U_{j-}^{-1})\phi_{jR}^{A\dagger} + \frac{\mu}{2}\phi_{jR}^{B\dagger} = 0.$$
(20)

Here we have obtained coupled linear equations. To solve the equations we seek solutions which are stationary in time and homogeneous in space in the sense that they satisfy
$$\phi_{jx}^A = \phi_{jx}^B \equiv \phi_{jx}, \qquad \chi = R, L. \tag{21}$$
Then the equations split into two sets of coupled equations. If we further choose the parameters such that $\mu = 2(2\omega a^3 - h)$, where ω is the sum of frequencies of ϕ and U, they turn to the equations:
$$\nabla_\pm \phi_{jL}^\dagger = c_\pm E_\pm \phi_{jL}^\dagger,$$
$$\tilde{\nabla}_\pm \phi_{jR}^\dagger = c_\pm \tilde{E}_\pm \phi_{jR}^\dagger.$$

Here we have used the notations of the previous section through the identification $j = (x, y, z) = (l, m, n)$.

We have just arrived at the point discussed in the previous section. Namely both $\phi_R^\dagger$ and $\phi_L^\dagger$ are solutions of HBDE only if the potential U solve the same equation.

Here a remark is in order. Equations for ϕ_{jR} and ϕ_{jL} derived from (19) are exactly the same with those for $\phi_{jL}^\dagger$ and $\phi_{jR}^\dagger$, respectively. Therefore the same arguments hold for these fields too. In other words all fields $\phi_{jL}, \phi_{jR}, \phi_{jL}^\dagger, \phi_{jR}^\dagger$ and U_j obey the same equation (1), provided we assume $\phi_j^A = \phi_j^B$ and $\phi_j^{A\dagger} = \phi_j^{B\dagger}$.

4. Summary

We have discussed the possibility that the behaviour of holes and spins of La_2CuO_4 compounds are described by HBDE when small amount of impurities are doped.

Our model is based on the modified Hubbard model. This, however, does not have any correspondence with one dimensional Hubbard model which has been known integrable. In fact, in our model, the spin correlation part of the Lagrangean is considered only implicitly to prohibit double occupation of $d_{x^2-y^2}$ orbit on copper. On the other hand the hopping term is not as simple as it appears, but determines the behaviour of holes and spins through the slave boson representation. The spin part links the holes. Taking into account the properties of real materials we found that the wave function of holes obeys the linearized equations of HBDE, hence HBDE itself, if we require that the behaviour of holes on the A- and B-sublattice is determined by a single wave function.

The duality relation of the wave function and the gauge potential manifests the strong correlation between them. The spin flow plays the role of gauge potential which determines the flow of holes and vice-versa.

HBDE is a three dimensional integrable system and explicit form of solutions has been known in terms of Riemann theta function[10]:
$$f(l, m, n) = \prod_{i,j=1}^{3}\left(\frac{E(z_i, z_j)}{z_i - z_j}\right)^{\frac{1}{2}p_i p_j}\theta\left(\zeta + \sum_{j=1}^{3}p_j\int^{z_j}\omega\right). \tag{23}$$

Here
$$p_1 = m + n - \frac{1}{2}, \quad p_2 = -m - \frac{1}{2}, \quad p_3 = l - n - \frac{1}{2},$$
z_1, z_2, z_3 are three complex parameters defined on a Riemann surface, $E(z_i, z_j)$ is the prime form, ζ is a constant and ω is the Abel differential of first kind. It is interesting to notice that this function also appears as a correlation function of three particles in the string theory. There p_j's represent momentum of three external particles. The solutions are distinguished by number of genus of the Riemann surface. It is expected that this topological feature will characterize the behaviour of holes and spins.

Acknowledgment

The author would like to thank the organizers of the conference for the invitation and hospitality. She also likes to thank Professor Bo-Yu Hou, Professor Han-Ying Guo, Professor Chucen Ping Chen and Professor Peizhu Luo for their kind hospitality and various support during her stay at their Institutes and to S.Saito for discussions.

References

[1] R.Hirota, J. Phy. Soc. Jpn. **51** (1981) 3485.
[2] T. Miwa, Proc. Japan Acad. **58A** (1982) 9.
[3] M. Sato and Y. Sato, Proc. U.S. - Japan Seminar, Tokyo 1982, North Holland (1983) pp. 259-271.
[4] S. Saito and N. Saitoh, Phys. Lett. **120A** (1987) 332.
[5] S. Saito and N. Saitoh, J. Math. Phys. **28** (1987) 1052.
[6] N. Saitoh and S. Saito, J. Phys. Soc. Jpn. **56** (1987) 1664.
[7] G. Baskaran and P. W. Anderson, Phys. Rev. **B37** (1988) 580.
[8] P. W. Wiegmann, Phys. Rev. Lett. **60** (1988) 821.
[9] N. Saitoh, J. Phys. Soc. Jpn. **55** (1986) 435.
[10] S. Saito, Phys. Rev. Lett. **59** (1987) 1798.
[11] N. Saitoh and S. Saito, Integrable Lattice Gauge Model Based on Soliton Theory, submitted to J. Phys. A:
[12] Y. Nambu and G. Jona-Lasinio, Phys. Rev. **122** (1961) 345.; **124** (1961) 246.
[13] P. W. Anderson, Mater. Res. Bull. **8** (1973) 153.
[14] Z. Zou and P. W. Anderson, Phys.Rev **B37** (1988) 627.
[15] G. Shirane et. al., Phys. Rev. Lett. **59** (1987) 1613.; T. Freltot et. al., Phys. Rev **B36** (1987) 826.; K. Yamada et. al., Solid State Commun. **64** (1987) 753.

Soliton Phenomena in a Porous Medium

D. Takahashi[1], J.R. Sachs[2], and J. Satsuma[1]

[1]Department of Applied Physics, University of Tokyo, Bunkyo-ku, Tokyo 113, Japan
[2]Department of Biomedical Engineering, Northwestern University, Evanston, IL 60208, USA

I. INTRODUCTION

In the mantle of the Earth, rock is partially melted. Melt phase (melt) and solid phase (matrix) convolute each other complicatedly. Since the density of melt is smaller than that of matrix, melt migrates through the solid interstices. The flow becomes like a porous flow because of this phase mixing. Several researchers studied such a flow and proposed model equations. In 1984, Scott and Stevenson [1] proposed a set of equations which describes the evolution of vertical distribution of melt phase and is equivalent to the following dimensionless equation;

$$(1) \qquad u_t = [u^n\{(u^{-m}u_t)_x - 1\}]_x ,$$

where u is the volume fraction of melt (porosity), t is the time, and x is the vertical space coordinate. Parameters n and m denote the dependency of a matrix permeability k and an effective viscosity η characterizing the rate of matrix compaction and distension as $k \propto u^n$ and $\eta \propto u^{-m}$, respectively. The reasonable values of n and m are $2 \sim 5$ and $0 \sim 1$, respectively. We call (1) the magma equation. Scott and Stevenson numerically showed that pulse-like solitary waves interact one another like solitons.

The magma equation appears in another simple physical situation. It is the flow of two kinds of fluid of which densities are different. If a low-density fluid is injected continuously from a tube at a bottom of a tank filled with a high-density fluid, the former migrates through the latter and forms a thin pipe. If the flux of low-density fluid is controlled at the entrance, the shape of pipe becomes hump-like. Scott, Stevenson and Whitehead [2] studied the interaction of the humps experimentally and found that they behave like solitons. A set of equations which describes such a flow is also proposed by them. It reduces to the following dimensionless equation;

$$(2) \qquad u_t + \alpha u u_z + \beta(u_t u_{zz} - u u_{zzt}) = 0 ,$$

where u is the horizontal cross section of the low-density fluid, t is the time, z is the vertical space coordinate and α, β are constants. Equation (2) is equivalent to the magma equation for $n = 2$ and $m = 1$.

In a preceding paper [3], we have shown explicit travelling wave solutions of (1) for some particular choices of the parameters n and m, and discussed the existence of weak solutions with compact support. Moreover, we proposed a modified version of (1), which reduces to the Korteweg-de Vries (KdV) equation by means of a variable transformation.

In this paper, we give a brief summary of the preceding paper and present some further results on the magma equation. A detailed account will be given in a forthcoming paper [4].

II. EXPLICIT ANALYTICAL TRAVELLING WAVE SOLUTIONS

In this section, we give explicit travelling wave solutions of (1). Here and hereafter, we confine ourselves to the case $n = 3$ and $m = 0$ for simplicity. Namely we consider

(3) $$u_t = [u^3(u_{tx} - 1)]_x .$$

Then, substituting $u = u(z)$, $z = x - ct$ into (3) and integrating twice with respect to z, we obtain an ordinary differential equation governing the travelling wave solutions,

(4) $$\frac{c}{2}u_z^2 = -\frac{1}{u^2}(u^3 - Bu^2 + cu - A) ,$$

where c is the wave velocity, and A, B are integration constants. If we introduce a transformation of the independent variable from z into ζ,

(5) $$\zeta = \int^z u^{-1} dz ,$$

then, (4) is reduced to

(6) $$\frac{c}{2}u_\zeta^2 = -(u^3 - Bu^2 + cu - A) \equiv -f(u) ,$$

the solutions of which are expressed in terms of the Jacobian elliptic functions. Let us assume that $f(u)$ is factorized as

(7) $$f(u) = (u - u_1)(u - u_2)(u - u_3) , \quad u_1 \leq u_2 \leq u_3 .$$

Then a regular solution of (6) is given by

(8) $$u = u_2 + (u_3 - u_2)\,\mathrm{cn}^2 p\zeta ,$$

where $p = \sqrt{(u_3 - u_1)/2c}$ and cn is the Jacobian elliptic function with the modulus $k = \sqrt{(u_3 - u_2)/(u_3 - u_1)}$. The inverse transformation of (5) is given by

(9) $$z = \int^\zeta u\, d\zeta .$$

Substituting (8) into (9), we get

(10) $$z = u_1 \zeta + \sqrt{2c(u_3 - u_1)}\, E(s; k) ,$$

where $s = \mathrm{sn}\, p\zeta$ and $E(s; k)$ is the elliptic integral of the second kind defined as $E(x; k) \equiv \int_0^x \sqrt{(1 - k^2 t^2)/(1 - t^2)}\, dt$. Equations (8) and (10) give a periodic wave solution for $u_2 > 0$. The period L is calculated as $L = \sqrt{8c/(u_3 - u_1)}\,\{K(k) + (u_3 - u_1)E(k)\}$, where $K(k)$ and $E(k)$ are the complete elliptic integrals of the first and the second kind, respectively.

If we take a limit of $k \to 1$ or $u_1 \to u_2$ in (8) and (10), we obtain a solitary wave solution

(11) $$u = u_2 + (u_3 - u_2)\,\mathrm{sech}^2 p\zeta ,$$
(12) $$z = u_2 \zeta + \sqrt{2c(u_3 - u_2)}\,\tanh p\zeta .$$

These solutions may correspond to the numerical solutions obtained by Scott and Stevenson [1].

There exists another variable transformation by which solutions can be expressed explicitly in terms of the Jacobian elliptic functions. Let us transform z into ζ as

(13) $$\zeta = \int^z u^{-3/2} dz .$$

Then (4) is reduced to

(14) $$\frac{c}{2} u_\zeta^2 = -u(u^3 - Bu^2 + cu - A),$$

from which we obtain a singular periodic wave solution

(15) $$u = \frac{u_1 u_3 \mathrm{sn}^2 p\zeta}{u_3 - u_1(1 - \mathrm{sn}^2 p\zeta)},$$

(16) $$z = \frac{u_1 \sqrt{u_2}}{p}\left[F(s;k) - \left(1 - \frac{u_3}{u_1}\right)\left(E(s;k) + \frac{\sqrt{(1-s^2)(1-k^2 s^2)}}{s}\right)\right],$$

where $u_1 > 0$, $p = \sqrt{u_2(u_3 - u_1)/2c}$, $s^2 = 1-(1-\mathrm{sn}^2 p\zeta)u_1/u_3$, $k^2 = (u_3 - u_2)/(u_3 - u_1)$, and F is the elliptic integral of the first kind defined as $F(x;k) = \int_0^x 1/\sqrt{(1-t^2)(1-k^2 t^2)}\, dt$. The maximum and the minimum values of the above periodic solution are u_1 and 0, respectively, and the gradient of $u(z)$ is infinite where $u = 0$. The configuration of $u(z)$ is shown in Figure 1.

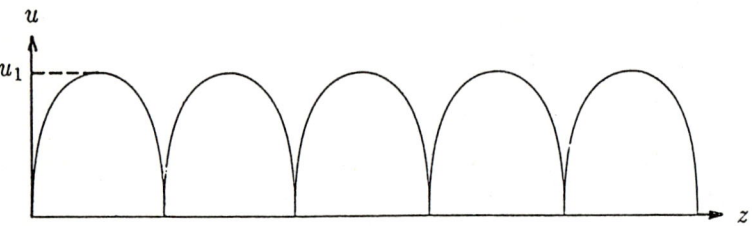

Figure 1. Singular periodic wave solution $u(z)$ described by (15) and (16).

III. COMPACT SUPPORT SOLUTIONS

It is formally possible to cut off one hump of the singular periodic wave given in the preceding section. If we force u to be zero outside the hump, we obtain a weak solution as shown in Figure 2. It satisfies the magma equation except the feet at $z = z_1$ and z_2. We here consider the validity of the existence of such a weak solution.

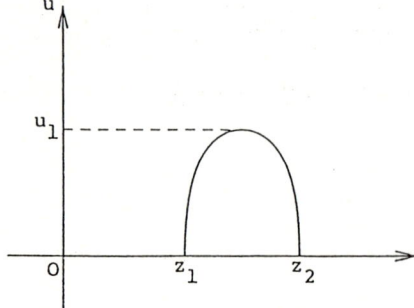

Figure 2. Solitary wave solution with compact support constructed from the solution of Figure 1.

Let us assume that there exists the weak solution as shown in Figure 2, which satisfies (4) on $z_1 < z < z_2$. Then

(17) $$-cu = u^3(-cu_{zz} - 1) - 2A ,$$

holds on $z_1 < z < z_2$. We note that the integration constant A can not be zero because of the balance of the both sides of (4). On the other hand, in the outside region, $z < z_1$ or $z > z_2$,

(18) $$-cu = u^3(-cu_{zz} - 1) ,$$

holds because $u \equiv 0$ there. Hence, the solitary wave solution satisfies

(19) $$u_t = [u^3(u_{tx} - 1)]_x + 2A\{\delta(x - ct - z_2) - \delta(x - ct - z_1)\} ,$$

where $\delta(x)$ is the Dirac's delta function. The last two terms of the right-hand side may be interpreted as source terms for u. Equation (19) equation conserves the total mass, $\int_{-\infty}^{\infty} u dx$. Hence, if the solitary wave solution is applied to (3) which has no source terms, the balance of mass u breaks at the feet of the hump and the solution should change its shape or diverge.

It is suggested from the above argument that solitary wave solutions with compact support are not possible for (3). However, it is probable that such waves exist in real physical systems. We believe that the basic equation should be modified in order to describe them. Still then it is an interesting problem to investigate how the initial data with compact support develops in time for (3).

We here show a result of numerical computation for the initial condition of the single hump given by Figure. 2. The time evolution is calculated with a finite-difference scheme using a potential of u. Figure 3 shows the time development of the profile.

The mass moves to the right more rapidly than the right hand endpoint, and, since the area under the profile must remain constant to conserve mass, the wave steepens as a result of this motion. As the profile steepens and moves to the right, a single solitary wave separates off moving with a constant velocity and profile. This process then repeats as the mass remaining behind the solitary wave(s) splits into another solitary wave (each smaller than its predecessor) and a remainder. At approximately $t = 10$, the profile contains a peak which is, apparently, as steep as can be represented by the numerical grid. Therefore, there exists a possibility that the current results would be of the numerical damping due to the use of finite-difference differentiation. We calculated the same initial value problem with a different number of grid points or with radically different numerical scheme. The results coincide with the above qualitatively. Therefore, it may be suggested that the given numerical results should correspond well to the behavior of a physical system modeled by (3).

The second observation on these results is that the leading solitary wave resembles an analytical solitary wave with very low constant base which is described by (11) and (12). Figure 4 shows a comparison of the leading solitary wave obtained numerically with an analytical solitary wave solution described by (11) and (12); the solid line shows the former and the dotted one shows the latter. Though they are different especially near the x-axis because their origins are quite different, their profiles are very similar to each other as a whole. This argument may be valid for other smaller solitary waves released from the initial hump with compact support. From above discussions, we conclude that an initial data with compact support breaks into some quasi-stable waves which are very similar to solitary waves with low constant base described by (11) and (12), with the help of a damping or a diffusing effect undescribed here.

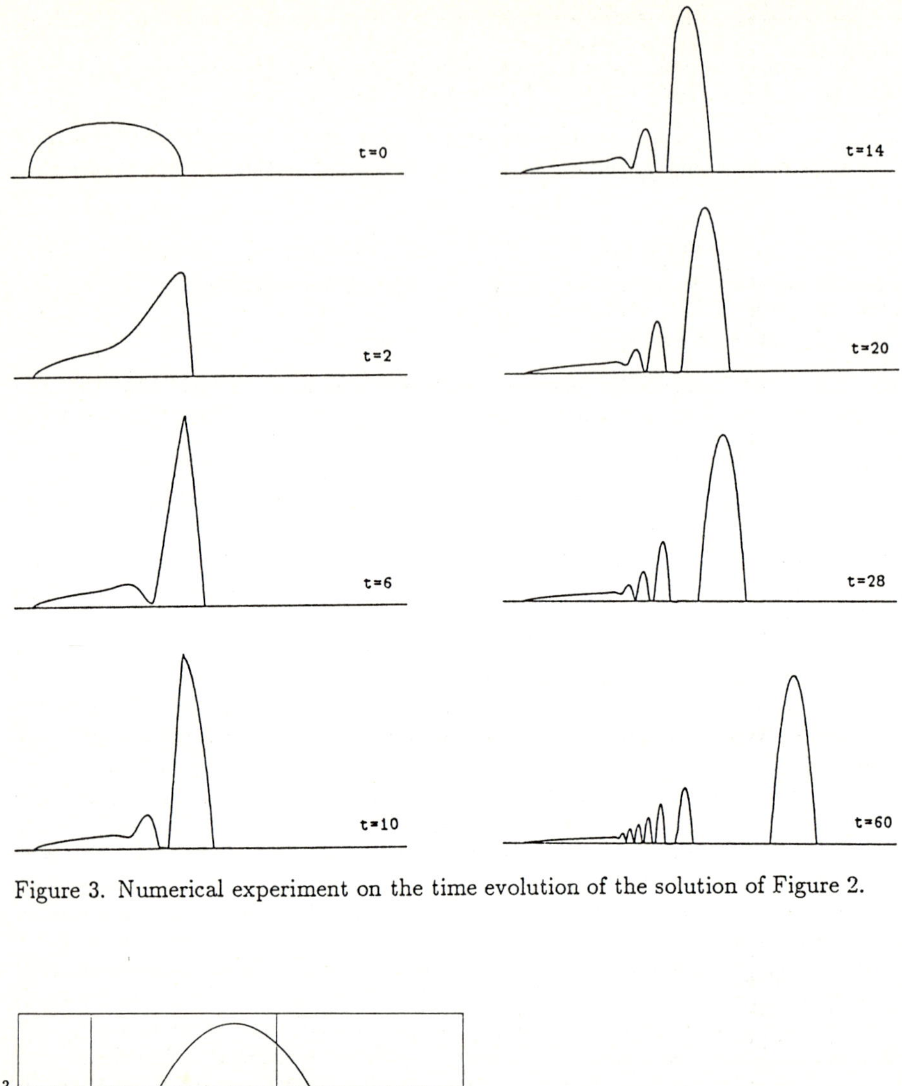

Figure 3. Numerical experiment on the time evolution of the solution of Figure 2.

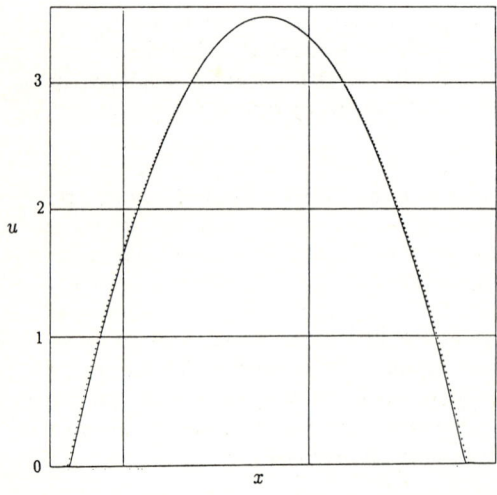

Figure 4. Comparison of the leading solitary wave with an analytical solitary wave described by (11) and (12). The straight line indicates the former and the dotted line the latter.

IV. MODIFIED MAGMA EQUATION

The magma equation has solutions which behave like solitons. Moreover, the results of numerical computation suggests that it may not be integrable. In this section we propose a modified equation which is related to (3) and integrable. The equation is written as

(20) $$u_t + [u^3(3u_{xx} + 1)]_x = 0 .$$

We call this the modified magma equation. It is noted that (20) has been introduced by Ito and Kako [5] as an example which possesses higher order conserved quantities. In the long wave and small amplitude approximation, both (3) and (20) have the linear dispersion relation of the type

(21) $$\omega = \alpha k - \beta k^3 .$$

If we introduce new independent variables ξ and τ by

(22) $$\xi = \int^x u^{-1} dx ,$$

(23) $$\tau = t ,$$

then (20) reduces to the KdV equation,

(24) $$u_\tau + 6uu_\xi + 3u_{\xi\xi\xi} = 0 .$$

By using this fact, we can construct the solitary wave solution with finite constant base of (20),

(25) $$u = c + \frac{3}{2}p^2 \text{sech}^2 \frac{1}{2}(p(\xi + 6c\tau) - 3p^3\tau) ,$$

(26) $$x = c\xi + 3p\tanh\frac{1}{2}(p(\xi + 6c\tau) - 3p^3\tau) ,$$

where c is the height of constant base. The solution is quite similar to (11) and (12). Since the KdV equation has N-soliton solution, (20) also has N-soliton solution which describes the interaction of solitary waves of the type (25) and (26).

It is also possible to construct solutions with compact support of (20) from the soliton solutions of the KdV equation. Let us take the 1-soliton solution of (24),

(27) $$u(\xi, \tau) = \frac{3}{2}p^2 \text{sech}^2 \frac{1}{2}(p\xi - 3p^3\tau) ,$$

where p is an arbitrary parameter. Through the variable transformation (22) and (23), (27) reduces to

(28) $$u(x, t) = \begin{cases} \frac{3}{2}p^2 - \frac{x^2}{6} , & |x| \leq \sqrt{3}p , \\ 0 , & |x| > \sqrt{3}p , \end{cases}$$

which is a stationary solution with compact support. We note that the incline of u at the feet is finite. If we take the 2-soliton solution of (24) which vanish as $\xi \to \pm\infty$, we get a nonstationary solution with compact support of (20). Figure (5) shows a typical example of time evolution of such a solution.

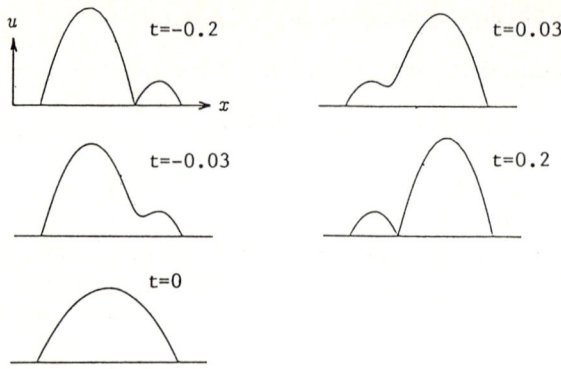

Figure 5. Time evolution of a solution for (20) constructed from the 2-soliton solution of the KdV equation.

Furthermore, if we take the N-soliton solution of (24), we obtain a nonstationary solution describing the interaction of N humps. We note that exact solutions describing this type of interaction of pulses with compact support have been obtained for a nolinear diffusion equation [6].

By using the fact that (20) reduces to the KdV equation through the transformation (22) and (23), we can show the modified magma equation has an infinite number of conserved densities. Let a conserved density of the KdV equation (24) be denoted by I, which satisfies $\frac{\partial}{\partial \tau} \int_{-\infty}^{\infty} I d\xi = 0$. Then, by transforming the variables ξ and τ into x and t, we find that the corresponding conserved density of the modified magma equation (20) is I/u satisfying $\frac{\partial}{\partial t} \int_{x_1}^{x_2} \frac{I}{u} dx = 0$, where x_1 and x_2 are positions of both ends of the positive part of solution. Since one of the conserved densities of the modified magma equation is 1 corresponding to the conserved density u of the KdV equation, we find that the width of compact support is conserved.

Although we do not know any physical relevance of (20) yet, we hope that it plays a role as a clue to the analysis of the systems exhibiting the behavior shown by the above solutions.

REFERENCES

[1] D. R. SCOTT AND D. J. STEVENSON, Magma solitons, Geophys. Res. Lett. <u>11</u> (1984), 1161-1164.
[2] D. R. SCOTT, D. J. STEVENSON AND J. A. WHITEHEAD JR., Observations of solitary waves in a viscously deformable pipe, Nature <u>319</u> (1986), 759-761.
[3] D. TAKAHASHI AND J. SATSUMA, Explicit solutions of magma equation, J. Phys. Soc. Jpn. <u>57</u> (1988), 417-421.
[4] D. TAKAHASHI, J. R. SACHS AND J. SATSUMA, ??? title etc.???.
[5] M. ITO AND F. KAKO, A REDUCE program for finding conserved densities of partial differential equations with uniform rank, Comp. Phys. Comm. <u>38</u> (1985), 415-419.
[6] N. C. FREEMAN AND J. SATSUMA, Exact solutions describing an interaction of pulses with compact support in a nonlinear diffusive system, preprint (1989).

Two-Dimensional Chiral Gauge Theories on a Lattice

Ma Zhongshui[1] and Guo Shuohong[2]

[1]Department of Physics, Zhongshan University, Guangzhou,
People's Rep. of China
[2]Center of Theoretical Physics, CCAST (World Lab.), Beijing, and
Department of Physics, Zhongshan University, Guangzhou,
People's Rep. of China

I. Introduction

Chiral gauge theories in two dimensions have been receiving much attention recently. Though quantum field theories with gauge field coupling to chiral fermion suffer from the loss of gauge invariance, it can be consistently quantized[1]. A proper quantum treatment of the gauge degrees of freedom in chiral gauge theories uncovers the presence of the Wess-Zumino effective action exactly[2]. In view of these interests in chiral gauge theories, the study of two-dimensional chiral gauge model on the lattice should prove useful. It is well-known that there will appear the species dubling when we naively discretize a fermion on the lattice[3]. In our approach we shall use Wilson fermion[4] which copes the problem[5]. The Wilson term, however, will break not only the chiral invariance but also the gauge invariance when the chiral gauge field is introduced. Aoki[6] had redrived the nonabelian anomaly with consideration of chiral gauge theory on the lattice and thought the existence of the anomaly may be related with the gauge symmetry breaking due to the Wilson term. It will be shown in follows that the Wess-Zumino effective action and the vector-boson mass as the anomalous generation terms appear due to a nontrivial change in the Wilson term under the chiral transformation in fermion.

II. Chiral Schwinger Model

The action of the chiral Schwinger model[7] with Wilson fermion is given by

$$S = S_G + S_F \tag{1}$$

where S_G is gauge action, the chiral fermion action is

$$S_F(\psi,\bar{\psi},U_\mu) = \sum_{xy} \bar{\psi}(x) D_{xy} \psi(y)$$

$$D_{xy} = \sum_\mu \gamma^\mu [\nabla^+_\mu - \nabla^-_\mu] + \sum_\mu \gamma^\mu P_L [U_\mu(x)\nabla^+_\mu - \nabla^-_\mu U_\mu(x)]$$

$$-r\sum_\mu [\nabla^+_\mu + \nabla^-_\mu - \frac{1}{a}] \tag{2}$$

$(\nabla^{\pm}_{\mu})_{xy}=(1/2a)\delta_{y,x\pm\mu}$, $P_{L\atop R}=(1\pm\gamma_5)/2$, a is the lattice spacing, and the gauge link is parameterized as $U_\mu(x)=\exp iag\Lambda_\mu(x)$. We employ the notations, $h^{\mu\nu(\pm)}=\frac{1}{2}(g^{\mu\nu}\pm\epsilon^{\mu\nu})$, $g_{00}=-g_{11}=1$, $\epsilon_{01}=-\epsilon_{10}=1$, and separate $\gamma^\mu(U_\mu(x)-1)$ into $\gamma_\mu[bg^{\mu\nu}-(1-b)\gamma_5\epsilon^{\mu\nu}](U_\nu(x)-1)$ by consideration of the dual identity $\gamma^\mu\gamma_5=\epsilon^{\mu\nu}\gamma_\nu$. So the operator in the left-fermion part of the action for the naive one becomes as

$$\frac{1}{2a}\sum_{\mu\nu}\gamma_\mu P_L (h^{\mu\nu(-)}+bh^{\mu\nu(+)})[(U_\nu(x)-1)\nabla^+_\nu - \nabla^-_\nu(U_\nu(y)-1)] \qquad (3)$$

here we have introduced an arbitrary parameter b. Formally the terms with parameter b in eq.(3) equal identically to zero due to the dual identity, so that it seems the arbitrary parameter b does not exist in the action. However, because of the chiral anomaly, and theory with γ_5 interaction must be defined carefully with a proper regularization[8] scheme. Therefore we retain the b terms in the action and will show in the follows that these terms do give nontrivial contribution to the effective action in our approach. After separating the fermion field into the left- and right- handed chiral fermions ψ_L and ψ_R, and using the shift $\psi_R \to \psi_R' = \psi_R - M_2^{-1}N\psi_L$, $\bar\psi_R \to \bar\psi_R' = \bar\psi_R - \bar\psi_L NM_2^{-1}$, the fermion part of the functional can be written as

$$\int D\psi_L D\bar\psi_L \exp\bar\psi_L(x)[M_1-NM_2^{-1}N]_{xy}\psi_L(y) \int D\psi_R' D\bar\psi_R' \exp\bar\psi_R'(x)M_{2xy}\psi_R'(y)$$

$$M_{1xy}=\sum_\mu \gamma_\mu [g^{\mu\nu}(\nabla^+_\nu-\nabla^-_\nu)+(h^{\mu\nu(-)}+bh^{\mu\nu(+)})[(U_\nu(x)-1)\nabla^+_\nu-\nabla^-_\nu(U^+_\nu(y)-1)]] \qquad (4)$$

$$M_{2xy}=\sum_\mu \gamma_\mu[g^{\mu\nu}(\nabla^+_\nu-\nabla^-_\nu)], \qquad N_{xy}=r\sum_\mu[g^{\nu\nu}(\nabla^+_\nu+\nabla^-_\nu)-\frac{1}{a}]$$

The path integration of ψ_R' is det M_2 which can be absorbed in the normalization constant. In order to define the integration of ψ_L meaningfully, we assume there exists the left and right eigenfunctions of the operator $(M_1-NM_2^{-1}N)$, φ_n and φ_n' ($\varphi_n'\neq\varphi_n$), satisfying orthogonality and completeness conditions

$$(M_1-NM_2^{-1}N)\varphi_n=\lambda_n\varphi_n, \qquad \varphi_n'^+(M_1-NM_2^{-1}N)=\lambda_n\varphi_n'^+$$
$$\sum_x \varphi_n'^+(x)\varphi_m(x)=\delta_{m,n}, \qquad \sum_n \varphi_n'^+(x)\varphi_n(y)=\delta_{xy} \qquad (5)$$

We then expand the left-handed fermion with the chiral orthonormal set

$$\psi_L(x)=\sum_{\lambda_n\geq 0} a_n \varphi_n^L(x), \qquad \bar\psi_L(x)=\sum_{\lambda_n\geq 0} b_n \varphi_n'^R(x)$$

$$\varphi_n^L(x)=(1+\gamma_5)/\sqrt{2}\varphi_n(x) \qquad (\lambda>0),$$
$$=(1+\gamma_5)/2\varphi_n(x) \qquad (\lambda=0), \qquad (6)$$

$$\varphi_n'^R(x)=(1-\gamma_5)/\sqrt{2}\varphi_n'(x) \qquad (\lambda>0),$$
$$=(1-\gamma_5)/2\varphi_n'(x) \qquad (\lambda=0),$$

a_n and $\bar{b}_n$ are elements of Grassmann algebra. By using these relations the left-handed fermion integral becomes

$$\int \prod da_n d\bar{b}_n \, \exp \sum_{nxy} \bar{b}_n \varphi_n'^+(x)(M_1 - NM_2^{-1}N)_{xy} \varphi_n(y) a_n \tag{7}$$

To get the anomalous generation of the theory, we make a chiral change of fermion variables

$$\psi_L(x) \to V^+(x,\alpha)\psi_L \quad \bar{\psi}_L(x) \to \bar{\psi}_L(x) V(x,\alpha), \quad V(x,\alpha) = \exp i\alpha\phi(x) \tag{8}$$

the α is a real parameter varied from 0 to 1. The measure of lattice fermion is invariant under this change and now the variation of the effective action is

$$\delta W_{eff}(\alpha) = \sum_x \sum_n \varphi_n'^+ \{D(U_\mu)\}^{-1} \hat{GD}(\alpha,\phi,U_\mu)\}\varphi_n(x)$$

$$D(U_\mu) = M_1 - NM_2^{-1}N \tag{9}$$

$$\hat{GD}(\alpha,\phi,\mu) = -i\alpha[\phi(x)\gamma_5(M_1 - NM_2^{-1}N) + (M_1 - NM_2^{-1}N)\gamma_5\phi(y)]$$

Using the weak coupling expansion for $U_\mu(x)$, $U_\mu(x) = 1 + iagA_\mu(x)$, we can get the result by a tedious calculation

$$\delta W_{eff} = -g \frac{1}{4\pi} \int d^2 x \phi \epsilon_{\mu\nu} g^{\nu\delta} \partial_\delta (h^{\mu\lambda(-)} + bh^{\mu\lambda(+)}) A_\lambda(x) \tag{10}$$

It is seen that the result does not depend on the Wilson parameter r. This is the anomalous generation term which contributes as a manifestation of the Schwinger mechanism[7].

In two dimensions, the gauge field A_μ can be written as

$$A_\mu = \frac{1}{g}(\partial_\mu \sigma + \epsilon_{\mu\nu}\partial^\nu \rho)$$

It is easy to see that if the phase ϕ of the chiral transformation eq.(8) is taken to be $\phi = \sigma + \rho$, then the fermion field ψ is decoupled completely from the gauge field in the continuum limit[9]. With the expression of A_μ in terms of σ and ρ, we can get a compact expression for the effective action (adding $F_{\mu\nu}^2/4$)

$$W_{eff} = \frac{1}{2g^2} \rho \Box \Box \rho + \frac{1}{8\pi}[(1+b)\rho \Box \rho + (1-b)\sigma \Box \sigma + 2\sigma \Box \rho)] \tag{12}$$

which gives the equation of ρ in form

$$\Box(\Box + \frac{g^2}{4\pi}(\frac{b^2}{b^2 - 1}))\rho = 0 \tag{13}$$

This indicates that the chiral Schwinger model consists of a massive boson ρ with the mass $m^2 = g^2 b^2/4\pi(b-1)$, for $b>1$, which is the same as demonstrated by Jackiw and Rajaraman[11].

223

III. Wess-Zumino Effective Action in Two Dimensions

In this section we consider the chiral $SU(N)_L \times SU(N)_R$ fermion model, with chiral gauge field and Goldstone field added. It is shown below that Wess-Zumino effective action as an anomalous term also can be derived with the lattice regularization. The lattice functional integral considered by us is

$$Z[U_\mu, V] = \int \psi D\bar{\psi} \exp S_F \qquad (14)$$

where S_F is the fermion action

$$S_F(\psi, U_\mu, V) = \sum_{xy} \bar{\psi}(x) D_{xy} \psi(y)$$

$$D_{xy} = \sum_\mu \gamma^\mu [\nabla_\mu^+ - \nabla_\mu^-] P_R + \sum_\mu \gamma^\mu [U_\mu(x)\nabla_\mu^+ - \nabla_\mu^- U_\mu(y)] P_L - m[V(x) P_R + V^+(x) P_L]$$

$$-r \sum_\mu [\nabla_\mu^+ + \nabla_\mu^- - \frac{1}{a}] \qquad (15)$$

m is mass of fermion, and V is Goldstone field. With the fermionic local chiral transformations

$$\psi(x) = (V(x) P_L + P_R) \psi'(x), \quad \bar{\psi}(x) = \bar{\psi}'(x)(V^+(x) P_R + P_L) \qquad (16)$$

$V^{-1}(x)V(x)=1$, the fermion action becomes as

$$S_F'(\psi', U_\mu, V) = \sum_{xy} \bar{\psi}'(x) D'_{xy} \psi'(y)$$

$$D'_{xy} = (V^+(x) P_R + P_L) D_{xy}(V(x) U_\mu(x) V^+(x+\mu), V(x)V(x))$$

$$(V(y) P_L + P_R) \qquad (17)$$

Under this transformation, the Wilson fermion will suffer a nontrivial change and make the functional integral to be a new one with a manifestation of the correct anomaly. We define a new function $V(x,\tau)$, on a three-dimensional manifold with the space-time as boundary, to interpolate between well-defined space-time fields by introducing a continuum parameter $\tau \in [0,1]$,

$$V(x,0) = 1, \quad V(x,1) = V(x) \qquad (18)$$

The effective action $W_{eff}[U_\mu, V]$ is

$$W_{eff}[U_\mu, V] = -\ln Z[U_\mu, V] = -\ln \det D \qquad (19)$$

Then the variation of it can be written as

$$\delta W_{eff} = \int_1^0 d\tau \hat{G}W = -N_c \sum_x \int_1^1 d\tau \mathrm{Tr}^x [D^{-1}(\hat{G}D)]$$

$$\hat{G}D = D' - D \qquad (20)$$

In the consideration of infinitesimal chiral transformation $\hat{G}$ on the chiral symmetry broken part of the fermionic action with

$$\hat{G}V(x,\tau) = \bar{g}V(x,\tau), \quad \hat{G}V^{+}(x,\tau) = -V^{+}(x,\tau)\bar{g}, \quad \bar{g} = (\partial_\tau V)V^{+} \qquad (21)$$

the trace terms can be calculated by introducing a complete set of the plane waves. Following the same procedure taken in former section, the effective action in continuum limit then is given by

$$\delta W_{eff} = \frac{Nc}{4\pi}\int_0^1 d\tau \int d^2x \epsilon^{\mu\nu} tr \mathcal{L}_\mu \mathcal{L}_\nu \mathcal{L}_\tau + \frac{Nc}{4\pi} ig \int d^2x \epsilon^{\mu\nu} tr A_\mu \mathcal{L}_\nu ,$$

$$\mathcal{L}_\mu = (\partial_\mu V)V^{+} \qquad (22)$$

Here we have integrated out in the second term with the boundary condition $\mathcal{L}_\mu(\tau=0)=0$. For the first term in eq.(22), we can identify $d\tau d^2 x \epsilon^{\mu\nu}$, up to a constant which we would like to set as 1, with the volume element $\frac{1}{3}d\Sigma^{ijk}$ of a three-dimensional manifold. The effective action can finally be written in following form

$$\delta W_{eff} = \frac{Nc}{12\pi^2}\int d\Sigma^{ijk} tr \mathcal{L}_i \mathcal{L}_j \mathcal{L}_k + \frac{ig}{4\pi}Nc \int d^2x \epsilon^{\mu\nu} tr A_\mu \mathcal{L}_\nu \qquad (23)$$

This just is the Wess-Zumino effective action in two dimensions.

IV. Conclusion

We have shown the change in the fermion action under the chiral transformation and stressed the generation of vector boson mass and Wess-Zumino effective action on a lattice in two dimensions. it is found that the change of effective action, due to the nontrivial change of Wilson's chiral symmetry breaking term associated with the chiral transformation of fermionic variable, has been transferred to the anomalous action. Our results in continuum limit coincide with that obtained by the methods beyond the lattice regularization. Let us comment briefly on the consequences of our results for studying the chiral gauge theories on a lattice. In this approach we hope further to show the anomaly cancellation in our model based on the corresponding work done in continuous case[10]. Besides, it is that the lattice calculation is a possible nonperturburbative approach in a realistic four dimensional world. We have shown the mass-independent Wess-Zumino effective action arises in four-dimensional case in lattice method[11].

Reference

1. R. Jackiw, R. Rajaraman, Phys. Rev. Lett. 54(1985)1219.
2. J. Wess, B. Zumino, Phys. Lett. 37B(1971)95.
3. H.B. Nielsen, N. Ninomiya, Nucl. Phys. B185(1981)20, B193(1981)173.

4. K.G. Wilson, in New phenomena in subnuclear physics, ed. A.Zichichi, Erice, 1975 (plenum, New York, 1977).
5. L.H. Kasten, J. Smit, Nucl. Phys. B183(1981)103.
6. S. Aoki, Phys. Rev. D35(1987)1435.
7. J. Schwinger, Phys. Rev. 82(1951)664.
8. A. Andrianov, Nucl. Phys. B233(1984)232,247.
9. K. Harada, H. Kabota, I. Tsutsui, Phys. Lett. 173B(1986)77.
10. L.D. Faddeev, S.L. Shatashvili, Phys. Lett. 167B(1986)225.
11. Z.S. Ma, S.H. Guo, Preprint, 1988.

Transformation for the Solutions of the Two-Dimensional Toda Lattice

Liu Qiming

Department of Mathematics, Shanghai University of Science and Technology, Shanghai, People's Rep. of China

In the development of soliton theory, various exact methods were found for solving nonlinear evolution equations. Among them, the bilinear transformation method [1] initiated by Hirota is a powerful tool. In the bilinear formalism, a given nonlinear equation is first transformed into a bilinear form through a dependent variable transformation. Then the N-soliton solutions, the Bäcklund transformations and an infinite number of conservation laws of this bilinear equation can be derived in a systematic way.

Recently Nakamura [2] proposed a type of bilinear N-soliton solutions of the two-dimensional Toda lattice:

$$f_n(\partial_x \partial_s f_n) - (\partial_x f_n)(\partial_s f_n) - f_{n+1} f_{n-1} + f_n f_n = 0 \tag{1}$$

which is the determinant of the matrix whose matrix elements are in the integral form. This type of solutions is proved by the purely algebraic procedure and the knowledge of the Gel'fand-Levitan-Marchenko integral equation is not required. To simplify this procedure the "four-operators" bilinear form of eq.(1) was introduced [3]:

$$f_n(L_a L_{a'} - L_b L_{b'}) f_n - (L_a f_n)(L_{a'} f_n) + (L_b f_n)(L_{b'} f_n) = 0 , \tag{2}$$

where

$$L_a = \partial_x, \quad L_{a'} = \partial_s, \quad L_b = \exp(\partial_n) - 1, \quad L_{b'} = 1 - \exp(-\partial_n). \tag{3}$$

The propose of this paper is to prove a transformation formula of the solutions of eq.(1) by the use of Nakamura's method.

Theorem If g_n satisfies eq.(1) and $\varphi_n^i, \bar{\varphi}_n^i$ ($1 \leq i \leq N$) satisfy

$$\begin{cases} \partial_x \varphi_n^i = u_n \varphi_{n+1}^i , \\ \partial_s \varphi_n^i = -\varphi_{n-1}^i + v_n \varphi_n^i , \end{cases} \tag{4}$$

$$\begin{cases} \partial_x \bar{\varphi}_n^i = -u_{n-1} \bar{\varphi}_{n-1}^i , \\ \partial_s \bar{\varphi}_n^i = \bar{\varphi}_{n+1}^i - v_n \bar{\varphi}_n^i , \end{cases} \tag{5}$$

where $u_n = \dfrac{g_n g_{n+2}}{g_{n+1}^2}$ and $v_n = \dfrac{\partial_s g_n}{g_n} - \dfrac{\partial_s g_{n+1}}{g_{n+1}}$. Then $f_n = g_{n+1} \det H_n$ also satisfies eq.(1), where $(H_n)_{ij} = c_{ij} + \sum\limits_{m=-\infty}^{n} \bar{\varphi}_m^i \varphi_m^j$ ($1 \leq i, j \leq N$).

Proof Since

$$f_n(L_a L_{a'} - L_b L_{b'})f_n - (L_a f_n)(L_{a'} f_n) + (L_b f_n)(L_{b'} f_n)$$
$$= [g_{n+1}(L_a L_{a'} - L_b L_{b'})g_{n+1} - (L_a g_{n+1})(L_{a'} g_{n+1}) + (L_b g_{n+1})(L_{b'} g_{n+1})](\det H_n)^2$$
$$+ g_{n+1}^2 [H_n(L_a L_{a'} - u_n L_b L_{b'})H_n - (L_a H_n)(L_{a'} H_n) + u_n(L_b H_n)(L_{b'} H_n)] ,$$

it is sufficient to prove that

$$H_n(L_a L_{a'} - u_n L_b L_{b'})H_n - (L_a H_n)(L_{a'} H_n) + u_n(L_b H_n)(L_{b'} H_n) = 0 .$$

In order to calculate the quantities including the four linear operators $L_a, L_{a'}, L_b$ and $L_{b'}$, first we see the operations of these operators on the matrix element $(H_n)_{ij}$. Using eqs.(4) and (5), we have

$$L_a(H_n)_{ij} = u_n \bar{\varphi}_n^i \varphi_{n+1}^j , \quad L_{a'}(H_n)_{ij} = \bar{\varphi}_{n+1}^i \varphi_n^j ,$$
$$L_a L_{a'}(H_n)_{ij} = u_n(\bar{\varphi}_{n+1}^i \varphi_{n+1}^j - \bar{\varphi}_n^i \varphi_n^j) ,$$
$$L_b(H_n)_{ij} = \bar{\varphi}_{n+1}^i \varphi_{n+1}^j , \quad L_{b'}(H_n)_{ij} = \bar{\varphi}_n^i \varphi_n^j ,$$
$$L_b L_{b'}(H_n)_{ij} = \bar{\varphi}_{n+1}^i \varphi_{n+1}^j - \bar{\varphi}_n^i \varphi_n^j .$$

Next we consider the operations of these operators on $\det H_n$. Using the above equations and the following matrix identity formula

$$\sum_{k=1}^{N} \det A_k = - \begin{vmatrix} a_{11} & a_{12} & \cdots & a_{1N} & x_1 \\ a_{21} & a_{22} & \cdots & a_{2N} & x_2 \\ \vdots & \vdots & \ddots & \vdots & \vdots \\ a_{N1} & a_{N2} & \cdots & a_{NN} & x_N \\ y_1 & y_2 & \cdots & y_N & 0 \end{vmatrix} ,$$

where

$$(A_k)_{ij} = \begin{cases} a_{ij} & (i \neq k) , \\ x_i y_j & (i = k) , \end{cases}$$

we have

$$L_a H_n = -u_n \begin{vmatrix} & & \bar{\varphi}_n^1 \\ & H_n & \vdots \\ & & \bar{\varphi}_n^N \\ \varphi_{n+1}^1 & \cdots & \varphi_{n+1}^N & 0 \end{vmatrix} , \quad L_{a'} H_n = - \begin{vmatrix} & & \bar{\varphi}_{n+1}^1 \\ & H_n & \vdots \\ & & \bar{\varphi}_{n+1}^N \\ \varphi_n^1 & \cdots & \varphi_n^N & 0 \end{vmatrix} ,$$

and

$$L_b H_n = - \begin{vmatrix} & H_n & & \bar{\varphi}^1_{n+1} \\ & & & \vdots \\ & & & \bar{\varphi}^N_{n+1} \\ \varphi^1_{n+1} & \cdots & \varphi^N_{n+1} & 0 \end{vmatrix}, \quad L_{b'} H_n = - \begin{vmatrix} & H_n & & \bar{\varphi}^1_n \\ & & & \vdots \\ & & & \bar{\varphi}^N_n \\ \varphi^1_n & \cdots & \varphi^N_n & 0 \end{vmatrix}.$$

Using the following matrix identity formula

$$\sum_{\substack{k,k'=1 \\ (k \neq k')}}^{N} \det A_{kk'} = \begin{vmatrix} a_{11} & a_{12} & \cdots & a_{1N} & x_{11} & x_{12} \\ a_{21} & a_{22} & \cdots & a_{2N} & x_{21} & x_{22} \\ \vdots & \vdots & & \vdots & \vdots & \vdots \\ a_{N1} & a_{N2} & \cdots & a_{NN} & x_{N1} & x_{N2} \\ y_{11} & y_{12} & \cdots & y_{1N} & 0 & 0 \\ y_{21} & y_{22} & \cdots & y_{2N} & 0 & 0 \end{vmatrix},$$

where

$$(A_{kk'})_{ij} = \begin{cases} a_{ij} & (i \neq k, k'), \\ x_{i1} y_{1j} & (i = k), \\ x_{i2} y_{2j} & (i = k'), \end{cases}$$

we have

$$(L_a L_{a'} - u_n L_b L_{b'}) H_n = -u_n \begin{vmatrix} & H_n & & \bar{\varphi}^1_n & \bar{\varphi}^1_{n+1} \\ & & & \vdots & \vdots \\ & & & \bar{\varphi}^N_n & \bar{\varphi}^N_{n+1} \\ \varphi^1_n & \cdots & \varphi^N_n & 0 & 0 \\ \varphi^1_{n+1} & \cdots & \varphi^N_{n+1} & 0 & 0 \end{vmatrix}.$$

Thus

$$H_n (L_a L_{a'} - u_n L_b L_{b'}) H_n - (L_a H_n)(L_{a'} H_n) + u_n (L_b H_n)(L_{b'} H_n)$$

$$= -u_n \left(|H_n| \begin{vmatrix} & H_n & & \bar{\varphi}^1_n & \bar{\varphi}^1_{n+1} \\ & & & \vdots & \vdots \\ & & & \bar{\varphi}^N_n & \bar{\varphi}^N_{n+1} \\ \varphi^1_n & \cdots & \varphi^N_n & 0 & 0 \\ \varphi^1_{n+1} & \cdots & \varphi^N_{n+1} & 0 & 0 \end{vmatrix} \right.$$

$$\left. - \begin{vmatrix} & H_n & & \bar{\varphi}^1_n \\ & & & \vdots \\ & & & \bar{\varphi}^N_n \\ \varphi^1_{n+1} & \cdots & \varphi^N_{n+1} & 0 \end{vmatrix} \begin{vmatrix} & H_n & & \bar{\varphi}^1_{n+1} \\ & & & \vdots \\ & & & \bar{\varphi}^N_{n+1} \\ \varphi^1_n & \cdots & \varphi^N_n & 0 \end{vmatrix} + \begin{vmatrix} & H_n & & \bar{\varphi}^1_{n+1} \\ & & & \vdots \\ & & & \bar{\varphi}^N_{n+1} \\ \varphi^1_{n+1} & \cdots & \varphi^N_{n+1} & 0 \end{vmatrix} \begin{vmatrix} & H_n & & \bar{\varphi}^1_n \\ & & & \vdots \\ & & & \bar{\varphi}^N_n \\ \varphi^1_n & \cdots & \varphi^N_n & 0 \end{vmatrix} \right),$$

which is identically zero due to the Jacobi formula. Therefore we have proved that f_n actually satisfies eq.(1).

If we take $g_n=1$, then we get the cylindrical N-soliton solutions[4]

$f_n = \det H_n$, where $(H_n)_{ij} = c_{ij} + \sum_{m=-\infty}^{n} \bar{\varphi}_m^i \varphi_m^j$ ($1 \leq i, j \leq N$) and the quantities φ_n^i and $\bar{\varphi}_n^i$ ($1 \leq i \leq N$) satisfy

$\partial_x \varphi_n^i = \varphi_{n+1}^i$, $\partial_s \varphi_n^i = -\varphi_{n-1}^i$; $\partial_x \bar{\varphi}_n^i = -\bar{\varphi}_{n-1}^i$, $\partial_s \bar{\varphi}_n^i = \bar{\varphi}_{n+1}^i$.

The author would like to express his sincere thanks to Professor Ben-Yu Guo for continual encouragement.

References

[1] Y. Matsuno, Bilinear Transformation Method, Academic Press, New York, 1984.
[2] A. Nakamura, Exact cylindrical solitons of the sine-Gordon equation, the sinh-Gordon equation and the periodic Toda equation, J. Phys. Soc. Japan 57(1988), 3309-3322.
[3] A. Nakamura, A bilinear N-soliton formula for the KP equation, J. Phys. Soc. Japan 58(1989), 412-422.
[4] A. Nakamura, Exact Bessel type solution of the two-dimensional Toda lattice equation, J. Phys. Soc. Japan 52(1983), 380-387.

Part V

Other Topics

Some Ideas on Nonlinear Evolution Equations

F. Calogero

Dipartimento di Fisica, Università di Roma "La Sapienza", I-00185 Roma, Italy, and Istituto Nazionale di Fisica Nucleare, Sezione di Roma, Italy

0. This is a terse review of some ideas and recent results helpful to understand nonlinear evolution equations. The discussion is mainly limited to problems in 1+1 dimensions.

1. <u>S-integrability and C-integrability</u> ([1]). S-integrable nonlinear evolution equations are those solvable (linearizable) via the Spectral transform (or the inverse Scattering technique); typical examples are the KdV equation, the Nonlinear Schroedinger equation, the Sine-Gordon equation. C-integrable equations are those solvable (linearizable) via an appropriate Change of variable (generally a change of the dependent variable); typical examples are the Burgers equation, the equation $u_t = u_{xxx} + 3u_{xx}u^2 + 9u_x^2 u + 3u_x u^4$ ([2]), the Eckhaus equation ([3]). The requirement to be C-integrable is generally more stringent than that to be S-integrable: generally C-integrable equations are also S-integrable (but not viceversa). But there are C-integrable equations that are not S-integrable (if S-integrability is meant to imply the existence of an infinity of <u>local</u> conservation laws); one such instance is the Eckhaus equation ([3]).

2. <u>C-integrable equations displaying a solitonic behaviour</u>. Since C-integrable equations can be linearized by a simple change of variable, they are generally much easier to study than S-integrable equations. It is therefore interesting to know that some such equations display typical solitonic features, such as the presence of solitons that collide purely elastically. One such example is provided by the Eckhaus equation, that in fact displays a very rich solitonic phenomenology ([3]).

3. **Boundary/initial value problems for certain nonlinear evolution equations (on the semiline, on a finite interval).** Certain C-integrable equations can be solved non only on the whole line, but also on the semiline or finite interval. One such example is the Eckhaus equation ([4]); another interesting example (treated only recently) is the Burgers equation ([5]) (which is however solvable only if the unknown function itself is assigned on the boundary; but cannot be fully linearized if the space derivative of the unknown function is instead assigned on the boundary ([6])).

4. **Why are certain nonlinear evolution equations both widely applicable and integrable?** One might well take the view that this question is metascientific; or, equivalently, one might quote Galileo: "this great book that stands always open before our eyes (I mean the universe)...is written in mathematical language" ([7]). We submit that there exist a less methaphysical (if somewhat heuristic) answer: "certain `universal' nonlinear evolution PDEs can be obtained, by a limiting procedure involving rescalings and an asymptotic expansion, from very large classes of nonlinear evolution equations; for instance, from the class of autonomous nonlinear evolution equations whose linear part is dispersive but otherwise arbitrary, and whose nonlinear part depends in an analytic but otherwise arbitrary manner on the dependent variable and its derivatives. Because this limiting procedure is the correct one to evince nonlinear effects, the universal model equations obtained in this manner (of which the nonlinear Schroedinger equation in 1+1 dimensions is the prototype) show up in many, disparate, applicative contexts; they are widely applicable. Because this limiting procedure generally preserves integrability, these universal model equations are likely to be integrable, since in order for this to happen it is sufficient that the very large class from which they are obtainable contain just one integrable equation; indeed, while the fact that an arbitrarily given equation turns out to be integrable may be seen as an exceptional event, the fact that a very large class of equations contain at least one integrable

equation may be considered normal, i. e. not exceptional; hence a universal model equation that is obtainable by a limiting procedure from (all!) the equations of a large class is likely to be integrable, provided the limiting procedure preserves integrability" (⁸).

5. **Necessary conditions for integrability of nonlinear PDEs**. The argument we have just given "may also be run backwards: if a universal model equation, obtainable via a limiting procedure that preserves integrability from all the equations of a large class, turns out not to be integrable, then none of the equations contained in the large class is integrable; hence this approach also yields necessary conditions for integrability of wide applicability" (⁸,⁹).

6. **A novel phenomenon: solitons that change their shape upon collision**. Besides explaining "what had hitherto appeared to us a puzzling miracle, namely the fact that certain nonlinear PDEs appear in many applications and are applicable" (¹), the approach outlined above "provides moreover a powerful, if heuristic, methodology, to understand the integrability of known equations, to test the integrability of new equations and to obtain novel integrable equations likely to be applicable" (⁸). This methodology has already been widely applied (also to problems in more than 1+1 dimensions (¹⁰)), and is currently under further investigation. An interesting result in this context has been the study of the nonlinear PDEs describing the interaction of N dispersive waves (¹¹). A particularly amusing phenomen, discovered in the context of certain (C-integrable) nonlinear PDEs describing the nonresonant interaction of N dispersive waves, is the occurrence of solitons that, whenever they collide, change their shape (in a remarkable way, described by a universal formula); the justification for calling these objects "solitons" is because the effect of the multiple collisions, that occur over time whenever more than two solitons are present, turns out to be independent of the order in which they occur (¹²).

7. _Outlook_. A goal is to identify and study interesting multidimensional C-integrable equations (if any).

References

[1] F. Calogero and W. Eckhaus: "Nonlinear evolution equations, rescalings, model PDEs and their integrability. I & II". Inverse Problems **3**, 229-262 (1987) & **4**, 11-33 (1988).

[2] F. Calogero: "The evolution PDE $u_t = u_{xxx} + 3 (u_{xx} u^2 + 3 u_x^2 u) + 3 u_x u^4$". J. Math. Phys. **28**, 538-555 (1987).

[3] F. Calogero and S. De Lillo: "The Eckhaus PDE i $psi_t + psi_{xx} + 2 (\psi\^2)_x$ psi $+ \psi\^4$ psi $= 0$ ". Inverse Problems **3**, 633-681 (1987).

[4] F. Calogero and S. De Lillo: "Cauchy problem on the semiline and on a finite interval for the Eckhaus equation". Inverse Problems **4**, L33-L37 (1988).

[5] F. Calogero and S. De Lillo: "The Burgers equation on the semi-infinite and finite intervals". Nonlinearity **2**, 37-43 (1989).

[6] F. Calogero and S. De Lillo: (to be published).

[7] Galileo Galilei: _Il Saggiatore_, 1623. [See: _Opere di Galileo Galilei_, Edizione nazionale, Barbera, Firenze, 1890-1909, vol. VI, p. 232].

[8] F. Calogero: "Why are certain nonlinear PDEs both widely applicable and integrable?". Rome preprint n. 582, January 1988. To be published in: V. E. Zakharov (editor): _What is integrability (for nonlinear PDEs)?_, Springer, 1989 (in press).

[9] F. Calogero and W. Eckhaus: "Necessary conditions for integrability of nonlinear PDEs". Inverse Problems **3**, L27-L32 (1987).

[10] F. Calogero and A. Maccari: "Equations of nonlinear Schroedinger type in 1+1 and 2+1 dimensions, obtained from integrable PDEs". In: P. C. Sabatier (editor): _Inverse Problems: an Interdisciplinary Study_ (Proceedings of the meeting on Inverse Problems held in Montpellier, November 1986), Advances in Electronics and Electron Physics, **19**, Academic Press, London & New York, 1987, pp. 463-480.

[11] F. Calogero: "Universality and integrability of the nonlinear PDEs describing N-wave interactions". J. Math. Phys. **30**, 28-40 (1989).

[12] F. Calogero: "Solutions of certain integrable nonlinear evolution PDEs describing nonresonant N-wave interactions". J. Math Phys. (in press).

Some Problems of the Generalized Kuramoto-Sivashinsky Type Equations with Dispersive Effects

Guo Boling

Center for Nonlinear Studies, Institute of Applied Physics
and Computational Mathematics, P.O. Box 8009, Beijing, People's Rep. of China

The Kuramoto-Sivashinsky equation has been studied by many authors [1-4]. In this note, firstly we consider the following initial value problem for the generalized KS type equations with the dispersive effects

$$u_t + f(u)_x + \alpha u_{xx} + \phi(u)_{xx} + \beta u_{xxx} + \gamma u_{xxxx} = g(u), \tag{1}$$

$$u(x,0) = u_0(x), \quad x \in R^1, \tag{2}$$

where α, β and γ are constants, $f(x)$, $\phi(x)$ and $g(x)$ are known functions. In $n(n \geq 2)$ dimensional case, the following initial value problem for the system of generalized K-S type equations has been studied:

$$\vec{u}_t + \sum_{i=1}^{n} \frac{\partial}{\partial x_i} grad\phi(u_1, \ldots, u_N) + \alpha \Delta \vec{u} + \beta \Delta^2 \vec{u} = \vec{g}(\vec{u}), \tag{3}$$

$$\vec{u}(x,0) = \vec{u}_0(x), \quad x \in R^n, \ n \geq 2, \tag{4}$$

where $\vec{u}(x,t) = (u_1(x,t), \ldots, u_N(x,t))$ is a N dimensional unknown functional vector. $\varphi(s_1, \ldots, s_N)$ is a scalar function of its variables. Under some conditions the existence, nonexistence and asymptotic behaviour as $t \to \infty$ for the global smooth solutions of the problem (1) (2) and problem (3) (4) are obtained.

Next, by using the qualitative theory of ODE, we discuss the structure of the travelling waves solution for the equation

$$u_t + f(u)_x + \phi(u)_{xx} = \alpha u_{xx} + \beta u_{xxx} + \gamma u_{xxxx}. \tag{5}$$

Finally, by using the Lie's infinitesimal transformations, the similarity solution for the equation

$$u_t + uu_x + \alpha u_{xx} + \beta u_{xxx} + \gamma u_{xxxx} = 0 \tag{6}$$

are obtained.

(I) Existence and Nonexistence

By using a priori estimations and the energy method, we can prove that

Theorem 1. Suppose that the following conditions are satisfied:

(1) $f(u) \in C^{m+1}$, $|f(u)| \leq A|u|^p$, $0 \leq p < 7$, $A = const > 0$
(2) $\phi(u) \in C^{m+2}$, $\phi'(u) \leq 0$, $|\phi'(u)| \leq B|u|^q$, $0 \leq q < 4$, $B = const > 0$
(3) $g(u) \in C^m$, $g'(u) \leq b$, $g(0) = 0$, $b = const > 0$
(4) $u_0(x) \in H^m(R^1)$, $m \geq 1$.

Then there exists a global smooth solution of the initial value problem (1) (2), $u(x,t) \in L^\infty(o, T, H^m(R^1))$.

Theorem 2. Assume that $f(u) \in C^2$, $\phi(u) \in C^3$, $g(u) \in C^1$. Then the smooth solution of the initial value problem (1) (2) is unique.

Theorem 3. Suppose that the conditions of Theorem 2 are satisfied, and assume that $\alpha \leq 0$, $b \leq 0$. Then for the solution of problem (1)(2), we have

$$\lim_{t \to \infty} \|u(.,t)\|_{L^\infty(R^1)} = 0 \tag{7}$$

Theorem 4. If the conditions of Theorem 2 are satisfied, then the smooth solution of problem (1) (2) is continuously dependent of the initial data, i.e., for any given $\epsilon > 0$, there is $\delta > 0$, such that

$$\|u_0 - v_0\|_{L_2(R^1)} \leq \delta \tag{8}$$

we have

$$\|u(.,t) - v(.,t)\|_{L_2(R^1)} \leq \epsilon, \quad 0 \leq t \leq T, \tag{9}$$

where u(x,t) and v(x,t) are two solutions of problem (1) (2) which take the initial data $u_0(x)$ and $v_0(x)$ respectively.

By using the method of the convex function, we have

Theorem 5. Suppose that there exists a local smooth solution u(x,t) of problem (10) (11),

$$u_t + \alpha u_{xx} + \beta u_{xxxx} + cu = g(u), \quad \alpha > 0, \ \beta > 0 \tag{10}$$

$$u(x,0) = u_0(x), \quad x \in R^1 \tag{11}$$

and assume that
  (1) $2\beta \geq \alpha > 0$, $c \geq \beta > 0$,
  (2) g(0)=0, and there is a constant $\delta > 0$, such that

$$2(\delta + 1)G(u) \leq (u, g(u)) \tag{12}$$

where $G(u) = \int_0^1 (g(\rho u), u) d\rho$.
  (3) The initial valued function $u_0(x)$ satisfies

$$G(u_0) > \frac{1}{2}[c(u_0, u_0) - \alpha(u_{0x}, u_{0x}) + \gamma(u_{0xx}, u_{0xx})]. \tag{13}$$

Then the solution of problem (10) (11) "blow up", i.e., there is a constant

$$T \leq T_{\gamma_0 \delta} = \frac{(2\delta+1)(u_0, u_0)}{2\delta^2(\delta+1)}\{G(u_0) - \frac{1}{2}[c(u_0,u_0) - \alpha(u_{0x}, u_{0x}) + \gamma(u_{0xx}, u_{0xx})]\}^{-1} \tag{14}$$

such that

$$\lim_{t \to T_{\gamma_0 \delta}^-} Sup_{\{0 < \tau < t\}}(u(.,\tau), u(.,\tau)) = +\infty. \tag{15}$$

Theorem 6. Suppose that there exists a local smooth solution of problem (16) (17),

$$u_t + \alpha u_{xx} + \beta u_{xxx} + f(u)_x = g(u) \tag{16}$$

$$u(x,0) = u_0(x), \quad x \in R^1 \tag{17}$$

and assume that the following conditions are satisfied:
  (1) $\alpha > 0$
  (2) $(u, g(u)) \geq c(u,u)^{1+\delta}$, $c = const > 0$, $\delta > 0$.
  (3) $\|u_0(x)\|_{L_2(R^1)} > 0$.
Then the solution of problem (16) (17) "blow up".

In $n(n \geq 2)$ dimensional case, we have

Theorem 7. Suppose that the following conditions are satisfied:
(1) $\phi(\vec{u}) \in C^{m+2}$,

$$|grad\phi(\vec{u})| \leq A|\vec{u}|^p, \quad 1 \leq p \leq \frac{6}{n}+1, \quad A = const > 0. \tag{18}$$

(2) $\vec{g}(\vec{u}) \in C^m$, and the Jacobi derivative matrix $\vec{g}_u(\vec{u})$ is semibounded, i.e., there is a constant b, such that

$$\vec{\xi} \cdot \vec{g}_u(\vec{u})\vec{\xi} \leq b|\vec{\xi}|^2, \quad \forall \vec{\xi} \in R^n. \tag{19}$$

(3) $\vec{u}_0(x) \in H^m(R^n)$, $m \geq [\frac{n}{2}]+1$.
Then there exists a global smooth solution $\vec{u}(x,t)$ of problem (3) (4),

$$\vec{u}(x,t) \in L^\infty(0,T,H^m(R^n)).$$

Theorem 8. Suppose that the conditions of Theorem 7 are satisfied. Assume that $\alpha \to 0$, then the global smooth solution of problem (3) (4) tend to the solution for the corresponding problem of equations

$$\vec{u}_t + \sum_{i=1}^n \frac{\partial}{\partial x_i} grad\phi(\vec{u}) + \beta\Delta^2 \vec{u} = \vec{g}(\vec{u}), \tag{20}$$

$$\vec{u}(x,0) = \vec{u}_0(x). \tag{21}$$

(II) The Travelling Waves Solutions

Now we consider the travelling waves solutions for the equation (3). Letting $\xi = x - Dt$, $D = const > 0$, we have

$$-\gamma u_{\xi\xi\xi} - \beta u_{\xi\xi} - \alpha u_\xi + \phi(u)_\xi + f(u) - Du = A.$$

Letting A=0, and $\frac{du}{d\xi} = y$, $\frac{dy}{d\xi} = z$, it follows

$$\begin{cases} \dfrac{du}{d\xi} = y, \\ \dfrac{dy}{d\xi} = z, \\ \dfrac{dz}{d\xi} = \dfrac{1}{\gamma}(f(u) - Du) + (\phi'(u) - \alpha)y - \dfrac{\beta}{\gamma}z. \end{cases} \tag{22}$$

Obviously, (0,0,0) is a singular point of the system of ODE (22) as f(0)=0, and $(u_1,0,0)$ is an another singular point of the system (22), where $Du_1 - f(u_1) = 0$, $u_1 > 0$.

The characteristic equation of the linearized equation of the system (22) at point (0,0,0) is

$$\lambda^3 + \frac{\beta}{\gamma}\lambda^2 + \frac{1}{\gamma}[\alpha - \phi'(0)]\lambda - \frac{1}{\gamma}(f'(0) - D) = 0 \tag{23}$$

Letting

$$\Delta \equiv \frac{1}{4\gamma^2}[\frac{2}{27}\frac{\beta^3}{\gamma^2} + \frac{\beta}{3\gamma}(\phi'(0) - \alpha) + (D - f'(0))]^2 + \frac{1}{27\gamma}(\alpha - \phi'(0) - \frac{\beta^2}{3\gamma}). \tag{24}$$

we can analyse the characters of the singular point as follows:

(a) $\Delta > 0$. Specially, $\alpha > \phi'(0) + \frac{\beta^2}{3\gamma}$ or $\alpha = \phi'(0) + \frac{\beta^2}{3\gamma}$, $D \neq f'(0) + \frac{1}{27}\frac{\beta^3}{\gamma^2}$.
Then the equation (23) have one real root, two complex roots (conjugate). There is a saddle-focus point. The picture is as follows

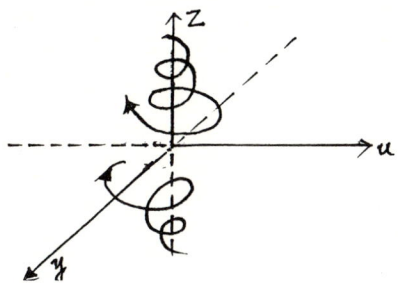

(b) $\Delta \equiv 0$. Specially, $\alpha = \phi'(0) + \frac{\beta^2}{3\gamma}$, $D = f'(0) + \frac{1}{27}\frac{\beta^3}{\gamma^2}$.
Then the equation (23) has three real roots. There is a saddle point. The picture is as follows

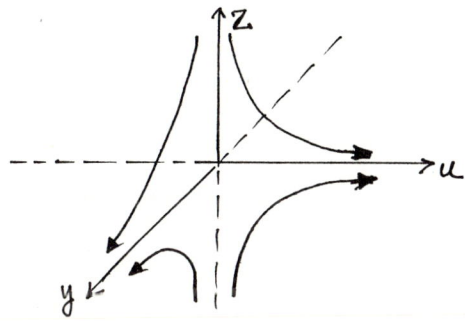

(c) $\Delta < 0$, $\beta^2 > 3(\alpha - \phi'(0))\gamma$.
Then the equation (23) has three roots as follows

$$\lambda_1 = 2\sqrt[3]{\gamma}\cos\frac{\theta}{3}, \quad \lambda_2 = -\sqrt[3]{\gamma}(\cos\frac{\theta}{3} + \sqrt{3}\sin\frac{\theta}{3}),$$

$$\lambda_3 = -\sqrt[3]{\gamma}(\cos\frac{\theta}{3} - \sqrt{3}\sin\frac{\theta}{3}),$$

where

$$\sin\theta = \frac{\sqrt{|\Delta|}}{\gamma}, \quad \cos\theta = \frac{-q}{\gamma}, \quad \gamma = \sqrt{\frac{q^2}{4} + |\Delta|},$$

$$q = 2(\frac{\beta}{3\gamma})^3 - (\alpha - \phi'(0))\frac{\beta}{3\gamma^2} - \frac{1}{\gamma}(f'(0) - D).$$

(a) If $\tan\frac{\theta}{3} \geq \frac{1}{\sqrt{3}}$, then $\lambda_3 \geq 0$, $\lambda_1 > 0$, $\lambda_2 < 0$. The singular point is a saddle point at $(0,0,0)$.
(b) If $\tan\frac{\theta}{3} < \frac{1}{\sqrt{3}}$, Then $\lambda_3 \leq 0$, $\lambda_1 > 0$, $\lambda_2 < 0$. The singular point is also a saddle point at $(0,0,0)$.

Then linearized equation of system (22) at singular point $(u_1, 0, 0)$ is

$$\begin{cases} \dfrac{du}{d\xi} = y \\ \dfrac{dy}{d\xi} = z \\ \dfrac{dz}{d\xi} = \dfrac{1}{\gamma}[f'(u_1) - D](u - u_1) + \dfrac{1}{\gamma}(\phi'(u_1) - \alpha) - \dfrac{\beta}{\gamma}z \end{cases} \quad (25)$$

Let $u' = u - u_1$, then the characteristic equation of the system (25) is

$$\lambda^3 + \frac{\beta}{\gamma}\lambda^2 + \left(\frac{\alpha}{\gamma} - \frac{\phi'(u_1)}{\gamma}\right)\lambda - \frac{1}{\gamma}(f'(u_1) - D) = 0. \quad (26)$$

It is similar to previous discussing cases at singular point (0,0,0).

(III) Similarity Transformations and Similarity Solutions.

We consider a one parameter ϵ Lie's group of infinitesimal transformations

$$\begin{aligned} x^* &= x + \epsilon\xi(x, t, u) + O(\epsilon^2) \\ t^* &= t + \epsilon\tau(x, t, u) + O(\epsilon^2) \\ u^* &= u + \epsilon\eta(x, t, u) + O(\epsilon^2) \end{aligned} \quad (27)$$

Assume that the equation

$$u_t + u u_x + \alpha u_{xx} + \beta u_{xxx} + \gamma u_{xxxx} = 0 \quad (28)$$

are invariant under the transformation (27), we get the following relations from the coefficient of the first order of ϵ

$$[\eta_t] + \alpha[\eta_{xx}] + \beta[\eta_{xxx}] + \gamma[\eta_{xxxx}] + \eta u_x + u[\eta_x] = 0, \quad (29)$$

where $[\eta_t]$, $[\eta_x]$, $[\eta_{xx}]$, $[\eta_{xxx}]$ and $[\eta_{xxxx}]$ denote the infinitesimals of $u_t, u_x, u_{xx}, u_{xxx}$ and u_{xxxx} respectively under the transformation (27).

Assuming $\xi = \xi(x,t)$, $\tau = \tau(x,t)$, $\eta = f(x,t) + g(x,t)u$, we get [5]

$$\begin{aligned} [\eta_t] &= \eta_t + (\eta_u - \tau_t)u_t - \xi_t u_x, \\ [\eta_x] &= \eta_x + (\eta_u - \xi_x)u_x - \tau_x u_t, \\ [\eta_{xx}] &= \eta_{xx} + (2\eta_{xu} - \xi_{xx})u_x - \tau_{xx}u_t + (\eta_u - 2\xi_x)u_{xx} - 2\tau_x u_{xt}, \\ [\eta_{xxx}] &= \eta_{xxx} + 3(\eta_{xu} - \xi_{xxx})u_x - \tau_{xxx}u_t + 3(\eta_{xu} - \xi_{xx})u_{xx} - 3\tau_{xx}u_{xt} \\ &\quad + (\eta_u - 3\xi_x)u_{xxx} - 3\tau_x u_{xxt}. \\ [\eta_{xxxx}] &= \eta_{xxxx} + (4\eta_{xxxu} - \xi_{xxxx})u_x - \tau_{xxxx}u_t + (6\eta_{xxu} - 4\xi_{xxx})u_{xx} - 4\tau_{xxx}u_{xt} \\ &\quad + (4\eta_{xu} - 6\xi_{xx})u_{xxx} - 6\tau_{xx}u_{xxt} + (\eta_u - 4\xi_x)u_{xxxx} - 4\tau_x u_{xxxt} \end{aligned} \quad (30)$$

Substituting (30) into (29), putting the coefficient of each term and noticing equation (28), we have a set of equations, solving the set, we can get that the infinitesimals ξ, τ and η are

$$\begin{aligned} \xi &= C_1 t + C_2, \\ \tau &= C_3, \\ \eta &= C_1, \end{aligned}$$

where $C_i (i = 1, 2, 3)$ are arbitrary real constants.

Solving the characteristic equation

$$\frac{dx}{\xi(x,t)} = \frac{dt}{\tau(x,t)} = \frac{du}{\eta(x,t,u)}$$

we obtain the following similarity variables and similarity form

$$\begin{aligned}\varsigma &= x - \frac{C_1}{2C_3}t^2 + \frac{C_2}{C_3}t, \\ f(\varsigma) &= u - \frac{C_1}{C_3}t.\end{aligned} \quad (31)$$

Substituting (31) into equation (28), we obtain the following ODE

$$\gamma f''''(\varsigma) + \beta f'''(\varsigma) + \alpha f''(\varsigma) + f(\varsigma)f'(\varsigma) - \frac{C_2}{C_3}f'(\varsigma) + \frac{C_1}{C_3} = 0. \quad (32)$$

Integrating (32) with respect to ς one time, we have

$$\gamma f'''(\varsigma) + \beta f''(\varsigma) + \alpha f'(\varsigma) + \frac{1}{2}f^2(\varsigma) - \frac{C_2}{C_3}f(\varsigma) + \frac{C_1}{C_3}\varsigma = 0. \quad (33)$$

If $\alpha = 0$, $\gamma = 0$, $\beta = 1$, $C_1 = 0$, then we have

$$f''(\varsigma) + \frac{1}{2}f^2(\varsigma) - \frac{C_2}{C_3}f(\varsigma) = 0 \quad (34)$$

The solution of KdV equation can be found

$$u(x,t) = \frac{3C_2}{C_3}\text{sech}^2\sqrt{\frac{C_3}{2C_2}}(x - \frac{C_2}{C_3}t) \quad (35)$$

Let $X_i (i=1,2,3)$ respect the infinitesimal operators corresponding to the parameters C_1, C_2, C_3 respectively. Then

$$\begin{aligned}X_1 &= t\frac{\partial}{\partial x} \\ X_2 &= \frac{\partial}{\partial x} + \frac{\partial}{\partial t} \\ X_3 &= \frac{\partial}{\partial u}\end{aligned} \quad (36)$$

It is easy to see that the closed Lie algebra of (36) is genarated by the operators X_1, X_2, X_3.

References

[1] Y. Kuramoto, Prog. Theo. Phys., 63(1980), 1885-1903.
[2] G. Sivashinsky, Acta Astronant, 4(1977), 1117-1206.
[3] T.Shlang and G. Sivashinsky, J. de Physique, 43(1982), 459-466.
[4] M.T. Aimar and P. Dener, Universite de TOULON et da VAR, preprint, 1982
[5] G.W. Bluman and J.D. Cole, Similarity Methods for Differential Equations, Springer, Berlin, 1974.

Standard Nonlinearities Associated with KdV-like Two-Soliton Solutions

F. Lambert and R. Willox

Theoretische Natuurkunde, Vrije Universiteit Brussel, Pleinlaan 2, B-1050 Brussel, Belgium

1. Introduction.

Shock-like solitary waves or kinks of the PKdV (or Burgers) type:

$$V_{sol}(x,t) = -2\partial_x \ln(1 + \frac{\lambda}{2k}\exp\theta), \quad \theta = -kx + \omega t, \quad \lambda, k > 0 \tag{1}$$

are among the simplest closed form solutions to NLPDE's which are of physical interest. They are the "potentials" from which the familiar KdV-like sech square waves can be derived.

The occurence of such solutions is not exceptional. The existence of a true soliton, however, requires an exceptional balance between the dispersive terms of the linear part $L(V)$ of the equation and the nonlinear part $K(V)$. A necessary (but not sufficient) condition is the "two-soliton balance": the equation must also have two-soliton (potential) solutions. This strong constraint happens to be automatically satisfied for equations which are derived from Hirota's bilinear equations of KdV-type [1]. Yet, it is not clear which precise conditions allow for the bilinearization of a given NLPDE. Nor has there been any attempt, so far, to characterize the various nonlinearities which balance a given $L(V)$, from a direct point of view (the equations of the KdV-hierarchy [2] are not expressible in terms of a single bilinear form)....

The present approach aims at expressing the two-soliton balance in terms of explicit conditions on the NLPDE for the original variable. The starting point is a set of lowest order consistency conditions which are *necessary* for the existence of PKdV-like two-soliton solutions. We restrict our discussion, for simplicity, to a particular class of L-operators including dispersive terms of the generalized KdV and RLW type. The requirement that the *two-soliton consistency conditions* be satisfied as *polynomial identities* opens several ways of realizing the two-soliton balance, three of which lead to standard hierarchies (the Hopf-Burgers hierarchy [3], the KdV-hierarchy and the bilinear hierarchies).

A striking feature of the two-soliton balance is the presence of simple combinatorial rules which characterize the standard nonlinearities of the linearizable Hopf-Burgers hierarchy as well as those of the bilinear hierarchies.

2. Necessary conditions for the existence of two-soliton solutions.

We consider dispersive NLPDE's for a field $V(x,t)$: $\quad L(V) = K(V) \quad$ (2)
where L stands for a linear differential operator with constant coefficients and where $K(V)$ denotes a polynomial nonlinearity in V and in its partial derivatives:

$$K(V) = K^{(2)}(V) + K^{(3)}(V) + \ldots \qquad (3)$$

$$K^{(2)}(V) = \sum_i \beta_i \, V_{x\ldots xt\ldots t}^{(p_1^i + q_1^i)} \, V_{x\ldots xt\ldots t}^{(p_2^i + q_2^i)} \,, \qquad (4)$$

$$K^{(3)}(V) = \sum_i \gamma_i \, V_{x\ldots xt\ldots t}^{(r_1^i + s_1^i)} \, V_{x\ldots xt\ldots t}^{(r_2^i + s_2^i)} \, V_{x\ldots xt\ldots t}^{(r_3^i + s_3^i)} \,, \qquad (5)$$

... with $\quad p_j^i$, q_j^i , r_j^i , s_j^i = integer or zero.

We wish to derive necessary conditions on $L(V)$ and $K(V)$ for the existence of two-soliton (potential) solutions:

$$V^{(2)} = -2\,\partial_x \ln\left[1 + \lambda\left(\frac{\exp\theta_1}{2k_1} + \frac{\exp\theta_2}{2k_2}\right) + \lambda^2 A_{12} \frac{\exp(\theta_1+\theta_2)}{4k_1 k_2}\right], \quad \theta_j = -k_j x + \omega_j t + \tau_j \,, \qquad (6)$$

which differ from the sum of two kinks by the presence of a two-soliton coupling factor (phaseshift parameter) $A_{12} \neq 1$.

Such conditions are easily obtained from a Rosales perturbation approach [4]. Setting $V = \lambda \bar{V}$ we expand the fraction $\bar{V}^{(2)}(x,t,\lambda)$ in powers of the nonlinearity parameter λ and require that $\bar{V}^{(2)}$ should solve the scaled equation:

$$L(\bar{V}) = \lambda^{-1} K(\lambda \bar{V}) = \lambda \, K^{(2)}(\bar{V}) + \lambda^2 K^{(3)}(\bar{V}) + \ldots \qquad (7)$$

to all orders in λ.

By applying this requirement to the first three orders in λ we obtain the following three consistency conditions:

$$\Phi(2k_j, 2\omega_j) = -2k_j \sum_i (-)^{p_1^i + p_2^i} \beta_i \, k_j^{p_1^i + p_2^i} \, \omega_j^{q_1^i + q_2^i} \,, \qquad (8)$$

$$\Phi(3k_j, 3\omega_j) = \sum_i (-)^{p_1^i + p_2^i + 1} \beta_i \left(2^{p_1^i + q_1^i + 1} + 2^{p_2^i + q_2^i + 1}\right) k_j^{p_1^i + p_2^i + 1} \, \omega_j^{q_1^i + q_2^i}$$

$$+ 4k_j^2 \sum_i (-)^{r_1^i + r_2^i + r_3^i} \gamma_i \, k_j^{r_1^i + r_2^i + r_3^i} \, \omega_j^{s_1^i + s_2^i + s_3^i} \,, \qquad (9)$$

$$k_1 (a_{112}^{(2)} + a_{112}^{(3)}) = -\frac{c_{12}}{\Phi(k_1+k_2, \omega_1+\omega_2)} \left[k_1 b_{112} + (2k_1+k_2) \frac{\Phi(2k_1+k_2, 2\omega_1+\omega_2)}{2(k_1+k_2)}\right] \qquad (10)$$

where we have introduced the dispersion polynomial:

$$\Phi(k,\omega) = \exp(kx-\omega t) L[\exp(-kx+\omega t)]. \qquad (11)$$

the polynomials c_{12}, b_{112} and $k_1 a_{112}^{(2)}$ which are determined by the quadratic part of the nonlinearity:

$$c_{12}(k_1, k_2; \omega_1, \omega_2) = \sum_i (-)^{p_1^i + p_2^i} \beta_i \left(k_1^{p_1^i} k_2^{p_2^i} \omega_1^{q_1^i} \omega_2^{q_2^i} + k_1^{p_2^i} k_2^{p_1^i} \omega_1^{q_2^i} \omega_2^{q_1^i} \right). \quad (12)$$

$$b_{112} = c_{12}(k_1, k_1+k_2; w_1, w_1+w_2) \quad , \quad k_1 a_{112}^{(2)} = -\frac{1}{2} c_{12}(2k_1, k_2; 2\omega_1, \omega_2). \quad (13)$$

and the polynomial $a_{112}^{(3)}$ which is determined by the cubic part of the nonlinearity:

$$a_{112}^{(3)} = \sum_i (-)^{r_1^i + r_2^i + r_3^i} \gamma_i \left(k_1^{r_1^i + r_2^i} k_2^{r_3^i} \omega_1^{s_1^i + s_2^i} \omega_2^{s_3^i} + k_1^{r_1^i + r_3^i} k_2^{r_2^i} \omega_1^{s_1^i + s_3^i} \omega_2^{s_2^i} \right.$$
$$\left. + k_1^{r_2^i + r_3^i} k_2^{r_1^i} \omega_1^{s_2^i + s_3^i} \omega_2^{s_1^i} \right) \quad (14)$$

The three conditions (8-10) must be satisfied for any two pairs of soliton parameters (k_j, ω_j), $j = 1,2$, which are related by the dispersion relation:

$$\Phi(k_j, \omega_j) = 0 \quad , \quad j = 1,2. \quad (15)$$

We also find that the two-soliton coupling factor A_{12} is determined by $\Phi(k,\omega)$ and by $K^{(2)}(V)$:

$$A_{12} = 1 + \left(\frac{2k_1 k_2}{k_1 + k_2} \right) \left[\frac{c_{12}}{\Phi(k_1+k_2, \omega_1+\omega_2)} \right]. \quad (16)$$

Higher order two-soliton consistency conditions imposing constraints on higher-order parts of the nonlinearity can be derived from equation (7) at higher orders in λ. It is worth noticing that a finite number of these conditions - the number depends on the order of L - constitutes a set of consistency conditions which are both necessary and sufficient for the existence of two-soliton solutions (6) (this follows from the polynomial character of K(V) and from the fact that $V^{(2)}(x,t,\lambda)$ is a rational fraction in λ).

3. Construction of "compensating" nonlinearities.

We shall now show that the consistency condition (10) can be used for constructing the quadratic and cubic parts of nonlinearities which realize the two-soliton balance for a hierarchy of linear dispersive terms.
We restrict this part of our discussion to the following class of L-operators:

$$L = \sum_\ell d_\ell \partial_x^{2m_\ell - 1} + \sum_j \bar{d}_j \partial_t \partial_x^{2m_j - 2}, \quad m_{\ell, j} = \text{integer}. \quad (17)$$

where it is assumed that the dispersion relation:

$$\Phi(k,\omega) \equiv \sum_j \bar{d}_j \omega k^{2m_j-2} - \sum_\ell d_\ell k^{2m_\ell-1} = 0 \tag{18}$$

leads to a non-trivial dispersion mode $\omega(k) = k\, v(k)$ with $\frac{d}{dk} v(k) \neq 0$.

A systematic construction of "compensating" nonlinearities $K^{(2,3)}(V)$ out of a given $L(V)$ follows from the requirement that the consistency conditions should be satisfied in such a way that each dispersive term of $L(V)$ be "compensated" by a specific nonlinearity and that a combination of various dispersive terms be compensated by the same combination of corresponding nonlinearities (linear superposition principle).

In order to obey this principle we must transform the above conditions (8-10) into equivalent ones wich may be satisfied as polynomial identities. It suffices to make use of the rel.(15) and to replace each of the dispersion polynomials by a "subtracted" polynomial which does no longer contain any linear contribution coming from possible first-order terms in $L(V)$:

$$\Phi_s(pk_j, p\omega_j) = \Phi(pk_j, p\omega_j) - p\,\Phi(k_j, \omega_j) \quad , \quad p=2 \text{ and } p=3 \, , \tag{19}$$

$$\Phi_s(k_1+k_2, \omega_1+\omega_2) = \Phi(k_1+k_2, \omega_1+\omega_2) - \Phi(k_1, \omega_1) - \Phi(k_2, \omega_2) \tag{20}$$

$$\Phi_s(2k_1+k_2, 2\omega_1+\omega_2) = \Phi(2k_1+k_2, 2\omega_1+\omega_2) - 2\Phi(k_1, \omega_1) - \Phi(k_2, \omega_2). \tag{21}$$

We then impose the following conditions:

$$\Phi_s(2k, 2\omega) \equiv 2 \sum_i (-)^{p_1^i+p_2^i+1} \beta_i\, k^{p_1^i+p_2^i+1}\, \omega^{q_1^i+q_2^i} \tag{22}$$

$$\Phi_s(3k, 3\omega) \equiv \sum_i (-)^{p_1^i+p_2^i+1} \left(2^{p_1^i+q_1^i+1} + 2^{p_2^i+q_2^i+1} \right) \beta_i\, k^{p_1^i+p_2^i+1}\, \omega^{q_1^i+q_2^i}$$
$$+ 4 \sum_i (-)^{r_1^i+r_2^i+r_3^i}\, \gamma_i\, k^{r_1^i+r_2^i+r_3^i+2}\, \omega^{s_1^i+s_2^i+s_3^i} \tag{23}$$

$$k_1 \left(a_{112}^{(2)} + a_{112}^{(3)} \right) \equiv - \frac{c_{12}\, F_{12}}{\Phi_s(k_1+k_2,\, \omega_1+\omega_2)} \quad , \tag{24}$$

with $\quad F_{12} = k_1 b_{112} + \dfrac{(2k_1+k_2)}{2(k_1+k_2)} \Phi_s(2k_1+k_2,\, 2\omega_1+\omega_2). \tag{25}$

It is easy to check that both $k_1 \left(a_{112}^{(2)} + a_{112}^{(3)} \right)$ and F_{12} are polynomials in k_j and ω_j, $j = 1,2$ (the substracted polynomial $\Phi_s(2k_1+k_2, 2\omega_1+\omega_2)$ contains a factor k_1+k_2). The identity (24) implies that $c_{12} F_{12}$ must contain the polynomial $\Phi_s(k_1+k_2, \omega_1+\omega_2)$ as a factor. This polynomial is a linear combination of various contributions, each of which corresponds to a linear dispersive term which is of the form $\partial_x^{2m-1} V$ or $\partial_t \partial_x^{2m-2} V$, $m \geq 2$.

Thus, there are essentially two ways to satisfy the above requirements.

The first way is to require that:

$$c_{12} \equiv -f(k_1, k_2) \Phi_s(k_1+k_2, \omega_1+\omega_2) \tag{26}$$

with an appropriate rational function $f(k_1, k_2)$, symmetric in the indices 1 and 2, and subject to the condition (in order to satisfy the condition (22)):

$$f(k_1, k_2)|_{k_1 = k_2} = \frac{1}{k_1} \tag{27}$$

This decouples the condition (24) into the identity (26) and the constraint:

$$k_1 \left(a_{112}^{(2)} + a_{112}^{(3)} \right) \equiv f(k_1, k_2) F_{12} . \tag{28}$$

It is easy to check that the conditions (26-28) imply the three identities (22-24). The only two choices for $f(k_1, k_2)$ which apply to any combination of dispersive terms of type $\partial_x^{2m-1} V$ are the following:

i) $\quad f(k_1, k_2) = \dfrac{k_1 + k_2}{2 k_1 k_2} \quad$ i.e.

$$\left[\begin{array}{l} c_{12} \equiv - \dfrac{(k_1+k_2)}{2 k_1 k_2} \Phi_s(k_1+k_2, \omega_1+\omega_2) \hfill (29) \\[1em] k_1 a_{112}^{(3)} \equiv \dfrac{(k_1+k_2) F_{12}}{2 k_1 k_2} - k_1 a_{112}^{(2)} \hfill (30) \end{array} \right.$$

ii) $\quad f(k_1, k_2) = \dfrac{2}{k_1 + k_2} \quad$ i.e.

$$\left[\begin{array}{l} c_{12} \equiv - \left(\dfrac{2}{k_1+k_2} \right) \Phi_s(k_1+k_2, \omega_1+\omega_2) \hfill (31) \\[1em] k_1 a_{112}^{(3)} \equiv \dfrac{2 F_{12}}{(k_1+k_2)} - k_1 a_{112}^{(2)} \hfill (32) \end{array} \right.$$

These two choices lead, respectively, to the linearizable equations of the Hopf-Burgers hierarchy [3] and to the potential equations of the KdV-hierarchy [2].

The second way is to require that the polynomial $\Phi_s(k_1+k_2, \omega_1+\omega_2)$ should be contained, up to a possible trivial factor, into the polynomial F_{12}. In fact, it is easy to check that F_{12} and $\Phi_s(k_1+k_2, \omega_1+\omega_2)$ are bound to contain some common non-vanishing terms, as a result of the identity (22), and that they differ at most by terms which involve at least two factors with the index 2. This means that the simplest nonlinearities which may be obtained in this way should satisfy the condition:

$$F_{12} \equiv \Phi_s(k_1+k_2, \omega_1+\omega_2) . \tag{33}$$

which decouples the identity (24) into the following two constraints:

$$\left[\begin{array}{l} k_1 b_{112} \equiv \Phi_s(k_1+k_2, \omega_1+\omega_2) - \dfrac{(2k_1+k_2)}{2(k_1+k_2)} \Phi_s(2k_1+k_2, 2\omega_1+\omega_2) \hfill (34) \\[1em] k_1 a_{112}^{(3)} \equiv - (c_{12} + k_1 a_{112}^{(2)}) \hfill (35) \end{array} \right.$$

It is easy to verify that the rel.(34,35) imply the rel.(22,23). On account of the rel.(18) one may rewrite the condition (34) in the more balanced form:

$$- 2 k_1 k_2 c_{12} \equiv (k_1+k_2) \Phi_s(k_1+k_2, \omega_1+\omega_2) + (k_1 - k_2) \Phi_s(k_1-k_2, \omega_1-\omega_2). \tag{36}$$

with $\quad \Phi_s(k_1 - k_2, \omega_1 - \omega_2) = \Phi(k_1 - k_2, \omega_1 - \omega_2) - \Phi(k_1, \omega_1) + \Phi(k_2, \omega_2).$ \hfill (37) which is precisely the condition that the two-soliton coupling factor A_{12} be expressible by the familiar formula which results from the bilinear formalism [1]:

$$A_{12} = -\frac{(k_1-k_2)\Phi_s(k_1,-k_2,\omega_1,-\omega_2)}{(k_1+k_2)\Phi_s(k_1,k_2,\omega_1,\omega_2)} \tag{38}$$

4. Hopf-Burgers hierarchy and KdV-hierarchy.

It follows from the rel. (16) that the first two constraints (29,30) imply degenerate two-soliton solutions (6) for which:

$$A_{12} \equiv 0 \tag{39}$$

By using the rel.(12-14), (18), (20) and (21) it is found that these constraints associate a particular quadratic nonlinearity $\tilde{K}_m^{(2)}(V)$ and cubic nonlinearity $\tilde{K}_m^{(3)}(V)$ with each dispersive term $\partial_x^{2m-1} V$. These nonlinearities are uniquely determined by the table of decompositions of the integer $2m-1$ as a sum of two and three integers.

Thus, setting $V = -2u_x$, it turns out that:

$$\tilde{K}_m^{(2)}(V = -2u_x) = 2 \sum_{r_1 \leq r_2} c_{2m-1}(r_1) u_{r_1 x} u_{r_2 x} , \qquad r_1 + r_2 = 2m-1 \tag{40}$$

$$\tilde{K}_m^{(3)}(V = -2u_x) = 2 \sum_{r_1 \leq r_2 \leq r_3} c_{2m-1}(r_1, r_2) u_{r_1 x} u_{r_2 x} u_{r_3 x} , \qquad r_1 + r_2 + r_3 = 2m-1 \tag{41}$$

where $u_{px} = \partial_x^p u$, and where each c-coefficient is equal to the combinatorial weight of the corresponding decomposition (number of ways of dividing $2m-1$ distinct elements into two or three boxes containing r_1 and r_2, or, r_1, r_2 and r_3 elements if each box is only characterized by the number of elements it contains). It should be noticed that the weights $c_r(r_1, r_2, \ldots r_{\ell-1})$ of the decompositions:

$$r = r_1 + r_2 + \ldots + r_{\ell-1} + r_\ell , \qquad r_1 \leq r_2 \leq \ldots \leq r_\ell , \tag{42}$$

which are defined as:

$$c_r(r_1, \ldots r_{\ell-1}) = \frac{r!}{s(\prod_{i=1}^{\ell} r_i!)} , \qquad s = \prod_j s_j! \tag{43}$$

where the integers s_j are the multiplicities of the different integers in $(r_1, r_2, \ldots r_\ell)$, are precisely the coefficients which appear in the explicit expression of

$$p_r(u_x, \ldots u_{rx}) \equiv \exp(-u) \partial_x^r \exp(u): \tag{44}$$

$$p_r(u_x, \ldots u_{rx}) = u_{rx} + \sum_{r=r_1+\ldots+r_\ell} c_r(r_1, \ldots r_{\ell-1}) u_{r_1 x} \ldots u_{r_\ell x} . \tag{45}$$

Hence, by extending the above rule (40,41) to all decompositions of 2m-1 as a sum of smaller integers on obtains all the nonlinear terms in the deriviatives of the field u(x,t) which appear in the equation:

$$\exp(-u) L_m (V = \exp u) = 0 \quad , \quad L_m(\partial_t, \partial_x) = \partial_t + \partial_x^{2m-1} \tag{46}$$

This shows that the nonlinearities $\tilde{K}_m^{(2)}(V)$ and $\tilde{K}_m^{(3)}(V)$ are, respectively, the quadratic and cubic nonlinearities which appear in the Hopf-Burgers hierarchy of nonlinear equations with linear part $L_m(V)$, which are linearized by the Cole-Hopf transformation [3]:

$$V = -2 \partial_x \ln f. \tag{47}$$

As to the second pair of constraints (31,32) it follows from the fomula (16) that they imply a two-soliton coupling of the KdV-form:

$$A_{12} = (\frac{k_1 - k_2}{k_1 + k_2})^2. \tag{48}$$

In this case, it turns out that a term $\partial_x^{2m-1} V$ is balanced by a nonlinearity $\hat{K}_m(V)$ the quadratic and cubic terms of which are those of the potential equations of the KdV-hierarchy [2] with linear part $L_m(V)$.

Thus, as m = 2 and 3 one gets from the rel.(20,21) the expressions:

$$\Phi_s^{(2)}(k_1+k_2, \omega_1+\omega_2) = -3k_1 k_2(k_1+k_2) \quad , \quad \Phi_s^{(2)}(2k_1+k_2, 2\omega_1+\omega_2) = -6k_1(k_1+k_2)^2. \tag{49}$$

$$\Phi_s^{(3)}(k_1+k_2, \omega_1+\omega_2) = -5k_1 k_2(k_1+k_2)(k_1^2+k_1 k_2+k_2^2),$$

$$\Phi_s^{(3)}(2k_1+k_2, 2\omega_1+\omega_2) = -10k_1(k_1+k_2)^2(3k_1^2+2k_1 k_2+k_2^2). \tag{50}$$

showing, through the formulas (12-14), (25), (31) and (32) that $\partial_x^3 V$ and $\partial_x^5 V$ are, respectively, balanced by the following nonlinearities:

$$\hat{K}_2(V) = \hat{K}_2^{(2)}(V) = 3V_x^2 \tag{51}$$

$$\hat{K}_3(V) = \hat{K}_3^{(2)}(V) + \hat{K}_3^{(3)}(V) = 10 V_x V_{3x} + 5V_{2x}^2 - 10V_x^3 \tag{52}$$

It also follows from the formulas (12), (20) and (31) that the x-derivative of $\hat{K}_m^{(2)}(V)$ is expressible by the combinatorial formula:

$$\partial_x \hat{K}_m^{(2)}(V) = 2 \sum_{r=1}^{m-1} \binom{2m-1}{r} V_{rx} V_{2m-2r-1} \tag{53}$$

5. Bilinear hierarchies.

It should be noticed that the above constraints (29) and (31) are somehow too strong as they impose a fixed two-soliton coupling for any combination of linear terms $\partial_x^{2m-1} V$, $m \geq 2$. On the other hand, it is clear from the formula (38) that the constraint (36) produces a two-soliton coupling which depends explicitly on the dispersive properties of L(V). By using the rel. (12-14), (20) and (37) it is found that the constraints (35) and (36) associate a specific quadratic and cubic nonlinearity with each dispersive term of L(V). These are given by a simple combinatorial rule for distributing a given number of x-derivatives (one more than those contained in the linear term) into two and three boxes. A generalized KdV-term $\partial_x^{2m-1} V$ is compensated by a nonlinearity $K_m(V)$ which contains as many quadratic (cubic) terms as there are different decompositions of the integer 2m as a sum of two (three) even integers:

$$2m = 2r+(2m-2r): \qquad K_m^{(2)}(V) = \sum c_{2m}(2r) \, V_{(2r-1)x} \, V_{(2m-2r-1)x} \qquad (54)$$

$$2m = 2r+2s+(2m-2r-2s): \quad K_m^{(3)}(V) = -\sum c_{2m}(2r, 2s) V_{(2r-1)x} V_{(2s-1)x} V_{(2m-2r-2s-1)x} \, . \quad (55)$$

A generalized RLW-term $\partial_t \partial_x^{2m-2} V$ is compensated by a nonlinearity $\bar{K}_m(V)$ the quadratic and cubic terms of which are similarly determined by the decompositions of 2m-1 as a sum of two and three integers of which only one is odd:

$$\bar{K}_m^{(2)}(V) = \sum c_{2m-1}(2r) \, V_{(2r-1)x} \, (V_{(2m-2r-2)x})_t \qquad (56)$$

$$\bar{K}_m^{(3)}(V) = -\sum c_{2m-1}(2r, 2s) V_{(2r-1)x} V_{(2s-1)x} (V_{(2m-2r-2s-2)x})_t \, . \qquad (57)$$

where $(V_{px})_t$ stands for $\partial_t \partial_x^p V$.

In deriving the cubic nonlinearities $K_m^{(3)}(V)$ and $\bar{K}_m^{(3)}(V)$ we have used the fact that the quadratic nonlinearities (54) and (56) contribute the following expressions to the polynomial: $T = -2k_1 k_2 [c_{12} + k_1 a_{112}^{(2)}]$

$$T^{(m)} = -2 \sum_{r=1}^{m-2} \binom{2m}{2r} (2^{2m-2r-2}-1) \, k_1^{2m-2r} \, k_2^{2r} \, , \qquad (58)$$

$$\bar{T}^{(m)} = -2 \sum_{r=1}^{m-2} (2^{2m-2r-2}-1) \left[\binom{2m-1}{2r} \omega_1 k_1^{2m-2r-1} k_2^{2r} + \binom{2m-1}{2r-1} k_1^{2m-2r} k_2^{2r-1} \omega_2 \right]. \qquad (59)$$

and that these expressions can be rewritten in the cyclic symmetric form:

$$T^{(m)} = -2 \sum_i c_{2m}(a_i, b_i) \left[k_2^{a_i} k_1^{b_i+c_i} + k_2^{b_i} k_1^{c_i+a_i} + k_2^{c_i} k_1^{a_i+b_i} \right], \qquad (60)$$

$$\bar{T}^{(m)} = -2 \sum_i c_{2m-1}(b_i, c_i) \left[k_2^{\alpha_i} \omega_2 k_1^{b_i+c_i} + k_2^{b_i} k_1^{c_i+\alpha_i} \omega_1 + k_2^{c_i} k_1^{\alpha_i+b_i} \omega_1 \right]. \qquad (61)$$

where the sum at the r.h.-side of equ.(60) is to be taken over all the decompositions of 2m as a sum of 3 even integers: $2m = a_i + b_i + c_i$, $a_i \leq b_i \leq c_i$, and that at the r.h.-side of equ.(61) is to be taken over all the decompositions of 2m-1 as a sum of two even integers ($b_i \leq c_i$) plus one odd integer α_i.

The simplicity of the combinatorial rule which governs the structure of $K_m^{(2,3)}(V)$ and $\bar{K}_m^{(2,3)}(V)$ suggests that it could be extended to higher orders in λ as a general recipe for the construction of a two-soliton generating nonlinearity associated with each dispersive term in L(V). This happens to be the case, for it turns out that the resulting nonlinearities $K_m(V)$ and $\bar{K}_m(V)$ are precisely those which appear in the following hierarchies of bilinear equations [1]:

$$D_x L_m(D_t, D_x)(f.f) \equiv D_x(D_t + D_x^{2m-1})(f.f) = 0 \tag{62}$$

$$D_x \bar{L}_m(D_t, D_x)(f.f) \equiv D_x(D_t + D_x + D_t D_x^{2m-2})(f.f) = 0 \tag{63}$$

by setting $f = \exp u$ and by rewriting the final equation in terms of the "potential field" $V = -2u_x$.

The point to be noticed is that the Hirota D-operators are precisely the tool needed for expressing the recipe in a compact form.

Indeed, it follows from the definition:

$$D_t^n . D_t^m (a.b) = (\partial_t - \partial_{t'})^n (\partial_x - \partial_{x'})^m a(x,t) b(x',t')|_{\substack{x'=x \\ t'=t}}, \tag{64}$$

that:

$$D_x^{2m}(\exp u, \exp u) = P_{2m}(u_x, ... u_{2mx}) \exp(2u), \tag{65}$$

with

$$P_{2m}(u_x, ... u_{2mx}) = 2\{p_{2m} + \sum_{r=1}^m (-)^r c_{2m}(r) p_r p_{2m-r}\} \tag{66}$$

To uncover the close link between $P_{2m}(q_x, ... q_{2mx})$ and the nonlinearities (54,55) it suffices to collect the quadratic and cubic terms of P_{2m}:

$$P_{2m}^{(2)} = 4 \sum_{r=1}^{E(m/2)} c_{2m}(2r) u_{2rx} u_{(2m-2r)x} \equiv K_m^{(2)}(V = -2u_x) \tag{67}$$

$$P_{2m}^{(3)} = 8 \sum c_{2m}(2r,2s) u_{2rx} u_{2sx} u_{(2m-2r-2s)x} \equiv K_m^{(3)}(V = -2u_x) \tag{68}$$

and to remark that the only linear term in P_{2m} is twice that of p_{2m}:

$$P_{2m}^{(1)} = 2u_{2mx} \equiv -\partial_x^{2m-1}(V = -2u_x). \tag{69}$$

It also follows from the definition (64) that:

$$D_t D_x^{2m-1}(\exp u . \exp u) = \bar{P}_{2m} \exp(2u) \tag{70}$$

with

$$\bar{P}_{2m} = 2\left\{\bar{p}_{2m} + \sum_{r=1}^{m-1}(-)^r\left[\binom{2m-1}{r}p_r\bar{p}_{2m-r} + \binom{2m-1}{r-1}p_{2m-r}\bar{p}_r\right] + (-)^m\binom{2m-1}{m}p_m\bar{p}_m\right\}. \quad (71)$$

where

$$\bar{p}_r = \exp(-u)\,\partial_t\,\partial_x^{r-1}\exp u. \quad (72)$$

The only term of $\bar{P}_{2m}$ which is linear in u is that of $\bar{p}_{2m}$:

$$\bar{P}_{2m}^{(1)} = 2(u_{(2m-1)x})_t \equiv -\partial_t\,\partial_x^{2m-2}(V = -2u_x). \quad (73)$$

By collecting the quadratic and cubic terms of $\bar{P}_{2m}$ it is also easily seen that they correspond precisely to the nonlinearities (56) and (57) written in terms of the u-variable:

$$\bar{P}_{2m}^{(2)} = 4\sum_{r=1}^{m-1}\binom{2m-1}{2r}u_{2rx}\,(u_{(2m-2r-1)x})_t \equiv \bar{K}_m^{(2)}(V = -2u_x) \quad (74)$$

$$\bar{P}_{2m}^{(3)} = 8\sum c_{2m-1}(2r,2s)\,u_{2rx}\,u_{2sx}\,(u_{(2m-2r-2s-1)x})_t \equiv \bar{K}_m^{(3)}(V = -2u_x). \quad (75)$$

Furthermore, one can show [5] that the quartic and higher order parts of P_{2m} and $\bar{P}_{2m}$ are expressible by a direct extension of the combinatorial rule which determines $K_m^{(2,3)}(V)$ and $\bar{K}_m^{(2,3)}(V)$:

$$P_{2m} = 2u_{2mx} + \sum_{e_1 \leq \ldots \leq e_\ell} c_{2m}(e_1,\ldots e_{\ell-1})\prod_{i=1}^{\ell}(2u_{e_i x}), \qquad \sum_{i=1}^{\ell}e_i = 2m \quad (76)$$

$$\bar{P}_{2m} = 2(u_{(2m-1)x})_t + \sum_{e_1 \leq \ldots \leq e_\ell} c_{2m-1}(e_1,\ldots e_\ell)(2u_{\alpha x})_t\prod_{i=1}^{\ell}(2u_{e_i x}), \qquad \alpha + \sum_{i=1}^{\ell}e_i = 2m-1. \quad (77)$$

References.

[1] Y. Matsuno, *Bilinear transformation method.* Acad. Press, Orlando (1984), 14.
[2] A. Newell, *Solitons in Mathematics and Physics.* SIAM, Philadelphia (1985), 128.
[3] D.V. Choodnovsky, G.V. Choodnovsky, *Pole expansions of nonlinear partial differential eqautions.* Nuovo Cimento, 40 B (1977), 339.
[4] R. Rosales, *Exact solutions of some nonlinear evolution equations.* Stud. Appl. Math. 59 (1987), 117.
[5] F. Lambert, R. Willox, *On the balance between dispersion and nonlinearity for a class of bilinear equations.* J. Phys. Soc. Japan, 58 (1989).

Complex Singularities and the Riemann Surface for the Burgers Equation

D. Bessis[1] *and J.D. Fournier*[2]

[1]Service de Physique Théorique de Saclay, Laboratoire de l'Institut de Recherche Fondamentale du Commissariat à l'Energie Atomique, F-91191 Gif-sur-Yvette Cedex, France

[2]Observatoire, Mont-Gros, BP 139, F-06003 Nice Cedex, France

The analytic structure of the solution of the Burgers equations is analysed: the viscous solution has an infinite number of complex poles. When the viscosity tends to zero, these poles condense, producing the inviscid singularities. A Riemann surface is attached to those non polar singularities. As a consequence, a shock appears to be the permutation of two Riemann sheets. This phenomenon can also be understood as a phase transition in a Curie-Weiss model.

Introduction

In this contribution, we are concerned with the space analytic structure of the Burgers' equation:

$$\frac{\partial u}{\partial t} + u\frac{\partial u}{\partial x} = \nu\frac{\partial^2 u}{dx^2} \qquad (I.1)$$

where ν, the viscosity is positive or zero and $u(x,t)$ is the velocity field at point x and time t.

We report on our previous work[1], incorporating some unpublished material, and we take the opportunity to add new features which have been pointed out to us by C. Newman[2].

This equation may be considered as a one dimensional version of an infinitely compressible Navier-Stokes fluid[3]. It is also the governing equations for the magnetization in a one dimensional Ising model, provided one makes the following identifications. The magnetization is identified with the velocity field, while the external magnetic field becomes the position $-x$, and the temperature β is associated to the time t. The infinite volume limit is obtained for the viscosity ν going to zero. A phase transition occurs in this model, because the coupling between neighbouring spin is inversly proportional to the volume (number of spins)[4].

In all cases, we are interested in analysing in details the behaviour for very small viscosity: is the small ν regime governed by the $\nu = 0$ limiting regime ? And what is the mathematical structure of the later.

The existence of the phase transition in the Ising-model at temperature β^*, reflects the existence of a shock at time $t_* = \beta^*$. This occurs of course, only in the infinite volume or zero viscosity limit. The lack of understanding of the delicate analytic structure of the Rieman surface in this limit, is the reason for paradoxes stressed in [5] concerning the high wave number behaviour of the velocity.

We have nothing to add to the nature of the singularities in the x complex plane. They are known to be poles[6][7] in the strictly positive viscosity case. In the inviscid case ($\nu = 0$), a discontinuity appears for real x (shock) after a time t_*. This discontinuity is announced at $t = t_*$ by a singularity in $x^{1/3}$. This last singularity itself results from the coalescence of two square root branch points moving in the complex plane before t_*.

Before ending this introduction, we must discuss the choice of the initial condition. Due to the fact that the residues of the poles are proportional to the viscosity, and that the number of poles is a constant of motion for a rational initial condition, the zero viscosity limit is trivial if the number of poles is finite. A simple way to generate an infinite number of poles, to get a non trivial inviscid limit is to start with an entire functions at time $t = 0$. The simplest choice is, as discussed in [1], the third degree polynomial:

$$u(x,0) = 4x^3 - \frac{x}{t^*} . \qquad (I.2)$$

which is obtained by truncations to third order of a generic entire function. It contains all the ingredient to produce the Maxwell catastrophy (shock).

II. Description of the Rieman surface in the inviscid case

In this case, the solution satisfies:

$$\begin{cases} u(x,t) = \dfrac{x}{t} + U(x,t) \\ 4t^3 U^3 + \dfrac{(t_* - t)}{t^*} U + \dfrac{x}{t} = 0 . \end{cases} \qquad (II.1)$$

Therefore the solution is an analytic function of x and is uniform on a three-sheeted Riemann surface.

For fixed positive t, $U(x,t)$ has three singularities: the point at infinity which a third order branch point plus two second order branch point x_s and $-x_s$ given by

$$x_s = \frac{i}{3\sqrt{3}} t_*^{-3/2} t^{-1/2} (t_* - t)^{3/2} \qquad (II.2)$$

Starting from $t = 0$ at infinity, they come down along the imaginary axis and meet at $t = t_*$, and then move away along the real axis. While before the time t_*, the moving singularities are on the physical sheet, after t_* both have moved to unphysical sheets. The

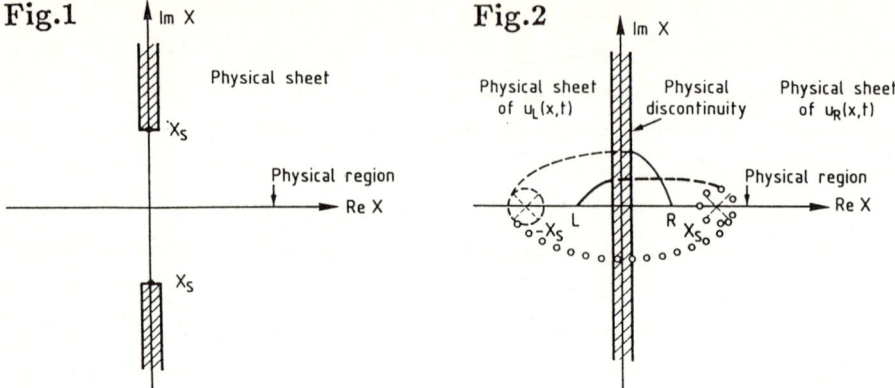

Fig.1-*The physical Riemann sheet and the corresponding singularities. For $0 < t < t_*$ in the inviscid case.*

Fig.2-*The physical Riemann sheets, for $t > t_*$ in the inviscid case. The square root singularities $\pm x_s$ are on other sheets. There is a connection between the right physical region and the left one through a path which partly belongs to a third sheet: this path is represented with the following symbols:—physical region; – – – left analytic continuation of the right physical region: ooo unphysical sheet; ---right analytic continuation of the left physical region.*

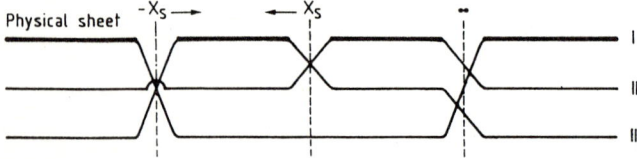

Fig.3- $0 < t < t_*$. *The singularity x_s is not present on sheet III, while $-x_s$ is absent on sheet II; they are both present on sheet I (physical sheet); the point at infinity is singular on all sheets.*

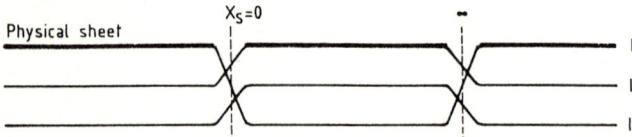

Fig.4-$t = t_*$.x_s *and $-x_s$ have collided on the real axis giving rise to a branch point of order 3 (preshock); this singularity as well as the one at infinity are present on all sheets.*

physical sheet after t_* is now the union of two separate domains. Physical sheets are defined as neighbouring domains of the complex plane surrounding the physical regions. Figs 1,2,3,4 and 5 illustrate the scenario.

In this description the shock appears as the permutation of two Riemann sheet.

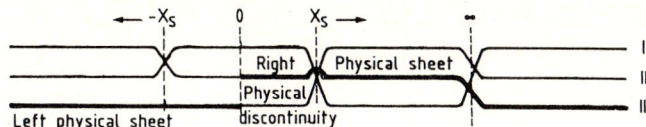

Fig.5-$t > t_*$. The singularities $-x_s$ and x_s are now non-physical, the physical sheets are sheets II and III, which are connected locally through sheet I; see the path on figure 4. Topology of the Riemann surface of the solution in the inviscid case.

III. The locus of the poles in the viscous case

We map the Burgers' equation into the heat equation[8], setting:

$$U(x,t) = -2\nu \frac{\partial}{\partial x} \ln \theta(t,x) \qquad (III.1)$$

on gets:

$$\frac{\partial \theta}{\partial t} = \nu \frac{\partial^2 \theta}{\partial \lambda^2} \qquad (III.2)$$

The solution of which with our initial condition (I.2) reads:

$$\theta(t,x) = (4\pi\nu t)^{-1/2} \exp[-x^2/4\nu t] E(x,t) \qquad (III.3)$$

with

$$E(x,t) = \int_{-\infty}^{+\infty} dy \, \exp\left[-\frac{1}{2\nu}\left[y^4 + \frac{1}{2}\left(\frac{1}{t}-\frac{1}{t^*}\right)y^2 - \frac{1}{t}xy\right]\right] \qquad (III.4)$$

It can be shown that $E(x,t)$ has the following properties for any positive time t.
(i) It is an entire functions of x.
(ii) Its order is $4/3$.
(iii) By Hadamard's theorem it has therefore an infinite number of zeros.
(iv) Being a symmetric real function, the zeroes appear in symmetric conjugate paires.
(v) By a theorem due to Polya[9], the zeroes can be shown to belong to the pure imaginary axis[2].

For the reader convenience, we state the Polya theorem:

$$\int_{-\infty}^{+\infty} e^{-ay^{4q}+by^{2q}+cy^2+izy} dy \qquad (III.5)$$

$$a > 0, \qquad b \text{ real}, \qquad c \geq 0$$

has only real zeros !

Combining (i) to (v), we see that it is possible to write, using Hadamard decomposition theorem for $\theta(x,t)$, and taking the logarithmic derivative of θ

$$u_\nu(x,t) = \frac{x}{t} - 4\nu x \sum_{n=1}^{\infty} \frac{1}{x^2 + \beta_n^2(t)}, \qquad \beta_n(t) > 0, \qquad t > 0 \qquad (III.6)$$

where $i\beta_n(t)$ is the nth ordered zero of $E(x,t)$ on the positive imaginary axis. It is easily derived, following[6], that $\beta_n(t)$ satisfies a Calogero-type[10] equation:

$$\overset{o}{\beta}_n = \sum_{\substack{\ell \neq n \\ \ell \geq 1}} \frac{2\nu}{\beta_n - \beta_\ell} + \frac{\beta_n}{t} \qquad (III.7)$$

IV. Pole condensation in the zero viscosity limit

We apply the saddle point method to the integral (III.4) to derive the location of the poles at small viscosity. At fixed positive time and in the limit $\nu \downarrow 0_+$, the distance between near by poles tends to zero proportionally to ν. Each pole being weighted by its residue, also proportional to the viscosity, this allows for a compensation which permit to define a limiting density:

$$\rho(y,t) = \lim_{\substack{\nu \longrightarrow 0_+ \\ n \longrightarrow \infty}} \frac{2\nu}{\beta_{n+1}(t,\nu) - \beta_n(t,\nu)}\bigg|_{y=\beta_n} \qquad (IV.1)$$

which enters in the Cauchy integral version of (III.6)

$$U(x,t) = \frac{x}{t} - 2x \int_0^\infty \frac{\rho(y,t)}{x^2 + y^2} \qquad (IV.2)$$

This limiting density characterizes the pole condensation process. As a function of y it has three different regimes.

For $0 < t < t_*$.
The density is zero for

$$0 \leq y < |x_s| \qquad (IV.3)$$

and behaves as $(y - |x_s|)^{1/2}$ for y near $|x_s|$.

For $t = t_*$.

$$\rho(y, t_*) = (2\pi)^{-1} 3^{1/2} (4t_*)^{-4/3} y^{1/3} \qquad (IV.4)$$

while for $t > t_*$.

$$\rho(y,t) \simeq (2\pi)^{-1} t_*^{-1/2} (t - t_*)^{1/2} t^{-3/2} \qquad (IV.5)$$
$$y \longrightarrow 0$$

Following ref. [11], one expects the discontinuity of the function accross the cut to be related to the local density of poles by:

$$\Delta_{x=0}(t) = U_L(0_-, t) - U_R(0_-, t) = -2i\pi\, \rho(0, t) \qquad (IV.6)$$

It is clear from the preceding discussion that this discontinuity is non zero only for $t > t_*$, and then using the odd parity of U and (IV.5) one gets:

$$U(0,t) = -\frac{1}{2}t^{-3/2}(t-t_*)^{1/2}t_*^{-1/2} \qquad (IV.7)$$

which is the correct value obtained directly from (II.1).

V. Conclusion

This pole condensation mechanism provides a new interpretation of the known features of a shock. In the Ising model analogy, it just describes the phase transition and the appearance of a spontaneous magnetisation. Furthermore, the detailed analysis of the dynamics of the Riemann surface in the inviscid case has cleared various previous paradoxes. In particular, when resumming the Taylor series expansion near $t=0$, of the Fourier transform $\hat{U}(k,t)$ of $U(x,t)$, one gets for $|k| \longrightarrow \infty$ a behaviour in $|k|^{-3/2}$, for all times. The absence of change in behaviour from $k^{-3/2}$ to k^{-1} after t_*, is due to fact that the analytic continuation in t, ends on the wrong sheet of $U(x,t)$.

References

[1] D. Bessis and J.D. Fourier, J.Phys.Let. **45**, L833-841 (1984)

[2] C.M. Newman, private communications

[3] J.A. Burgers, A.K. Verhaud-Kon, Nederl. Weterschappen Afd. Natuurkunde, Eerste Sertie, Vol.17 (1939) p.1-5
J.M. Burgers, "The Non-linear Diffusion equation" D. Reidel Publ. (1974)

[4] C.M. Newman, Schock waves and linear field bound, Talk given at Rutgers Statistical Mechanic Meeting, Dec. 1981.

[5] J.D. Fournier and U. Frisch, J.Mec.Th.Appl. **2** (1983) n°5, 689

[6] D.V. Choodnovsky and G.V. Choodnovsky, "Pole expansions of Non-linear Partial Differential equations" Nuovo Cimento **40B**, n°2, p.339-353 (1977)

[7] J. Weiss, M. Tabor, G. Carnevale, "The Painlevé Property for Partial Differential Equations", J.Math.Phys. **24**, (3) p.522-526 (1983)

[8] E. Hopf, Comm.Pure Appl.Mech. **3**, 201 (1950).
J.D. Cole, Quart. Appl.Math. **9**, 225 (1951)

[9] G. Polya, Uber trigonometische Integrale mit nur reellen Nullstelleir, Z. Reine Angew. Math. **158**, 6-18, 1927

[10] M.J. Ablowitz and H. Segur, Solitons and the inverse scattering transforms, Siam 1981

[11] T.D. Lee and C.N. Yang, Phys.Rev. **87**, 410 (1952).

From Soliton Theory to String Theory

S. Saito and H. Kato

Department of Physics, Tokyo Metropolitan University,
Setagaya-ku, Tokyo 158, Japan

A transformation of variables which relates soliton theory to string theory is discussed. Scattering amplitudes of bosonic strings with arbitrary momenta off fermionic (compactified bosonic) loops are shown to satisfy Hirota's bilinear difference equation, which is equivalent to infinite family of Kadomtsev-Petviashvili type of soliton equations.

1. The String Amplitudes

The string theory is a theory of elementary particles in which the fundamental object is not a point particle but a string. The supersymmetric version of this theory[1] is a unique candidate of consistent field theory which unifies all fundamental forces in nature including gravity. In this talk we are going to show how this theory is connected with the soliton theory. Particularly we will show that integrands of string amplitudes are characterized by a soliton equation.[2]

Let us begin with a brief introduction of the string theory. First we consider a point particle of momentum k_1^μ which propagates and emitts same kind of particles of momenta $k_2^\mu, k_3^\mu, ..., k_N^\mu$ at time $\tau = \tau_2, \tau_3, ..., \tau_N$, successively. The scattering amplitude of this process is given by

$$< 0|V(k_N)e^{-i\int_{\tau_{N-1}}^{\tau_N} H d\tau} V(k_{N-1})e^{-i\int_{\tau_{N-2}}^{\tau_{N-1}} H d\tau} V(k_{N-2})$$
$$\cdots e^{-i\int_{\tau_2}^{\tau_3} H d\tau} V(k_2)|k_1>, \tag{1}$$

where $H = \frac{1}{2}p^2$ and $|k_1>$ is the momentum eigenstate

$$p^\mu|k_1> = k_1^\mu|k_1>, \qquad p^\mu|0> = 0. \tag{2}$$

If the interaction is point like $V(k)$ is of the form

$$V(k) = g e^{ikx}, \qquad g \quad constant. \tag{3}$$

In the first quantized level only x^μ and p^μ are quantized quantities satisfying

$$[x^\mu, p^\nu] = i g^{\mu\nu}. \tag{4}$$

Using (4), we can rewrite (1) as

$$< 0|V(k_N, \tau_N)V(k_{N-1}, \tau_{N-1}) \cdots V(k_1, \tau_1)|0> \tag{5}$$

where

$$V(k,\tau) = g : e^{ik(x+p\tau)} : \equiv g e^{ikp\tau} e^{ikx}. \tag{6}$$

The extention to the propagation of a string can be done simply by considering the proper-time τ being a complex number. For this purpose we introduce a complex coordinate

$$z = e^{-i(\tau+i\sigma)}, \qquad (7)$$

and the string coordinate

$$X^\mu(z) = x^\mu + ip^\mu \ln z + \sum_{n=1}^{\infty}(a_n^\mu z^n + a_n^{\mu\dagger} z^{-n})/\sqrt{n}, \qquad (8)$$

where a_n^μ and $a_n^{\mu\dagger}$ satisfy

$$[a_m^\mu, a_n^{\nu\dagger}] = g^{\mu\nu}\delta_{mn}. \qquad (9)$$

$X^\mu(z)$ is a scalar under the conformal transformations which are generated by

$$H(z) = \frac{\partial X^\mu}{\partial \tau}\frac{\partial X_\mu}{\partial \tau} + \frac{\partial X^\mu}{\partial \sigma}\frac{\partial X_\mu}{\partial \sigma}$$

$$T(z) = \frac{\partial X^\mu}{\partial \tau}\frac{\partial X_\mu}{\partial \sigma}.$$

The integrand of the scattering amplitude of N-ground state particles is now given by

$$<0|\prod_{j=1}^{N} V(k_j, z_j)|0>, \qquad (10)$$

where

$$V(k, z) = g : e^{ikX(z)} :$$
$$\equiv g e^{ikX^+(z)} e^{ikX^-(z)}, \qquad (11)$$

$$X^{\mu+}(z) = ip^\mu \ln z + \sum_{n=1}^{\infty} \frac{1}{\sqrt{n}} a_n^{\mu\dagger} z^{-n},$$

$$X^{\mu-}(z) = x^\mu + \sum_{n=1}^{\infty} \frac{1}{\sqrt{n}} a_n^\mu z^n.$$

This represents a propagation of a string along which external ground state particles of momenta $k_1^\mu, k_2^\mu, ..., k_N^\mu$ are attached at $z_1, z_2, ..., z_N$. Integrating over z's the scattering amplitude is obtained. If the external particles themselves are strings we have to replace the verterx operator (11) by[3]

$$W(P, z) = g : e^{i \oint P(w)X(z(w))\frac{1}{w}dw} : . \qquad (12)$$

Here $P^\mu(z) = dY^\mu(z)/id\ln z$ denotes a distribution of momenta of the external string $Y^\mu(z)$.

A closed string consists of two independent coordinates $X^\mu(z)$ and $\bar{X}^\mu(\bar{z})$ where $\bar{z}$ denotes the complex conjugate of z. As it propagates it forms a sphere embedded in the space-time. Interaction of strings can be studied by using the three string vertex operator (12). Strings can split into pieces and join together. Then we obtain a Riemann surface with many holes on which external strings are attached. The string amplitude associated to this process turns out to be a modular function defined through the surface.[4]

2. The Soliton Equations

Now we recall that modular functions defined through the Riemann surfaces also appear as solutions to some soliton equations, known as quasi-periodic solutions.[5] Solitons are nothing but special limits of the solutions in which periods are taken infinity. The KP (Kadomchev-Petviashvili) -hierarchy[6] is a set of infinite number of soliton equations including the KdV equation, the KP equation, the sine-Gordon equation and so on. The dense set in a solution space of these equations are known explicitly in terms of modular functions defined through the Riemann surfaces. At this point it is apparent that there exists some correspondence between the soliton theory and the string theory.

To see the connection explicitly let us remind you some detail structure of solutions to the KP-hierarchy.[6] The solutions are most conveniently expressed in terms of free fermionic fields. Let ψ_n and $\psi_n^\dagger$ be fermionic operators satisfying

$$\{\psi_m, \psi_n^\dagger\} = \delta_{mn}, \quad \{\psi_m, \psi_n\} = \{\psi_m^\dagger, \psi_n^\dagger\} = 0, \quad m, n \in \mathbf{Z}. \tag{13}$$

The vacuum state is defined by

$$\begin{aligned} \psi_n|0> &=< 0|\psi_n^\dagger = 0, \quad n \geq 0, \\ < 0|\psi_n &= \psi_n^\dagger|0> = 0, \quad n < 0. \end{aligned} \tag{14}$$

We define a g-operator having and satisfying the following relations

$$\begin{aligned} g\psi_n &= \sum_{m \in \mathbf{Z}} \psi_m g a_{mn} \\ \psi_n^\dagger g &= \sum_{m \in \mathbf{Z}} a_{nm} g \psi_m^\dagger, \quad a_{mn} \in \mathbf{C} \end{aligned} \tag{15}$$

and the Hamiltonian

$$H(t) = \sum_{n=1}^{\infty} t_n \sum_{m \in \mathbf{Z}} : \psi_m^\dagger \psi_{m+n} :, \quad t_n \in \mathbf{C}, \quad n = 1, 2, 3, \ldots \tag{16}$$

Then the general solutions to the KP-hierarchy are given by

$$\tau(t) =< 0|e^{H(t)} g|0 > . \tag{17}$$

The proof goes as follows: If

$$\psi(z) = \sum_{n \in \mathbf{Z}} \psi_n z^n, \quad \bar{\psi}(z) = \sum_{n \in \mathbf{Z}} \psi_n^\dagger z^{-n-1}, \tag{18}$$

one can prove the identity

$$\oint \frac{dz}{2\pi i} < 1|e^{H(t)} \psi(z) g|0 >< -1|e^{H(t')} \bar{\psi}(z) g|0 >= 0 \tag{19}$$

for arbitrary operator g satisfying (15). We now notice

$$\begin{aligned} e^{H(t)} \psi(z) e^{-H(t)} &= e^{-\sum_{n=1}^{\infty} t_n z^{-n}} \psi(z) \\ e^{H(t)} \bar{\psi}(z) e^{-H(t)} &= e^{\sum_{n=1}^{\infty} t_n z^{-n}} \bar{\psi}(z). \end{aligned} \tag{20}$$

The same relations hold also as we replace $\psi(z)$ and $\bar\psi(z)$ by

$$\exp\left[\mp \sum_{n=1}^{\infty} \frac{z^{-n}}{n} \sum_{m\in\mathbf{Z}} :\psi_m^\dagger \psi_{m-n}:\right],$$

since

$$\sum_{n=1}^{\infty} \frac{z^{-n}}{n} \sum_{m\in\mathbf{Z}} [H(t), :\psi_m^\dagger \psi_{m-n}:] = \sum_{n=1}^{\infty} t_n z^{-n}. \tag{21}$$

This enables us to write (19) as

$$\oint \frac{dz}{2\pi i} e^{\sum_{n=1}^{\infty}(t'_n - t_n)z^{-n}} e^{\sum_{n=1}^{\infty} z^n(\partial_n - \partial'_n)/n} <0|e^{H(t)}g|0><0|e^{H(t')}g|0>$$

$$= \sum_{l=0}^{\infty} P_{l+1}(t'_n - t_n) P_l\left(\frac{1}{n}\partial_n - \frac{1}{n}\partial'_n\right) \tau(t)\tau(t') = 0, \tag{22}$$

where $\partial_n = \partial/\partial_n$, $\partial'_n = \partial/\partial t'_n$ and P_l $(l = 0, 1, 2, ...)$ are the Schur polynomial defined by

$$e^{\sum_{n=1}^{\infty} s_n z^n} = \sum_{l=0}^{\infty} P_l(s_n) z^l.$$

It is well known[6] that eq.(22) reduces to various Hirota bilinear differential equations for $\tau(t)$ as we truncate the dependence of τ on t_n in various ways in the limits of $t'_n \to t_n$. For instance it reduces to the KdV equation

$$\frac{\partial}{\partial t} u + 6u \frac{\partial}{\partial x} u + \frac{\partial^3}{\partial x^3} u = 0, \tag{23}$$

$$u = 2 \frac{\partial^2}{\partial x^2} \ln \tau,$$

when $t_1 = x, t_3 = t$, and $t_n = 0$ otherwise. The meaning of the parameters t_n is now apparent. They are real space-time variables of soliton equations.

3. Soliton String Correspondence

Our problem is to establish the correspondence between the string amplitudes of eq.(10) and the τ-functions of eq.(17). It can be done by the following steps. First we observe that the compactified one-dimensional piece of the string coordinate $X(z)$ behaves exactly the same way as

$$\int^z dw :\bar\psi(w)\psi(w):$$

under the commutation relations. This also implies that $\psi(z)$ and $\bar\psi(z)$ can be expressed by $X(z)$:

$$\psi(z) =: e^{iX(z)}:$$
$$\bar\psi(z) =: e^{-iX(z)}: \quad . \tag{24}$$

These are nothing but the boson-fermion equivalence.[7]

The second step is to relate the variables t_n of soliton equations with the external momenta p_j's of the string amplitude through the transformation[2,8]

$$t_n = \frac{1}{n} \sum_{j=0}^{\infty} p_j z_j^n. \qquad (25)$$

z_j are parameters of the transformations.

The third step is taken by substituting eq.(25) into the Hamiltonian

$$H(t) = H(p) \equiv \sum_{j=0}^{\infty} p_j X^-(z_j), \qquad (26)$$

where we have used $\sum_j p_j = 0$. This enables us to rewrite the generator of evolution of solitons $e^{H(t)}$ in terms of the vertex operators of strings[2]:

$$<0|e^{H(p)} = \frac{<0|\prod_j V(p_j, z_j)}{<0|\prod_j V(p_j, z_j)|0>}. \qquad (27)$$

When $p_j = 0, j \geq N+1$ the denominator is the tree amplitude in (10). The parameters z_j are nothing but the Koba-Nielsen variables which specify positions on the string where external particles are attached.

For the final step we have to interpret the operator g by the language of the string theory. The fermionic propagator on an arbitrary Riemann surface has been known[9]

$$<0|\psi(x)\bar{\psi}(y)g_F|0> = \frac{\theta\left(\zeta + \int_y^x \omega\right)}{\theta(\zeta)E(y,x)}. \qquad (28)$$

Here θ and $E(y,x)$ are the Riemann theta function and the prime form and ω denotes an Abel differential of first kind. From this one can derive an explicit form of g which is associated with multiloop configuration of fermionic strings [10]. It is given by

$$g_F = \exp\left[\frac{1}{4\pi^2} \oint dx \oint dy B(x,y)\bar{\psi}(x)\psi(y)\right], \qquad (29)$$

where

$$B(x,y) = <0|\psi(x)\bar{\psi}(y)g_F|0> - \frac{1}{y-x}. \qquad (30)$$

Combining above steps all together we can relate the τ-function, a solution to the soliton equations, to the string amplitude of N-external particles with many loops:

$$\tau(p) = \frac{<0|\prod_{j=1}^{N} V(p_j, z_j)g_F|0>}{<0|\prod_{j=1}^{N} V(p_j, z_j)|0>}. \qquad (31)$$

The string amplitude on the right hand side is a hybrid amplitude in the sense that bosonic strings are attached to the fermionic vacuum state. The fermionic vacuum state itself is also possible[2] to be expressed in terms of the bosonic fields defined over a compactified space, owing to the use of the boson-fermion equivalence (24).

What do we learn from the correspondence between the τ-functions and the string amplitudes ? From the string theoretical point of view it is remarkable that the totality of the compactified bosonic string amplitudes is given explicitly in terms of a modular function of period matrix τ irrespective to the number of loops. The general form of the solution to the KP-hierarchy is given by[10][11]

$$\tau(t) = \theta(\zeta + \sum_{n=1}^{\infty} A_n t_n) \exp\left[\sum_{m,n=1}^{\infty} t_m Q_{mn} t_n\right], \qquad (32)$$

where Q_{mn} and A_n are the coefficients of the Taylor expansion of the first Abel differential ω_k and $\ln \frac{E(x,y)}{x-y}$:

$$\omega_k = \sum_{n=1}^{\infty} A_{kn} z^{n-1} dz, \qquad k = 1, 2, ...h$$

$$\ln \frac{E(x,y)}{x-y} = 2 \sum_{m,n=1}^{\infty} \frac{x^m y^n}{mn} Q_{mn}.$$

Secondly we can characterize string amplitude by soliton equations.[2] In particular we can derive an equation satisfied by the string amplitudes in which momenta p_j of external bosonic particles are independent variables. If we had changed variables from t_n to p_j using (25) and substituted into : $\exp[\pm iX(z)]$: in the calculation of the left hand side of (19) we could derive Hirota's bilinear *difference* equation[12]:

$$\frac{z_0 - z_1}{(z_1 - z_2)(z_1 - z_3)} \tau(p_0, p_1+1, p_2, p_3)\tau(p_0-1, p_1, p_2+1, p_3+1)$$
$$+ \frac{z_0 - z_2}{(z_2 - z_3)(z_2 - z_1)} \tau(p_0, p_1, p_2+1, p_3)\tau(p_0-1, p_1+1, p_2, p_3+1)$$
$$+ \frac{z_0 - z_3}{(z_3 - z_1)(z_3 - z_2)} \tau(p_0, p_1, p_2, p_3+1)\tau(p_0-1, p_1+1, p_2+1, p_3)$$
$$= 0. \qquad (33)$$

In this equation (p_0, p_1, p_2, p_3) and (z_0, z_1, z_2, z_3) are arbitrary set of four momenta and the Koba-Nielsen variables among the N-external particles. This is an interesting equation which was derived by Hirota as a symmetric generalization of the Toda lattice to three discrete variables. It is satisfied by every solution of the KP-hierarchy and reproduces various soliton equations, such as the KdV equation, the Toda lattice, the KP equation, sine-Gordon equation etc. as we take certain limits of parameters and variables[12]. Moreover it has been shown[2] that it reduces to Fay's trisecant formula if we substitute the general form of the solution (32), after the variables are changed:

$$\tau(p) = \prod_{i,j} \left(\frac{E(z_i, z_j)}{z_i - z_j}\right)^{\frac{1}{2} p_i p_j} \theta\left(\zeta + \sum_{j=0} p_j \int_{z_0}^{z_j} \omega\right). \qquad (34)$$

There exist another form of solutions, i.e., soliton solutions.[8] They are implied as certain limits of the general solution (34). Nevertheless they must be useful to investigate some properties of string amplitudes since a simple compact expression of the soliton solution is known in the form of Casorati determinant[13]:

$$\tau^N(p) = \det |\phi_{mn}|, \qquad m, n = 1, 2, ..., N \qquad (35)$$

$$\phi_{mn} \equiv (1 - a_m z_0)^{-n} \prod_{j=0}^{\infty} (1 - a_m z_j)^{-p_j} + (1 - b_m z_0)^{-n} \prod_{j=0}^{\infty} (1 - b_m z_j)^{-p_j},$$

where a_m, b_m $(m = 1, 2, ..., N)$ are arbitrary parameters.

Finally we like to notice that the soliton-string correspondence leads to a generalization[14] of Fay's addition theorem[15] for Abel functions. Namely we have two expressions of the τ-functions, one of (31) is derived from the string consideration whereas another one of (34) comes from soliton equations. To compare these expressions more in detail we rewrite $<0|\prod_{j=1}^{N} V(p_j, z_j) g_F |0>$ of (31) as

$$<0| \exp\left[\frac{1}{4\pi^2} \oint dx \oint dy\, B(x,y) \prod_{j=1}^{N}\left(\frac{x-z_j}{y-z_j}\right)^{p_j} \bar{\psi}(x)\psi(y)\right] |0> .$$

Then we obtain

$$\theta\left(\zeta + \sum_{j=1}^{N} p_j \int_{z_0}^{z_j} \omega\right) = \prod_{i,j=1}^{N} \{E(z_i, z_j)\}^{-\frac{1}{2} p_i p_j}$$

$$\times \sum_{k=0}^{\infty} \frac{1}{k!} \prod_{l=1}^{k} \frac{1}{4\pi^2} \oint dx_l \oint y_l B(x_l, y_l) \prod_{j=1}^{N} \left(\frac{x_l - z_j}{y_l - z_j}\right)^{p_j} \det\left|\frac{1}{x_m - y_n}\right|. \quad (36)$$

When p's are integers the summation on the right hand side terminates at finite terms. In particular this reduces to Fay's addition formulae[15]:

$$\theta\left(\zeta + \sum_{j=1}^{N/2} \int_{z_j}^{w_j} \omega\right) = \frac{\prod_{i,j}^{N/2} E(z_i, w_j)}{\prod_{i<j}^{N/2} E(z_i, z_j) E(w_i, w_j)} \det\left|\frac{\theta\left(\zeta + \int_{w_j}^{z_i} \omega\right)}{\theta(\zeta) E(z_i, w_j)}\right|, \quad (37)$$

when p_j $(j = 1, 2, ..., N)$ can take values only ± 1. Therefore eq.(36) is a generalization[14] of Fay's addition theorem.

Acknowledgement

The authors would like to thank the organizers of this conference for the invitation and warm hospitality. One of the authors(S.S) thanks Professor Bo-Yu Hou and Professor Han Ying Guo for their kind hospitality and various support during his stay at their Institutes. The Yamada Science Foundation is also acknowledged for the financial support of travels.

References

[1] M. B. Green and J. M. Schwarz, Nucl. Phys. **B181** (1981) 502; J. M. Schwarz, "Superstring Theory" Phys. Rep. **89C** (1982) 223.
[2] S.Saito, Tokyo Metropolitan Univ. preprint TMUP-HEL-8813 (1986); Phys. Rev. **D36**(1987) 1819; Phys. Rev. Lett. **59** (1987) 1798; Phys. Rev. **D37** (1988) 990; K.Sogo, J.Phys. Soc. Jpn. **56** (1987) 2291.
[3] A.Della Selva and S.Saito, Lett. Nuov. Cim. **4** (1970) 689; see, for recent development, P.Di Vecchia, R.Nakayama, J.L.Petersen and S.Sciuto, Nucl. Phys. **B282** (1987) 103; U. Carow-Watamura, Z. F. Ezawa and S. Watamura, Nucl. Phys. **B315** (1989) 166.
[4] A.A.Belavin and V.Knizhnik, Phys. Lett. **168B** (1986) 201; Yu.I.Manin, Phys. Lett. **177B** (1986) 184, G.Moore, Phys. Lett. **176B** (1986) 369.
[5] I.M.Krichever, Russian Math. Surveys **32** (1977) 185.

[6] M.Sato and Y.Sato, Publ. Res. Inst. Math. Sci. (Kyoto Univ.) **388** (1980) 183; *ibid.* **414** (1981) 181; M.Sato, *ibid.***433** (1981) 30; E.Date, M.Jimbo, M.Kashiwara and T.Miwa, J.Phys. Soc. Jpn. **50** (1981) 3806,3813; "Nonlinear Integrable Systems" *ed.* M.Jimbo and T.Miwa (World Scientific, 1983) p.39.

[7] S.Coleman, Phys. Rev. **D11** (1975) 2088; S.Mandelstam, *ibid.* 3025.

[8] T.Miwa, Proc. Jpn. Acad. **58A** (1982) 9; E.Date, M.Jimbo and T.Miwa, J. Phys. Soc. Jpn. **51** (1982) 4116, 4125.

[9] M. A. Namazie, K. S. Narain and M. H. Sarmadi, Phys. Lett. **177B** (1986) 329, T.Eguchi and H. Ooguri, Phys. Lett. **187B** (1987) 127; E. Verlinde and H. Verlinde, Nucl. Phys. **B288** (1987) 357.

[10] C.Vafa,Phys. Lett. **190B** (1987) 47; L.Alvarez-Gaume, C.Gomez and C.Reina, Phys. Lett. **190B** (1987) 55; N.Ishibashi, Y.Matsuo and H.Ooguri, Mod. Phys. Lett.**A2**(1987) 119.

[11] M.Mulase, J.Diff. Geom. **19** (1984) 403; T.Shiota, Invent. Math. **83** (1986) 333.

[12] R.Hirota, J.Phys. Soc. Jpn. **50** (1981) 3785; "Nonlinear Integrable Systems" *ed.* M.Jimbo and T.Miwa (World Scientific, 1983) p.17.

[13] N.Saitoh and S.Saito, TMUP-HEL-8809 (1988).

[14] H.Kato and S.Saito, Lett. Math. Phys. **18** (1989) to be published.

[15] J. D. Fay, "Theta Functions on Riemann Surfaces" *ed.* A. Dold and B. Eckmann, Lect. Notes in Math. (Springer-Verlag, 1973) **Vol.352**.

Non-Linear Equations from a String-Theoretical Point of View

H. Kato and S. Saito

Department of Physics, Tokyo Metropolitan University,
Setagaya-ku, Tokyo 158, Japan

Non-linear equations are investigated from the string theoretical point of view. As a result a new hierarchy of non-linear integrable equations is constructed from open bosonic string.

1. Introduction

In soliton theory it was known that an infinite series of non-linear integrable differential equations called KP-hierarchy exists. This series includes KP equation, KdV equation and many known non-linear differential equations. The mathematical structure of such an interesting object "KP-hierarchy" was revealed by Kyoto school [1] and many analogous hierarchies[†] were constructed by them [2] and other authors [3].

While one of the most charming subject in a recent developement in particle physics is string theory because it is conjectured that string theory unifies all the forces in nature [4]. An interesting feature of string theory is that scattering amplitudes of the theory can be calculated by integrating some kind of vacuum expectaion values of a special type of 2-dimensional field theory. This theory is called conformal field theory (CFT) [5] which is defined by the feature that the action of the theory is invariant under 2-dimensional conformal transformation (=analytic mapping in 2-dim.). Thus the "space-time" for CFT is not a real 4-dimensional manifold but a complex 1-dimensional manifold. If we restrict the manifold to be compact (which is the case of interest in string theory), it becomes famous closed Riemann surface whose rich mathematical structure has been deeply investigated by a number of mathematitians. Hence the interests of particle physicists reduce to CFT on Riemann surfaces. CFT is also an interesting objet for statistical mehanics because the critical behavior of two dimensional statistical models are known to be described by some kind of CFT [5].

The main subject we would like to treat in this talk is that intimate connection between CFT's and non-linear integrable systems. That connection was originally discovered by Ref.[6] and revived and further developed by the authors of Refs.[7][8] inspired by the fever of string theory. In Ref.[9] we have treated two kinds of conformal field theories which are spin 1/2, charged (Weyl) fermion and spin 0, **R**/**Z**-valued boson both of which are free theories. The two are actually known to be equivlent theories (chiral bosonization)[10][11][12]. The equivalence between the fermionic theory and the bosonic theory is a very special featue in dim.=2. The bosonic CFT mentioned above describes the special sector (compactifed dimension) [4] of string theory because the field is **R**/**Z**-valued. Then

---

† In this talk by the word hierarchy we mean an infinite series of non-linear differential, difference or mixed equations.

we would like to consider the generalization to **R**-valued case because this generalization corresponds to an usual sector (uncompactified dimension) in string theory.

2. From CFT to hierarchy

In this section we quickly review how we could get a hierarchy from a CFT in the case of spin 1/2, fermion [6][2]. Then we give the generalization to the case of spin 0, **R**-valued boson.

In general the procedure of associating a hierarchy to a CFT consists of two steps. The first step is to prove so called bilinear identity (explained below) for vacuum expectation values of a given CFT. The second step is to seek so called vertex operators having some required properties. These two steps combine to produce a hierarchy associated to the CFT.

*Spin 1/2 fermion case (= Spin 0 **R** /**Z** -valued boson)*

First we prepare a 2-dimensional charged free fermion system[*] whose action is

$$S = \frac{1}{\pi} \int d^2z\, b(z,\bar{z})\bar{\partial}c(z,\bar{z})$$

where $z \in \mathbf{C}$, a bar denotes the complex conjugation, $\bar{\partial} = \partial/\partial\bar{z}$ and $b(z), c(z)$ are anticommuting fields of conformal spin 1/2 [5].

Normal mode expansion of these fields are

$$b(z) = \sum_{r \in \mathbf{Z}+1/2} b_r z^{r-1/2}, \quad c(z) = \sum_{r \in \mathbf{Z}+1/2} c_r z^{r-1/2}.$$

Canonical quantization and the definition of the vacuum are

$$[b_r, c_s]_+ = \delta_{r+s,0} \quad \text{otherwise zero} \tag{2.1}$$

$$b_r|0> = c_r|0> = 0 \quad \text{for } r > 0. \tag{2.2}$$

In fact this theory has an infinite number of conserved currents

$$J_n^{bc}(z) = -:b(z)c(z): z^n \quad n = 0, 1, 2, \cdots. \tag{2.3}$$

because the condition of conservation $\partial_\mu J^\mu = 0$ now reduces to $\bar{\partial}J = 0$. Conserved charges corresponding to these are

$$\alpha_n = \oint \frac{dz}{2\pi i} J_n^{bc}. \tag{2.4}$$

---

[*] We use here the notation which are popular in CFT to clarify the correspondence though those do not always agree with Ref.[9].

With these charges we can have an infinite number of "time" so we define so called Hamiltonian as

$$H(t) = -\sum_{n\geq 1} t_n \alpha_n. \tag{2.5}$$

Next we define so called τ-function as complex valued function of $(t_1, t_2, t_3, \cdots)$

$$\tau(t) = \tau(t_1, t_2, t_3, \cdots) = <0|e^{H(t)}g|0> \tag{2.6}$$

where

$$g = \exp[\oint \frac{dx}{2\pi i} \oint \frac{dy}{2\pi i} B(x,y) b(x) c(y)] \tag{2.7}$$

whith $B(x,y)$: antisymmetric analytic function. We remark here that $[\alpha_m, \alpha_n] = m\delta_{m+n}$, so α_n's are bosonic oscillators[†] . Now we take the first step which is to prove

$$\oint dz <1|e^{H(t)}b(z)g|0><-1|e^{H(t')}c(z)g|0>= 0 \tag{2.8}$$

where $|m> (m \in \mathbf{Z})$ are defined to satisfy (2.8) and $\alpha_0|m>= m|m>$ [6]. (2.8) is the bilinear identity mentioned previously. The proof is found in [6].

The second step is to find the operators like

$$\begin{aligned}\Lambda(t,z) <0|e^{H(t)} &=< 1|e^{H(t)}b(z) \\ \Lambda^\dagger(t,z) <0|e^{H(t)} &=< -1|e^{H(t)}c(z).\end{aligned} \tag{2.9}$$

We simply give the answer here [6][3]

$$\begin{aligned}\Lambda(t,z) &= \exp(-\sum_{n\geq 1} t_n z^{-n}) \exp(\sum_{n\geq 1} \frac{1}{n} z^n \frac{\partial}{\partial t_n}) \\ \Lambda^\dagger(t,z) &= \exp(\sum_{n\geq 1} t_n z^{-n}) \exp(-\sum_{n\geq 1} \frac{1}{n} z^n \frac{\partial}{\partial t_n}).\end{aligned} \tag{2.10}$$

Combining two steps we have

$$\oint dz \Lambda(t,z) <0|e^{H(t)}g|0> \Lambda^\dagger(t',z) <0|e^{H(t')}g|0>= 0. \tag{2.11}$$

If we make variable transformation [3]

$$t \to t+s, \quad t' \to t-s \tag{2.12}$$

and z-integration this reduces to

$$\sum_{m=0}^{\infty} P_{m+1}(-2s) P_m(\tilde{D}_t) \exp[\sum_{n=1}^{\infty} s_n D_{t_n}] \tau \cdot \tau = 0 \tag{2.13}$$

where

[†] Different notation $a_n = i\frac{1}{\sqrt{n}}\alpha_n$, $a_n^\dagger = -i\frac{1}{\sqrt{n}}\alpha_{-n}$ for $n \geq 1$ is used in Ref.[9].

$$\boldsymbol{s} = (s_1, \cdots, s_n, \cdots), \quad \exp[\sum_{n\geq 1} s_n z^n] = \sum_{m\geq 0} P_m(\boldsymbol{s}) z^m$$

$$D_{t_n}\tau \cdot \tau = \frac{\partial}{\partial s_n}\tau(t_1, t_2, \cdots, t_n + s_n, \cdots)\tau(t_1, t_2, \cdots, t_n - s_n, \cdots)\big|_{s_n=0} \quad (2.14)$$

$$\tilde{D}_t = (\cdots, \frac{1}{n}D_{t_n}, \cdots).$$

D_{t_n} above is called Hirota derivative [3].

This generates the KP-hierarchy

$$(D_1^4 + 3D_2^2 - 4D_1 D_3)\tau \cdot \tau = 0$$
$$(D_1^3 D_2 + 2D_2 D_3 - 3D_1 D_4)\tau \cdot \tau = 0 \quad (2.15)$$

R-valued spin 0 chiral scalar (open bosonic string)

First we define the model precisely.
The action is

$$S = \frac{1}{\pi}\int d^2 z \frac{1}{2}\partial X^\mu(z,\bar{z})\bar{\partial} X_\mu(z,\bar{z}) \quad (2.16)$$

where μ denotes the Lorentzian index running from $\mu = 0$ to $\mu = D - 1$ with D : (uncompactified) space-time dimension [4].

Equation of motion of the system is $\partial\bar{\partial} X^\mu = 0$.

So the normal mode expansion reads

$$X^\mu = x^\mu + ip^\mu \ln z + ip^\mu \ln \bar{z} + i\sum_{n\neq 0}\frac{1}{n}(\alpha_n^\mu z^n + \bar{\alpha}_n^\mu \bar{z}^n). \quad (2.17)$$

This field is known to describe closed string in bosonic string theory [4]. Since in this theory holomorhic part and antiholmorphic part are mixed (such a theory is called non-chiral), some complexity arises. So we remove the anti-holomorphic part to get

$$X^\mu = x^\mu + ip^\mu \ln z + i\sum_{n\neq 0}\frac{1}{n}\alpha_n^\mu z^n. \quad (2.18)$$

This kind of theory is called chiral. This truncation is not artificial in the meaning that, (2.18) is the field which describes the open string in bosonic string theory [4]. We define the decomposition of X^μ for later convenience as follows[†]

$$X_+^\mu(z) = x^\mu + i\sum_{n\geq 1}\frac{1}{n}\alpha_n^\mu z^n, \quad X_-^\mu(z) = ip^\mu \ln z + i\sum_{n\leq -1}\frac{1}{n}\alpha_n^\mu z^n. \quad (2.19)$$

Oscillators are canonically quantized as

$$[x^\mu, p^\nu] = ig^{\mu\nu}, \quad [\alpha_m^\mu, \alpha_n^\nu] = mg^{\mu\nu}\delta_{m+n,0} \quad (2.20)$$

where $g^{\mu\nu}$ is Minkovski metric. This concludes the definition of the model.
To this CFT we now associate a hierarchy.

[†] Remark that this definition is slightly different from that of Ref.[9].

Define a τ-function in this case as

$$\tau(t) = \tau(t_1^0, \cdots, t_1^{D-1}, t_2^0, \cdots, t_2^{D-1}, \cdots) = <0|e^{H(t)}g|0> \qquad (2.21)$$

with $H(t)$ defined to be a natural extension of the previous case (2.5)

$$H(t) = H(\{t_n^\mu\}) \stackrel{\text{def}}{=} -\sum_{n\geq 1} \alpha_n^\mu t_{n\mu} \qquad (2.22)$$

where summation of μ is understood.

As for g we define it to be

$$g = \exp[\frac{-1}{2}\oint \frac{dx}{2\pi i}\oint \frac{dy}{2\pi i}F(x,y)\partial X^\mu(x)\partial X_\mu(y)] \qquad (2.23)$$

where $F(x,y)$: analytic symmetric function.

Now it is the time to take the first step which is to prove

$$\oint dz <0|e^{H(t)}X^\mu(z)g|0><0|e^{H(t')}\partial X^\nu(z)g|0> = 0. \qquad (2.24)$$

From the form of g, following transformation is easily seen

$$\begin{aligned} g^{-1}X^\mu(z)g &= X^\mu(z) + \oint \frac{dy}{2\pi i}\partial_y F(z,y)X^\mu(y) \\ g^{-1}\partial X^\mu(z)g &= \partial X^\mu(z) - \oint \frac{dy}{2\pi i}\partial_z F(z,y)\partial X^\mu(y). \end{aligned} \qquad (2.25)$$

Making use of (2.25), (2.24) can be explicitly shown. Then take the second step. We define the differential operators as

$$\begin{aligned} \Gamma_\mu(t,z) &= -i\sum_{n\geq 1}[\frac{1}{n}z^n\frac{\partial}{\partial t_n^\mu} + z^{-n}t_{n\mu}] \\ \partial\Gamma_\mu(t,z) &= -i\sum_{n\geq 1}[z^{n-1}\frac{\partial}{\partial t_n^\mu} - nz^{-n-1}t_{n\mu}]. \end{aligned} \qquad (2.26)$$

It is clear that they satisfy

$$\begin{aligned} \Gamma_\mu(t,z)<0|e^{H(t)} &= <0|e^{H(t)}X_\mu(z) \\ \partial\Gamma_\mu(t,z)<0|e^{H(t)} &= <0|e^{H(t)}\partial X_\mu(z). \end{aligned} \qquad (2.27)$$

Combine the two steps to get

$$\oint dz\Gamma_\mu(t,z)\partial\Gamma_\nu(t',z)\tau(t)\tau(t') = 0. \qquad (2.28)$$

After carring out the z-integration around the origin we have

$$\sum_{n\geq 1}[t_{n\mu}\frac{\partial}{\partial t_n^{\prime\nu}} - t'_{n\nu}\frac{\partial}{\partial t_n^\mu}]\tau(t)\tau(t') = 0. \qquad (2.29)$$

This equation can be rewritten using Hirota derivative as

$$sym. \sum_{m \geq 1}(t_{m\mu}D_{l_m^\nu} - s_{m\mu}\frac{\partial}{\partial t_m^\nu}) \exp[\sum_{n \geq 1} s_n^\lambda D_{l_n^\lambda}]\tau \cdot \tau = 0 \qquad (2.30)$$

where $sym.$ denotes the symmetrization w.r.t. μ and ν. We have now found a new hierarchy associated to open bosonic string theory.

3. Discussion

In the previous section we have shown a new example of making a hierarchy from a CFT. This kind of arguments for the superghost system are treated in Refs. [13][14][15][16][17]. We consider that this knid of connection between CFT and hierarchy is more widely valid.

In fact the first step which is to make a bilinear identity is general enough. Actually assume that we find two conformal fields $A(z)$, $B(z)$ in the primary field content [5] of a given CFT such that

$$A(z)B(w) \sim \frac{1}{z-w} \quad \text{and} \quad B(w)A(z) \sim \frac{1}{z-w}(-1)^\epsilon \qquad (3.1)$$

where $\epsilon = 0$ or $\epsilon = 1$ corresponding to fields are bosonic or fermionic respectively. Then a similar argument as above leads to the bilinear identity of the type

$$\oint dz <a|A(z)g|0><a'|B(z)g|0> = 0 \qquad (3.2)$$

where $<a|$ and $<a'|$ are arbitrary bras.

But the second step needs the existence of an infinite number of conserved currents to make Hamiltonian $H(t)$ having infinite times t_n. The currents in the previous examples were (2.3) in the fermionic case (see (2.4) and (2.5)), and $J_n^X(z) = \partial X(z)z^n \quad n = 0, 1, 2, \cdots$ in the bosonic case (see (2.18) and (2.22)). So we expect the following :
To any chiral conformal field theory with infinite dimensional additional chiral currents, there exists a hierarchy associated to it. And in some case correlation function of the CFT is characterized by this hierarchy [7].

ACKNOWLEDGEMENTS

The authors greatly appreciate kind hospitality of the organizers of International Conference on Non-linear Physics.

References

[1] M.Sato and Y.Sato, Publ. Res. Inst. Math. Sci. (Kyoto Univ.) 388 (1980) 183; 414 (1981) 181;
M.Sato, *ibid.* 433, (1981) 30;
M.Sato and Y.Sato, "Nonlinear Partial Differential Equation in Applied Science", U.S.-Japan seminor Tokyo 1982, ed. P.D.Lax and H.Fujita
(North-Holland/Kinokuniya, 1982) p.259

[2] E.Date, M.Jimbo and T.Miwa, J. Phys. Soc. Jpn. 51 (1982) 4116; 4125;
ibid. 52 (1983) 388;761;766

[3] Nonlinear Integrable Systems, ed. M.Jimbo and T.Miwa
(World Scientific, 1983)

[4] M.B.Green J.H.Schwarz E.Witten, Superstring theory vol.1,2
(Cambridge University Press 1987)

[5] A.A.Belavin, A.M.Polyakov and A.B.Zamolodchikov,
Nucl. Phys. B241 (1984) 333

[6] E.Date, M.Jimbo, M.Kashiwara and T.Miwa,
Proc. Jpn. Acad. 57A (1981) 342, 387;
Physica 4D (1982) 343;
J.Phys. Soc. Jpn. 50 (1981) 3806, 3813

[7] S.Saito, Tokyo Metropolitan Univ. preprint TMUP-HEL 8613 (1986);
Phys. Rev. Lett. 59 (1987) 1798;
Phys. Rev. D36 (1987) 1819;
Phys. Rev. D37 (1988) 990;
K.Sogo, J. Phys. Soc. Jpn. 56 (1987) 2291

[8] N.Ishibashi, Y.Matsuo and H.Ooguri, Mod. Phys. Lett. A2 (1987) 119;
L.Alvarez-Gaumé, C.Gomez and C.Reina , Phys. Lett. 190B (1987) 55

[9] S.Saito and H.Kato, From soliton theory to string theory, TMUP-HEL-8906

[10] D.Friedan, D.Martinec, and S.H.Shenker, Nucl. Phys. B271 (1986) 93

[11] L.Alvarez-Gaumé, G.Moore, and C.Vafa, Comm. Math. Phys. 106 (1986) 40

[12] T.Eguchi and H.Ooguri, Phys. Lett. 187B (1987) 127;
E.Verlinde and H.Verlinde, Nucl. Phys. B288 (1987) 357

[13] M.A.Awada and A.H.Chamseddine, CERN-TH. 4980/88

[14] L.Alvarez-Gaumé, C.Gomez, P.Nelson, G.Sierra and C.Vafa, Nucl. Phys. B311 (1988) 253

[15] V.Kac and J.van de Leur, Ann. Inst. Fourier Grenoble 37(4) (1987) 99

[16] H.Nohara, Phys. Lett. B216 (1989) 103

[17] A.LeClair, Nucl. Phys. B314 (1989) 425

Painlevé Analysis and Integrability of the Evolution Equation $u_t = u_{xxx} + u^2 u_{xx} + 3uu_x^2 + 1/3 u^4 u_x$

M. Daniel and R. Sahadevan

Department of Physics, Bharathidasan University, Tiruchirapalli 620 024, India

It is shown that the evolution equation $u_t = u_{xxx} + u^2 u_{xx} + 3uu_x^2 + \frac{1}{3} u^4 u_x$ possesses the generalized Painlevé property. Also, it is demonstrated that linearization of the equation follows in a straightforward manner from the singularity structure analysis (Painlevé analysis).

1. INTRODUCTION

Nonlinear problems are in the form of evolution equations. They are often described by nonlinear ordinary and partial differential equations depending on whether the system is discrete or continuous in nature. There has been considerable progress during the past two decades in the understanding of a class of evolution equations leading to fascinating new concepts such as solitons, and complete integrability [1,2]. Of particular interest is the question whether a given system of nonlinear ordinary or partial differential equations is integrable or not. It is possible to identify integrable dynamical systems described by nonlinear ordinary/partial differential equations by looking at the singularity structure of the solutions in the complex plane/manifold which is widely referred to in the literature as Painlevé analysis [3-6].

The first application of these ideas is due to Kovalevskaya [7] in her work on the rigid body problem. Her approach was to determine the conditions under which the only movable singularities of the solutions to the equations of motion in the complex plane are ordinary poles. Also, Fuchs [8] earlier considered the classification of first order ordinary differential equations and concluded that out of all forms, the only equation which is free from movable critical points is the generalized Riccati equation. Following Kovalevskaya, came the extensive work of Painlevé and his contemporaries to find the class of second order equations whose only movable singularities are ordinary poles [9]. This analytic property of ordinary differential equations is now referred to as the Painlevé property. Later, while studying similarity reductions of known integrable or Inverse Scattering Transform (IST) solvable partial differential equations it was observed that they all led to ordinary differential equations of Painlevé type. This has in fact been verified for a large class of nonlinear partial differential equations by Lakshmanan and Kaliappan [10] using the underlying Lie symmetries. Following this, Ablowitz, Ramani and Segur (ARS) [11] formulated the following conjecture: every ordinary differential equation obtained by an exact reduction of a partial differential equation solvable by IST possesses the Painlevé property. Thus, integrability is related to the absence of movable critical points. The main virtue of ARS approach is that it proposes an algorithm by which it is possible to check whether a given system of ordinary differential equations does or does not satisfy necessary criteria (absence of movable critical points) for possessing the Painlevé property. In the cases where these criteria are found to be satisfied, further investigation is needed to prove integrability by constructing the full set of integrals or linearizing the equations or reducing them to one of Painlevé transcendental equations.

In Section 2, we present at some length the procedure to identify Painlevé property in partial differential equations. Then in the succeeding sections, we apply the procedure to the nonlinear partial differential equation under study and find that the equation passes the generalized Painlevé test. Also, we construct the Bäcklund transformation and bring out the fact that the linearization of the equation follows in a straightforward manner from the above analysis.

2. PAINLEVÉ PROPERTY FOR PARTIAL DIFFERENTIAL EQUATIONS

Though ARS conjecture has proven to be quite valuable there are certain drawbacks such as the difficulty of identifying all possible reductions of the partial differential equations to ordinary differential equations and the inability of the conjecture to provide further information about the actual solutions to the equations. In order to overcome these difficulties, Weiss, Tabor and Carnevale (WTC) [12] proposed a generalized Painlevé property directly applicable to partial differential equations wherein the solutions are required to be single valued around movable singular manifolds in order that they be integrable. The motivation for such a generalization comes from the fact that the singularities of solutions to partial differential equations are in general not isolated unlike that of ordinary differential equations but rather lie on a manifold determined by an equation of the form $\Phi(Z_1, Z_2, \ldots, Z_n) = 0$, where Φ is analytic in some neighbourhood of the singular manifold [12,13].

The method involves expanding the solution in a Laurent series about a singular manifold from which one can deduce Lax pairs and

Bäcklund transformations. Thus if $u = u(x,t)$ is a solution of the partial differential equation $u_t = K(u)$, the solutions must be single valued about the noncharacteristic movable singularity manifold $\Phi(x,t) = 0$, $\Phi_x(x,t) \neq 0$, $\Phi_t(x,t) \neq 0$, then we seek an expansion for the solution of the form

$$u(x,t) = \Phi^\alpha(x,t) \sum_{j=0}^\infty u_j(x,t) \Phi^j(x,t) , \qquad (1)$$

where $u_0 \neq 0$, $u_j(x,t)$ are expansion coefficients analytic in the neighbourhood of the singular manifold $\Phi(x,t) = 0$ and α is a negative integer in order that the Painlevé property holds. The procedure involves the following steps.

i) Determination of the leading order behaviour. The equation may have different leading orders.

ii) Finding the powers at which arbitrary functions enter called resonances. One resonance always occurs at $j = -1$ and corresponds to the arbitrariness of Φ.

iii) Verifying that sufficient number of arbitrary functions exist as required by the Cauchy-Kovalevskaya theorem. Integrability requires that all branches possess the Painlevé property. If at a resonance the associated u_j fails to be arbitrary, terms of the form $a_j + b_j \log \Phi$ must be included in the expansion. This makes the solution multivalued about the singular manifold and hence the Painlevé property is lost.

iv) Establishing the connection with other solution characteristics. Truncating expansion (1) at $O(\Phi^0)$

$$u = u_0 \Phi^{-n} + u_1 \Phi^{-n+1} + \ldots + u_n , \qquad (2)$$

putting in the evolution equation and on equating the coefficients of various powers of Φ, one obtains a system of equations involving Φ and u_j ($0 \leq j \leq n$) with u and u_n satisfying the same evolution equation. For integrable systems this approach yields a nontrivial auto-Bäcklund transformation with the equations involving Φ being transformable into the associated Lax pair. The Hirota formulation is also a straightforward consequence.

3. PAINLEVÉ ANALYSIS OF THE EVOLUTION EQUATION
$$u_t = u_{xxx} + u^2 u_{xx} + 3uu_x^2 + \frac{1}{3} u^4 u_x$$

Among the class of evolution equations

$$u_t = u_{xxx} + u^2 u_{xx} + auu_x^2 + bu^4 u_x \qquad (3)$$

obtained in the reduction problem of Calogero and Degasperis [14], the set corresponding to a = 3 and b = 1/3 was found to possess properties associated with complete integrability and displays explicit solutions. Thus it has become important to study the singularity structure of the evolution equation (3) (with a = 3 and b = 1/3) here using WTC method discussed in the previous section. consider the equation [15]

$$u_t = u_{xxx} + u^2 u_{xx} + 3uu_x^2 + \frac{1}{3} u^4 u_x .$$ (4)

Equation (4) is quite valuable and exceptional in the sense that it was studied extensively by several authors [16-18] and many properties associated with the complete integrability of the equation have been established. For instance, it was discovered that Eq. (4) possesses prolongation structure and nontrivial Lie-Bäcklund symmetries [16]. Also, it belongs to the class of equations possessing an infinite Lie-Bäcklund algebra [17]. Further, it was exactly linearized by an appropriate change of dependent variable and several explicit solutions have been found [18].

3.1 Leading Order Analysis

To carry out the Painlevé analysis, we consider the Laurent series expansion (1) around the singularity manifold. Assuming a leading singularity of the form $u = u_0 \phi^\alpha$ (first term of the expansion (1)) and balancing all the four terms on the right-hand side of the equation, we obtain

$$\alpha(\alpha - 1)(\alpha - 2) u_0 \phi_x^3 \phi^{\alpha-3} + u_0^3 \phi_x^2 [\alpha(\alpha - 1) + 3\alpha^2] \phi^{3\alpha-2}$$
$$+ \frac{1}{3} u_0^5 \alpha \phi_x \phi^{5\alpha-1} = 0 ,$$ (5)

which gives

$$\alpha = -1/2 , \quad C^2 = 9C - 45/4 ,$$ (6a)

where

$$u_0^2 = C\phi_x .$$ (6b)

Thus we have the following two leading order behaviours corresponding to C = 3/2 and C = 15/2 respectively.

(i) $\quad \alpha = -1/2 , \quad u_0^2 = \frac{3}{2} \phi_x ,$ (7a)

(ii) $\quad \alpha = -1/2 , \quad u_0^2 = \frac{15}{2} \phi_x .$ (7b)

Though the fractional value for α is an indication of the algebraic branch point singular manifold it is possible to transform it to poles using the transformation $w = u^2$.

3.2 Resonances

Using (7) into the Laurent series, we substitute the resultant expansion

$$u = u_0 \phi^{-\frac{1}{2}} + \sum_{j>0} u_j \phi^{j-\frac{1}{2}} \qquad (8)$$

into (4) and establish a recursion relation for the u_j's. We readily obtain

$$(j+1)(j-3)(j+C-5/2)\phi_x^3 u_j = F_j(\phi, u_0, u_1, \ldots, u_{j-1}), \qquad (9)$$

where F_j can be explicitly written down. Thus the resonances are $-1, 1, 3$ when $C = 3/2$ and $-5, -1, 3$ when $C = 15/2$. The resonance value -1 for both the choices represents the arbitrariness of the singularity manifold ϕ.

3.3 Arbitrary Functions

For the case corresponding to $C = 3/2$, the compatibility conditions at $j = 1, 3$ are satisfied identically. From the recursion relation, we find

$$j = 0 : u_0^2 = \tfrac{3}{2}\phi_x,$$

$$j = 2 : u_2 = \tfrac{1}{3}\phi_x^{-3}[\tfrac{1}{2}(\phi_t - \phi_{xxx})u_0 + \tfrac{3}{4}\phi_{xx}u_{0x}$$
$$- \tfrac{3}{2}\phi_x \phi_{xx} u_1 - \phi_x^2 u_0 u_1^2 - 3\phi_x^2 u_{1x}]. \qquad (10)$$

In a similar manner, we checked that the series representation (8) for the choice $C = 15/2$ admits exactly two arbitrary functions and does not introduce movable logarithmic singularity manifolds. Thus Eq. (4) possesses the Painlevé property.

3.4 Bäcklund Transformation

In view of constructing the Bäcklund transformation, the series representation (8) is truncated at the constant level term by setting the resonance coefficients u_1 and u_3 zero, that is, $u_j = 0$, $j > \tfrac{1}{2}$ and so we have

$$u = u_0 \phi^{-\frac{1}{2}} + u_{\frac{1}{2}}. \qquad (11)$$

In order to check that Eq. (11) is a Bäcklund transformation, we substitute it into Eq. (4). On equating the coefficients of different powers of Φ, we find that the function $u_{\frac{1}{2}}$ is a zero solution and the functions $u_0(x,t)$ and $\Phi(x,t)$ satisfy

$$u_{0t} = u_{0xxx} , \qquad (12a)$$

and

$$\Phi_t = \Phi_{xxx} - \frac{3}{4} \Phi_{xx}^2 / \Phi_x . \qquad (12b)$$

Thus the Bäcklund transformation reads

$$u = u_0 \Phi^{-\frac{1}{2}} = (3\Phi_x/2\Phi)^{\frac{1}{2}} . \qquad (12c)$$

One can check in a straightforward way that the nonlinear equation (4) can be linearized using the Bäcklund transformation (12). Recently, Calogero [18] has also linearized Eq. (4), but in a different way, by transforming the dependent variable using the same transformation as (12). It is interesting to note that Calogero's linearization follows in a natural way from our analysis and hence several explicit solutions can be constructed.

4. CONCLUSIONS

In this paper, we pointed out how the singularity structure analysis of solutions of nonlinear partial differential equations systematically provides an algorithmic procedure to identify integrable dynamical systems. We illustrated this idea by considering an interesting nonlinear partial differential equation for which several integrability properties have been studied extensively in the recent years. We reported that the equation possesses the Painlevé property and so satisfies a necessary condition to be integrable. Also, we constructed the Bäcklund transformation and pointed out its connection with the linearization, which enables one to construct various classes of solutions.

REFERENCES

[1] M. J. Ablowitz and H. Segur, **Solitons and the Inverse Scattering Transform** (SIAM, Philadelphia, 1981).
[2] M. Lakshmanan, ed., **Solitons: Introduction and Applications** (Springer, Heidelberg, 1988).
[3] A. Ramani, B. Dorizzi and B. Grammaticos, Phys. Rev. Lett. **49** (1982) 1539.
[4] M. Tabor, Nature **310** (1984) 277.
[5] M. Lakshmanan and R. Sahadevan, Phys. Rev. **A31** (1985) 861.
[6] K. M. Tamizhmani and M. Lakshmanan, J. Math. Phys. **27** (1986) 2257.

[7] S. Kovalevskaya, Acta Math. **12** (1889) 177.
[8] L. Fuchs, Sitz. Akad. Wiss., Berlin **32** (1884) 699.
[9] E. L. Ince, **Ordinary Differential Equations** (Dover, New York, 1956).
[10] M. Lakshmanan and P. Kaliappan, J. Math. Phys. **24** (1983) 795.
[11] M. J. Ablowitz, A. Ramani and H. Segur, J. Math. Phys. **21** (1980) 715.
[12] J. Weiss, M. Tabor and G. Carnevale, J. Math. Phys. **24** (1983) 522.
[13] J. Weiss, J. Math. Phys. **24** (1983) 1405.
[14] F. Calogero and A. Degasperis, Lett. Nuovo Cimento **16** (1970) 425; **23** (1978) 150.
[15] M. Daniel and R. Sahadevan, Phys. Lett. **130** (1988) 19.
[16] A. Roy Chowdhury and S. Basak, Phys. Scr. **33** (1986) 197.
[17] N. Kl. Ibragimov and A. B. Shabat, Funkcional Anal. Prilozen **14** (1980) 79.
[18] F. Calogero, J. Math. Phys. **28** (1987) 538.

Subject Index

Reference is made to the *first* page of relevant articles

AKNS hierarchy 42
Asymptotic behaviour 236

Bäcklund transformation 29, 35
Bilinearization 242
Blow up 236
Bosonic string 258, 266
Braid group 111, 152
Braid-monoid algebra 111

C-integrability 232
Cellular automata 166
Chaos 166
Chiral gauge field 214
Classical statistical mechanics 98
Complex poles 252
Confocal involutive system 68, 79
Conserved quantity 12
Constraint 79, 85
Coupled nonlinear oscillators 54

Darboux transformation 23, 42
Dispersion relations 98
Dynamical system 54

Eigenvalue problem 68
Exact solution 29
Exactly solvable model 111
Existence 236

Finite-dimensional Hamiltonian system 79
Finite-dimensional integrable system 68

Gauge equivalent equation 47
Generalized Hamiltonian structure 2
Graph-state models 111

Hamiltonian
— finite-dimensional 79
— generalized 2
— integrable 47
— involutive 92
— Lax equation 136
Hirota's bilinear difference equation 205, 258
Hubbard model 205

Infinite dimensional integrable system 2
Integrability 2, 54, 79, 92, 146, 273
Integrable Hamiltonian system 47
Integrable hierarchy 12
Integrable lattices 98
Integrable models 98
Integrable system 68, 136
Integrals of motion 85
Invariance analysis 54
Invariant subspace 85
Inviscid singularities 252
Involutive Hamiltonians 92
Involutive conserved integrals 68
Isospectral 35

K-P equation 190
Kauffman's state model 152
KdV equation 190
KdV-like soliton solution 242
Knots 111

Lagrangean 205
Lattice 98, 205, 221, 227
Lax algebra 12
Lax operators 12
Lax pair 23, 35, 79, 85
Lie algebra structure 47
Lie algebra, semisimple 136
Lie bialgebra 146
Linearization 273

Link polynomial   111, 152
Liouville integrability   2

Markov trace   111
Multi-dimension   23
Multiple Darboux transformation   42
Multiple pole solution   42

Non-isospectral   35
Nonabelian anomaly   221
Noncanonical recursion operator   12
Nonhereditary operator   47
Nonlinear dynamical system   54
Nonlinear equation   29, 42, 92, 166, 190
Nonlinearized Lax pair   79
Numerically perturbed versions   166

Painlevé analysis   273
Phase shifts   98
Phase transition   252
Poisson brackets   136
Porous medium   214
Potential   68
Pulse-like solitary waves   214

Quantum nonlinear Schrödinger   98

R-matrix   111, 136, 146
Recursion operator   12, 47, 85
Reduction   42
Riemann surface shock   252

S-integrability   232
Scattering amplitudes   266
Schrödinger model, quantum nonlinear   98
Semisimple Lie algebra   136
Similarity transformations   236
Soliton   166, 214, 227, 242
Solitonic behaviour   232
Spectral transformation   232
String, bosonic   258, 266
Surface amplitude   190
Symmetry   12, 54

Topological invariant   111
Transformation   227
Travelling waves   236
Tri-hamiltonian Lax equation   136
Two soliton balance   242

Vector fields   54
Viscous solution   252

Water waves   190
Wilson fermion   221
Wilson term   221

x-deformation   152

Yang-Baxter equation   152

Zero curvature equation   2

List of Participants

Professor M.J. Ablowitz
Program in Applied Mathematics
University of Colorado at Boulder
Campus Box 426
Boulder, Colorado 80309-0426
USA

Prof. Yu Arefeva
Steklov Math. Institute
Academy of USSR, Moscow
USSR

Prof. C.W. Cao
Department of Mathematics
Zhengzhou University
Shandong
China

Prof. F. Calogero
DIPARTIMENTO DI FISICA
Universita Degli Studi di Roma
"La Sapienza"
Piazzale Aldo Moro, 2
I-00185 Roma
ITALY

Dr. P.J. Caudrey
Department of Mathematics
U.M.I.S.T.
PO Box 88, Manchester M 60 1QD
England

Prof. R. Chatterjee
Faculty of Science
Department of Physics
The University of Calgary
2500 University Drive N.W.
Calgary, Alberta
Canada T2N 1N4
Canada

Associated Prof. Z.Y. Chen
Department of Physics
Institute of Technology
of Middle China
Wuhan, Hubei
China

Prof. Au Chi
Department of Mathematical Studies
Hong Kong Polytechnic
Hong Kong

Prof. P.L. Christiansen
The Technical University of Denmark
Laboratory of Applied Mathematics
Physics
DK-2800 LYNGBY
Denmark

Dr. M. Daniel
Department of Physics
Bharathidasan University
Palkalaiperur Campus
Tiruchirapalli-620024
Tamilnadu
India

Prof. M.L. Ge
Institute of Mathematics
Nankai University
Tianjing
China

Lecturer Z.Q. Gu
Basic Department
College of Shijiazhuang Railway
Hebei
China

Prof. K.Y. Guan
Institute of Applied Mathematics
College of
Beijing Aeronautical Engineering
Beijing
China

Researcher B.L. Guo
Institute of Applied Phys. &
Computational Math.
P.O. Box 8009
Beijing
China

Prof. B.Y. Guo
University of Sci. and Tech.
of Shanghai
Shanghai
China

Associated Prof. F.K. Guo
Department of Mathematics
Shandong Minal College
Shandong
China

Prof. H.S. Hu
Institute of Mathematics
Fudan University
Shanghai
China

Prof. L.D. Huang
Department of Mathematics
Tongji University
Shanghai
China

Prof. N.N. Huang
Department of Physics
University of Wuhan
Wuhan, Hubei
China

Dr. H. Kato
Department of Physics
Tokyo Metropolitan University
Fukazawa 2-1-1
Setagaya-ku
Tokyo 158
Japan

Prof. M. Kruskal
Program in Applied and
Computational Mathematics
Princeton University
Fine Hall, Washington Road
Princeton, New Jersey 08544
USA

Prof. M. Lakshmanan
Department of Physics
Bharathidasan University
Palkalaiperur Campus
Tiruchirapalli-620024
Tamilnadu
India

Prof. Franklin J. Lambert
Theoretische Natuurkunde
Vrije Universiteit Brussel
Pleinlaan 2, 1050 Brussel
Belgium

Prof. D. Levi
DIPARTIMENTO DI FISICA
Universita degli Studi di Roma
"La Sapienza"
Piazzale Aldo Moro, 2
I-00185 Roma
Italia

Lecturer L.S. Li
Basic Department of
Central Television University
Beijing
China

Lecturer Y. Li
Department of Applied Mathematics
Tongji University
Shanghai
China

Prof. Y.S. Li
Department of Mathematics
Univ. of Sci. & Tech. of China
Hefei
China

Prof. S.D. Lillo
DIPARTIMENTO DI FISICA
Universita degli Studi di Roma
"La Sapienza"
Piazzale Aldo Moro 2
I-00185 Roma
Italia

Lecturer C.S. Lin
Department of Physics
Nationality Normal College
of Inner-Mongulia
Inner-Mongulia
China

Prof. Q.Y. Liu
Department of Mathematics
University of North-west
Xian
China

Associated Prof. Z.J. Lu
Department of Mathematics
Shandou University
Guangdong
China

Associated Prof. Y.C. Ma
Department of Mathematics
Xian Jiaotong University
Xian, Saanxi
China

Dr. Z.S. Ma
Department of Physics
Zhongshan University
Guangdong
China

Prof. S.V. Manakov
L.D. Landau Institute
for Theoretical Physics
Moscow V 334
USSR

Lecture D.Z. Meng
Department of Applied Mathematics
University of Beijing Industry
Beijing
China

Researcher Assistant J.L. Mu
Institute of Physics
of Academia Sinica
P.O. Box 603 team 105 Beijing
Beijing
China

Dr. W. Oevel
FB 17 Mathematik
Universitat Paderborn
D-4790 Paderborn
W-Germany

Associated Prof. X.F. Pang
Department of Physics
Chengdu South-west
Nationality College
Chengdu, Sichuan
China

Prof. F.Q. Pu
Department of Applied Mathematics
University of Qinghua
Beijing
China

Prof. S. Saito
Department of Physics
Tokyo Metropolitan University
Setagaya-Ku Tokyo 158
Japan

Prof. N. Saitoh
Department of Applied Mathematics
Yokohama National University
Tokiwadai, Hodogaya-ku
Yokohama 240
Japan

Prof. Junkichi Satsuma
Department of Applied Physics
Faculty of Engineering
University of Tokyo
Bunkyo-ku, Tokyo 113
Japan

Prof. K.J. Shi
Institute of Morden Physics
University of North-west
Xian
China

Prof. F.A. Smirnov
Steklov Mathematical Institute
Fontanka 27
Leningrad, D-11
USSR

Prof. E. Taflin
UAP departement informatique
20 ter, rue de Bezons
92411 Courbevoie Cedex
France

Mr. Daisuke Takahashi
Department of Applied Physics
Faculty of Engineering
University of Tokyo
Bunkyo-ku, Tokyo 113
Japan

Prof. L.A. Takhtajan
Steklov Mathematical Institute
Fontanka 27
Leningrad, D-11
USSR

Researcher G.Z. Tu
Computational Centre of
Academia Sinica
Beijing
China

Prof. M. Wadati
Institute of Physics
College of Arts and Sciences
University of Tokyo
Komaba 3-8-1
Tokyo 153
Japan

Associated Prof. J.C. Wang
Department of Math. & Phys.
University of Wuhan Industry
Wuhan, Hubei
China

Lecturer L.R. Wang
Department of Mathematics
East-China Normal University
Shanghai
China

Associated Prof. M.L. Wang
Department of Mathematics
Lanzhou University
Lanzhou, Gansu
China

Associated Prof. X.Y. Wang
Department of Physics
Suzhou University
Jiangsu
China

Associated Prof. L.W. Xiang
Department of Applied Mathematics
University of Shanghai Jiaotong
Shanghai
China

Associated Prof. H.G. Xie
Department of Math. & Phys.
Zhejiang Educational College
Hangzhou
China

Associated Prof. B.Z. Xu
Department of Zhejiang
Hangzhou, Zhejiang
China

Prof. J.R. Yan
Department of Physics
Xiandan University
Hunan
China

Associated Prof. X.Q. Yan
Department of Mathematics
Xiantan Normal College
Hunan
China

Associated Prof. Y.B. Zeng
Department of Mathematics
Univ. of Sci. & Tech. of China
Hefei
China

Editor B.G. Zhang
Room 1
Beijing Science Press
Beijing
China

Associated Prof. D.W. Zhang
Department of Mathematics
University of Beijing
Beijing
China

Associated Prof. Z.S. Zhao
Department of Mathematics
University of Liaoning
Sengyan, Liaoning
China

Associated Prof. Z.W. Zhao
Department of Machinery
Shanghai Industrial University
Shanghai
China

Advanced Engineer K.J. Zheng
Tel. Science Institute
of Zhejiang
Zhejiang
China

Associated Prof. L.Y. Zhou
Basic Department
Kunming Industrial College
Kunming, Yunnan
China

Index of Contributors

Ablowitz, M.J. 166
Akutsu, Y. 111

Bessis, D. 252
Boling, Guo 236
Bullough, R.K. 98

Calogero, F. 232
Cewen, Cao 68
Chen, Yu-Zhong 98
Cheng, Yi 98
Chi, Au 29
Chou, Tian 35

Daniel, M. 273
Dawei, Zhuang 92
Deguchi, T. 111

Fournier, J.D. 252

Ge, M.L. 152
Guizhang, Tu 2

Herbst, B.M. 166

Kato, H. 258,266
Keiser, J.M. 166

Lakshmanan, M. 54
Lambert, F. 242
Levi, D. 190

Min, Qian 146

Oevel, W. 136

Piao, F. 152
Pilling, D.J. 98

Qiming, Liu 227

Sachs, J.R. 214
Sahadevan, R. 273
Saito, S. 258,266
Saitoh, N. 205

Satsuma, J. 214
Shuohong, Guo 221

Takahashi, D. 214
Timonen, J. 98

Wadati, M. 111
Wang, L.Y. 152
Wenxiu, Ma 79
Willox, R. 242

Xianguo, Geng 68
Xinshen, Gu 42
Xue, K. 152

Yi, Cheng 12,47
Yishen, Li 47,85
Youjin, Zhang 35
Yuanqu, Lin 92
Yunbo, Zeng 47,85

Zhangju, Liu 146
Zhongshui, Ma 221
Zixiang, Zhou 23

M. Lakshmanan, Bharathidasan University, Tiruchirapalli, Tamil Nadu (Ed.)

Solitons

Introduction and Applications

Proceedings of the Winter School, Bharathidasan University, Tiruchirapalli, South India, January 5–17, 1987

1988. IX, 367 pp. 29 figs. (Springer Series in Nonlinear Dynamics) Hardcover DM 98,– ISBN 3-540-18588-7

Contents: Introduction. – Mathematical Theory: IST, Symmetries, Singularity Structure and Integrability. – Lattice Solitons. – Statistical Mechanics and Quantum Aspects. – Applications: Physics and Biology. – Index of Contributors.

J. M. Combes, A. Grossmann, P. Tchamitchian, CNRS, Marseille (Eds.)

Wavelets

Time-Frequency Methods and Phase Space

Proceedings of the International Conference, Marseille, France, December 14–18, 1987

1989. IX, 315 pp. 88 figs. Hardcover DM 148,–
ISBN 3-540-51159-8

Time-frequency methods and phase space are as well known to most physicists, engineers and mathematicians as traditional Fourier analysis, which has recently found for many applications a competitor in the concept of wavelets. Crudely speaking a wavelet decomposition is an expansion of an arbitrary function into smooth localized contributions labeled by a scale and a position parameter.
The meeting recorded in this volume brought together people exploring and applying these concepts in an interdisciplinary framework. Topics discussed range from purely mathematical aspects to signal and speech analysis, seismic and acoustic applications, and wavelets in computer vision.

Springer-Verlag Berlin
Heidelberg New York London
Paris Tokyo Hong Kong

A. Hasegawa, AT&T Bell Laboratories, Murray Hill, NJ

Optical Solitons in Fibers

2nd enl. ed. 1990. XII, 79 pp. 25 figs. Softcover DM 48,–
ISBN 3-540-51747-2

(The first edition was published as hardcover edition of the Series "Springer Tracts in Modern Physics" Vol. 116)

Already after six months high demand made a new edition of this textbook necessary. The most recent developments associated with two topical and very important theoretical and practical subjects are combined: **Solitons** as analytical solutions of nonlinear partial differential equations and as lossless signals in dielectric **fibers**. The practical implications point towards technological advances allowing for an economic and undistorted propagation of signals revolutionizing telecommunications. Starting from an elementary level readily accessible to undergraduates, this pioneer in the field provides a clear and up-to-date exposition of the prominent aspects of the theoretical background and most recent experimental results in this new and rapidly evolving branch of science. This well-written book makes not just easy reading for the researcher but also for the interested physicist, mathematician, and engineer. It is well suited for undergraduate or graduate lecture courses.

P. C. Sabatier, University of Languedoc, Montpellier (Ed.)

Inverse Methods in Action

(r.c.p. 264, Montpellier, November 1989)

1990. Approx. 650 pp. 115 figs. 12 tabs. Hardcover, in prep.
ISBN 3-540-51994-7

The basic idea of inverse methods is to extract from the evaluation of measured signals the details of the object emitting them. The applications range from physics and engineering to geology and medicine (tomography). Although most contributions are rather theoretical in nature, this volume is of practical value to experimentalists and engineers and of interest to mathematicians.

The review lectures and contributed papers are grouped into ten chapters dedicated to tomography, distributed parameter inverse problems, spectral and scattering inverse problems (exact theory), wave propagation and scattering (approximations); miscellaneous inverse problems and applications and inverse methods in nonlinear mathematics.

Springer-Verlag Berlin
Heidelberg New York London
Paris Tokyo Hong Kong

Research Reports in Physics

The categories of camera-ready manuscripts (e.g., written in TEX; preferably hard plus soft copy) considered for publication in the **Research Reports** include:

1. Reports of meetings of particular interest that are devoted to a single topic (provided that the camera-ready manuscript is received within four weeks of the meeting's close!).
2. Preliminary drafts of original papers and monographs.
3. Seminar notes on topics of current interest.
4. Reviews of new fields.

Should a manuscript appear better suited to another series, consent will be sought from the author for its transfer to the other series.

Research Reports in Physics are divided into numerous subseries, e.g., nonlinear dynamics or nuclear and particle physics. Besides covering material of general interest, the series provides an opening for topics that are too specialized or controversial to be published within the traditional context. The implied small print runs make a consistent price structure impossible and will sometimes have to presuppose a financial contribution from the author (or a sponsor). In particular, in the case of proceedings the organizers are expected to place a bulk order and/or provide some funding.

Within **Research Reports** the timeliness of a manuscript is more important than its form, which may be unfinished or tentative. Thus in some instances, proofs may be merely outlined and results presented that will be published in full elsewhere later. Since the manuscripts are directly reproduced, the responsibility for form and content is mainly the author's, implying that special care has to be taken in the preparation of the manuscripts.

Springer-Verlag
Berlin Heidelberg New York
London Paris Tokyo Hong Kong

Research Reports in Physics

Manuscripts should be no less than 100 and no more than 400 pages in length. They are reproduced by a photographic process and must therefore be typed with extreme care. Corrections to the typescript should be made by pasting in the new text or painting out errors with white correction fluid. The typescript is reduced slightly in size during reproduction; the text on every page has to be kept within a frame of 16 × 25.4 cm ($6\frac{5}{16}$ × 10 inches). On request, the publisher will supply special stationary with the typing area outlined.

Editors or authors (of complete volumes) receive 5 complimentary copies and are free to use individual parts of the material in other publications later on.

All manuscripts, including proceedings, must contain a subject index. In the case of many-author books and proceedings an index of contributors is also required. Proceedings should also contain a list of participants, with complete addresses.

Our Instructions for the Preparation of Camera-Ready Manuscripts and further details are available on request.

Manuscripts (in English) or inquiries should be directed to

Dr. Ernst F. Hefter,
Physics Editorial 4,
Springer-Verlag, Tiergartenstrasse 17,
D-6900 Heidelberg, FRG,

Springer-Verlag
Berlin Heidelberg New York
London Paris Tokyo Hong Kong

(Tel. [0]6221-487495;
Telex 461723; Telefax 06221-43982).

NOV 0 2 1990